Birkhäuser

Modeling and Simulation in Science, Engineering and Technology

Sebastian Aniţa • Vincenzo Capasso •
Simone Scacchi

Mathematical Modeling and Control in Life and Environmental Sciences

Regional Control Problems

Birkhäuser

Sebastian Aniţa
Faculty of Mathematics
Alexandru Ioan Cuza University of Iaşi
Iaşi, Romania

Vincenzo Capasso
ADAMSS
Università degli Studi di Milano
"La Statale"
Milan, Italy

Simone Scacchi
Department of Mathematics
Università degli Studi di Milano
"La Statale"
Milan, Italy

ISSN 2164-3679 ISSN 2164-3725 (electronic)
Modeling and Simulation in Science, Engineering and Technology
ISBN 978-3-031-49970-8 ISBN 978-3-031-49971-5 (eBook)
https://doi.org/10.1007/978-3-031-49971-5

This book is published under the imprint Birkhäuser, www.birkhauser-science.com by the registered company Springer Nature Switzerland AG
The registered company address is: Gewerbestrasse 11, 6330 Cham, Switzerland

Preface

To the planet Earth: our Environment.

Control theory has been intensively used in Engineering, but recently has moved to other sciences, such as Medicine, e.g., for optimizing therapeutic regimens in oncology (see, e.g., Jarrett, A. J., et al., Optimal control theory for personalized therapeutic regimens in oncology: Background, history, challenges, and opportunities. *J. Clin. Med.* 2020, 9, 1314; doi:10.3390/jcm9051314; see also [140, 186]). The current COVID-19 pandemic has revealed the inadequacy of usual methods of control not supported by the required body of expertise integrating mathematics, biology, physics, engineering, logistics, etc. This has been the main motivation for producing a monograph addressing the mathematical problem of controlling a dynamical system modelled by ordinary and/or partial differential equations; a problem which is common to a variety of real-life problems. It is neither a tract nor a recipe book as such; rather, it aims to providing an account of fundamental concepts as they are required in relevant modern applications and in existing literature. A set of significant case studies provide a substantial motivation to the knowledge of the proposed mathematical and computational methods. The originality of the approach presented in this monograph is in its "leit motiv": regional control. In many real structured populations, it is practically impossible, due to anyhow limited resources, to control the relevant system by global interventions in the whole habitat. In geographical epidemiology (for human populations, or in agriculture), this might also be due to unavoidable logistic restrictions. This has motivated, about 20 years ago, two of the authors [SA and VC] to propose regional control methods, along the principle of "Think Globally, Act Locally." They have shown that, under suitable assumptions, it is possible to eradicate an epidemic in the whole habitat by interventions only in a bounded subset of it.

For age-dependent populations, a similar problem, arisen in connection with social distancing and vaccination policies, concerns the optimal selection of the age segments on which to act for a fast and effective eradication of an epidemic. This problem has been of crucial importance in connection with COVID-19 vaccination strategies.

In geographical economics, a recent key issue concerns the interplay of economic growth and the induced environmental pollution.

We make no pretense of its being complete. Indeed, we have omitted many results that we feel are not directly related to the main theme or that are available in easily accessible sources.

Consistent with the concept of "learning by doing," we intend to guide the reader on how to treat control problems, starting with modelling, and continuing with mathematical concepts and methods from control theory, including their computational aspects.

Special attention has been paid on case studies concerning the control of relevant examples of epidemic phenomena, and the impact of human production activities on environmental pollution.

We may compare our book with a couple of other excellent books in the literature devoted to the control in mathematical biology: Lenhart S. , Workman J.T., Optimal Control Applied to Biological Models, Chapman and Hall/CRC, 2007; Schättler H., Ledzewicz U., Optimal Control for Mathematical Models of Cancer Therapies, Springer, 2015. In addition, there are several very good monographs devoted to the general theory of optimal control or to controllability and stabilizability problems (see, e.g., [40] and references therein).

We have not restricted ourselves to usual optimal control, controllability, and stabilizability problems, but we have considered some nonstandard control problems as well.

Indeed, our plan is to guide the reader on the modelling of some realistic and affordable control problems, including the way one may derive information about the structure of the control. All supported by analytical (whenever possible) and computational experiments. A special feature of the book is its focusing on regional control problems for structured populations, in particular age and space structured populations, that is models described by partial differential equations.

The plan is to present optimal control, as well as stabilization and controllability problems. Preventing the spread of a pest population or of an epidemic via a control is another type of problem which has been discussed.

As already mentioned above, due to practical limitations/constraints, in real-world applications, controls usually act only on subsets of a habitat or on subsets of the relevant age interval; this is the main reason why we have paid particular attention to regional control problems, which means not only to address the magnitude of the relevant control, but also the location and geometry of the region where the controls acts.

We aim to addressing three main groups of readers: first, mathematicians working in a different field; second, other scientists and professionals from a private or academic background; third, graduate or advanced undergraduate students of a quantitative subject related to the analysis and applications of dynamical systems and their control. This topic has become increasingly popular in terms of current research output and applications; many pure as well as applied mathematicians have become interested in learning about the fundamentals of control theory and modern applications. We have tried to write this volume in a language that all addressed groups can understand and in its content and structure will allow them to learn the essentials profoundly and in a time-efficient manner. Other scientist-practitioners and academics from fields like biology, medicine, and economics might be very familiar with a less mathematical approach to their specific fields and thus be interested in the mathematical methods required for modelling their applications. Furthermore, this book would be suitable as a textbook accompanying a graduate or advanced undergraduate course or as secondary reading for students of mathematical or computational sciences.

The book is divided into four main parts. In Part I, comprising Chapters 1 and 2, we consider spatially structured epidemics, with a particular attention to man–environment epidemics and vector-borne epidemics. In particular, in Chapter 1, we introduce the main ideas related to our paradigm of "THINK GLOBALLY, ACT LOCALLY" which is the *leit motiv* of the whole book. General concepts about eradicability and control of a paradigmatic epidemic model are presented, with a view to the existence of endemic states, their stability, and the existence of travelling waves. Within this excursus, a challenging open problem has been raised, concerning the possibility to control the advance of a travelling front, for which some interesting computational experiments have been presented. Moreover it has been reported about the use of nonlocal diffusion operators as an alternative to the Laplace operator which models random diffusion; anyhow a diffusion approximation of nonlocal operators is presented too.

In Chapter 2, an extension of the spatially structured epidemic model discussed in the previous chapter is presented concerning malaria as an example of vector-borne epidemics. In this chapter, the regional control problem has been analyzed, with respect to both the parameters of the model, and with respect to the region of intervention. All has been supported by a set of numerical simulations, based on the gradient method.

In Part II, comprising Chapters 3 and 4, we investigate the regional control for some optimal harvesting problems related to population dynamics. We consider the problem of maximizing the profit for structured populations with respect to both the harvesting effort and the selected subregion where the control acts.

In Chapter 3, we treat the regional optimal harvesting problem for space-structured population dynamics with a logistic term (nonlocal as well as local). Chapter 4 concerns two regional control problems related to continuous models for age-dependent populations. The first problem is devoted to an optimal control problem when the harvesting effort acts only on an age subinterval. The second one is related to a nonlinear model in a time-periodic environment. The attention is focused on the determination of the position and length of the time-subinterval when the harvesting is allowed/prohibited.

In Part III, comprising Chapters 5 and 6, recent results concerning the possible control of a devastating epidemic in olive orchards have been presented. This is an additional case of vector-borne disease in which the role of the human population is taken by olive trees, and the vector is a pest insect. In Chapter 5, the main three steps are presented, i.e., (i) the mathematical modelling of the population dynamics of the ecosystem in presence of the infection, (ii) possible strategies for the eventual eradication of the disease, and (iii) computational experiments supporting the theoretical outcomes. In Chapter 6, the role of a predator is analyzed as a possible biocontrol agent of epidemics in agriculture. The limits of this strategy are discussed with respect to specialist and generalist predators.

In Part IV, comprising only Chapter 7, as an additional application of the methods presented in this volume, we introduce an environmental issue concerning the control of the pollution produced by human activities. Based again on a spatially structured reaction diffusion model, the optimal control problem concerning the role of taxation for reducing the amount of produced pollution by human activities has been analyzed. A set of challenging open problems are submitted to interested readers.

All chapters lead to challenging open problems for the interested reader.

The Appendix includes two chapters.

In Appendix A, we have collected and deduced some mathematical definitions and results concerning Rayleigh's formula and Krein–Rutman theorem and their applications related to the principal eigenvalue and eigenvectors of elliptic and integro-elliptic problems, as well as some comparison results for integro-parabolic equations.

In Appendix B, we introduce a non-expert reader to the field of numerical approximation of partial differential equations with low-order finite element methods and to their implementation. We do not enter into the theoretical details, since our focus is to provide a simple tutorial for the implementation of the algorithms adopted throughout the book. We assume that the reader has some basic notions of numerical linear algebra.

The chosen computational platform has been MATLAB (www.mathworks.com), that has successfully emerged as one of the most used programming and numerical computing platforms for scientific computing. The MATLAB codes of most of the numerical simulations are available at the web site:

https://github.com/sn-code-inside/Mathematical-Modeling-and-Control-in-Life-and-Environmental-Sciences

Chapter by chapter, it is also provided a direct link to the code used for the corresponding numerical simulations. In the computational section of each chapter, we have also included a comprehensive description of the numerical methods implemented in the codes used for the numerical simulations.

It is worth reporting here a statement (abridged) taken from a very recent paper [156] concerning the possible role of our investigations:

> a model is only an approximate interpretation of reality and it is always wrong in some small or relevant elements. The destiny of the model presented here is to be rapidly improved thanks to novel knowledge coming from new observations and better assumptions. The Authors hope that many and more brilliant minds will read the present pages, will identify and highlight putative mistakes, will get inspiration for their research, and will produce better, more complete, and useful models........ If the speculations presented here on implications for surveillance, control, and therapy will contribute, even only minimally, to save ..... [human life - environment]........., then the Authors have accomplished their small mission.

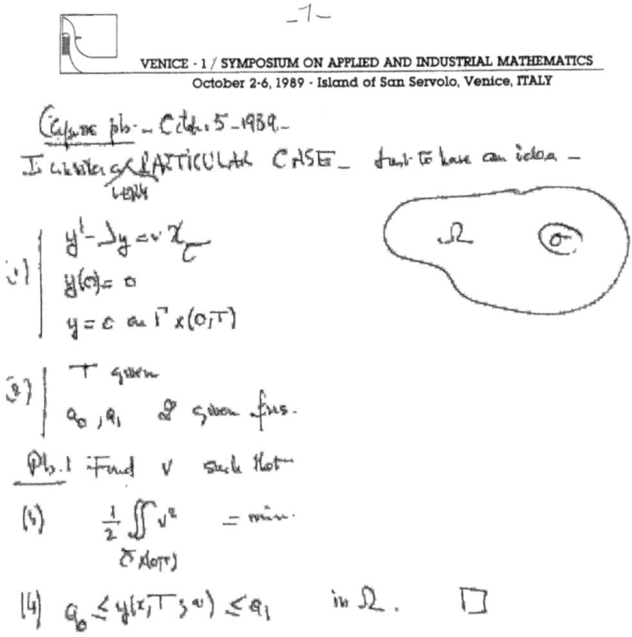

Fig. 1 Notes by Jacques L. Lions concerning a first discussion on regional control

We are indebted to many colleagues for discussions which have led to significant improvements of our results in the years.

But it is a particular duty to acknowledge the role played by the late Jacques L. Lions (see e.g. Figure 1) and by Viorel Barbu for having recognized the importance of regional control for reaction-diffusion systems, and for their consequent encouragement.

It is our duty and pleasure to acknowledge the editorial assistance of Dr. Rossana Caselli, to whom we are grateful in particular for the thorough revision of the bibliography.

Iaşi, Romania — Sebastian Aniţa
Milan, Italy — Vincenzo Capasso
Milan, Italy — Simone Scacchi
September, 2023

Contents

Nomenclature

"increasing" is used with the same meaning as "nondecreasing"; "decreasing" is used with the same meaning as "nonincreasing." In strict cases, "strictly increasing/strictly decreasing" is used.

$:=$	Equal by definition
$\equiv$	Coincide
$\emptyset$	Empty set
∇	Gradient
∂A	Boundary of a set A
$\square$	End of a proof
$\lvert a \rvert$	Absolute value of a number a, or modulus of a complex number a, or Euclidean norm on $\mathbb{R}^n$
$\lvert\lvert x \rvert\rvert$	Norm of a vector x
$\langle f, g \rangle$	Scalar product of two elements f and g in a Hilbert space
$\langle g, f \rangle_{V,V^*}$, $\langle f, g \rangle_{V^*,V}$	The duality between V and V^*
$\{x \in E : \mathcal{P}(x)\}$	set of elements of E, satisfying property $\mathcal{P}$
$\bar{A}$	Closure of a set A, depending upon context
A'	Transpose of matrix A, depending upon context
$A \setminus B$	Set of elements of A that do not belong to B
$C(A)$	Set of continuous functions from A to $\mathbb{R}$
$C(A; B)$	Set of continuous functions from A to B
$C^k(A)$	Set of functions from A to $\mathbb{R}$ with continuous derivatives up to order k

$L^p(\Omega)$	Set of equivalence classes of a.e. equal p-integrable functions with respect to the Lebesgue integral
$\\|\cdot\\|_p$ or $\\|\cdot\\|_{L^p(\Omega)}$	the usual norm in $L^p(\Omega)$
$L^\infty(\Omega)$	Set of equivalence classes of a.e. equal (with respect to the Lebesgue measure) essentially bounded functions
$\\|\cdot\\|_\infty$ or $\\|\cdot\\|_{L^\infty(\Omega)}$	the usual norm in $L^\infty(\Omega)$
$\mathbb{I}_A$	Characteristic function associated with a set A, i.e., $\mathbb{I}_A(x) = 1$, if $x \in A$, otherwise $\mathbb{I}_A(x) = 0$
a.e.	Almost everywhere
$\exp\{x\}$ or $\exp(x)$	Exponential function e^x
$f * g$	Convolution of functions f and g
f^-, f^+	Negative (positive) part of f, i.e., $f^- = \max\{-f, 0\}$ ($f^+ = \max\{f, 0\}$)
$\mathrm{sign}\{x\}$ or $\mathrm{sign}(x)$	Sign function: 1, if $x > 0$; 0, if $x = 0$; -1, if $x < 0$
$\mathrm{sgn}\{x\}$ or $\mathrm{sgn}(x)$	The multivalued sign function: 1, if $x > 0$; $[-1, 1]$, if $x = 0$; -1, if $x < 0$
$\mathbb{C}$	Complex plane
$\mathbb{N}$	Set of natural nonnegative integers
$\mathbb{N}^*$	Set of natural (strictly) positive integers
$\mathbb{R}^n$	n-dimensional Euclidean space
$\mathbb{R}_+$	Set of positive (nonnegative) real numbers
$\mathbb{R}_+^*$	Set of (strictly) positive real numbers
$\mathcal{A}$	Infinitesimal generator of a semigroup
$\mathcal{D}_A$	Domain of definition of an operator A
Δ	Laplace operator
div_x	divergence with respect to the variable x
δ_x	Dirac delta function localized at x

Part I
Regional Control of Spatially Structured Epidemics

Chapter 1
Regional Control for a Class of Spatially Structured Epidemics: *Think Globally, Act Locally*

1.1 Man–Environment Epidemics: A Nonlocal Reaction–Diffusion System

A widely accepted model for the spatial spread of epidemics in a habitat Ω, via environmental pollution produced by an infective population, e.g., via the excretion of pathogens in the environment, is the following one, as proposed in [66, 67] (see also [69] and the references therein). The model below is a more realistic generalization of a previous model proposed in [73, 75, 76] to describe fecal-orally transmitted diseases (cholera, typhoid fever, infectious hepatitis, etc.) which are typical for the European Mediterranean regions; anyhow it can be applied to other infections and other regions, which are propagated by similar mechanisms (see, e.g., [90]); schistosomiasis in Africa is a typical additional example [160].

$$\textbf{(E)} \quad \begin{cases} \dfrac{\partial u_1}{\partial t}(x,t) = d_1 \Delta u_1(x,t) - a_{11}u_1(x,t) + \displaystyle\int_\Omega k(x,x')u_2(x',t)dx', \\ \dfrac{\partial u_2}{\partial t}(x,t) = -a_{22}u_2(x,t) + g(u_1(x,t)), \end{cases} \tag{1.1}$$

in $\Omega \subset \mathbb{R}^N$ $(N \geq \mathbb{N}^*)$, a nonempty bounded domain with a sufficiently smooth boundary $\partial\Omega$; for $t \in (0, +\infty)$, where $a_{11} \geq 0$, $a_{22} \geq 0$, and $d_1 > 0$ are constants.

In this volume by "domain" we have meant a nonempty open and connected set; by the wording "sufficiently smooth" we have meant that the required regularity is satisfied, depending on the context.

- $u_1(x,t)$ denotes the concentration of the pollutant (pathogen material) at a spatial location $x \in \overline{\Omega}$ and a time $t \geq 0$.
- $u_2(x,t)$ denotes the spatial distribution of the infective population.
- The terms $-a_{11}u_1(x,t)$ and $-a_{22}u_2(x,t)$ model natural decays.

S. Aniţa et al., *Mathematical Modeling and Control in Life and Environmental Sciences*, Modeling and Simulation in Science, Engineering and Technology,
https://doi.org/10.1007/978-3-031-49971-5_1

- The total susceptible population is assumed to be sufficiently large with respect to the infective population, so that it can be taken as constant.

As an example, for fecal-orally transmitted infectious diseases, the infectious agent is multiplied by the infective human population and then sent to the environment (sea, lakes, ponds, etc.) mainly through sewage; because of the peculiar habits of the population of these regions, the agent may return via some diffusion-transport mechanism to any point of the habitat Ω, where the infection process is further activated; thus the nonlocal operator

$$\int_{\Omega} k(x, x')u_2(x', t)dx'$$

expresses the fact that the pollution produced at any point $x' \in \Omega$ of the habitat is made available at any other point $x \in \Omega$; when dealing with human pollution, this may be due to either malfunctioning of the sewage system or improper dispersal of sewage in the habitat. The linearity of the above integral operator is just a simplifying option.

The Laplace operator takes into account the pure random dispersal of the infectious agent in the habitat Ω, due to uncontrolled additional causes of dispersion (we take a constant diffusion coefficient to avoid purely technical complications).

As a further simplification, we assume that the infective population does not diffuse; possible extensions of the above model might include a more general nonlocal dispersion mechanism (see Section 1.6).

Under a suitable definition of its modelling ingredients, Model (E) in System (1.1) can be adopted as a good model for the spatial propagation of an infection in agriculture and forests, as well (see the case of *Xylella* in Chapter 5).

Concerning the local "incidence rate" of the human population at point $x \in \Omega$ and time $t \geq 0$, it is given by

$$(i.r.)(x, t) = g(u_1(x, t)),$$

depending upon the local concentration of the pollutant.

1.1.1 The "Force" of Infection

Based on the discussion originated in [77] concerning the functional response g on the *force of infection* due to the existing infective population, in [67] the following generalized form has been proposed for the force of infection due to the environmental pollution (see also [69, 127]):

$$g(z) = \frac{k_0 z^p}{\alpha_0 + \beta_0 z^q}, \qquad k_0, \alpha_0 > 0, \ \beta_0 \geq 0, \ p, q > 0. \tag{1.2}$$

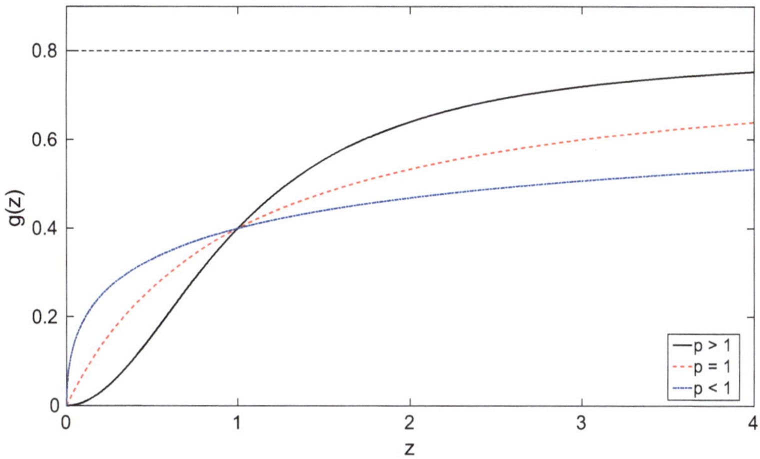

Fig. 1.1 Nonlinear forces of infection [77], according to Equation (1.2), with $k_0 = 0.8$, $\alpha_0 = \beta_0 = 1$, $p = q$, and $p = 2$ (continuous black line), $p = 1$ (dashed red line), and $p = 0.5$ (dashed-dotted blue line)

Particular cases are

$$g(z) = k_0 z^p, \qquad p > 0.$$

For the case $p = q$ we have the behaviors described in Figure 1.1.

Additional shapes of $g(z)$, as proposed in [77], which may decrease for large values of z ($p < q$ in (1.2)) may be interpreted as "awareness" effects in the contact rates (see Figure 1.2). Significant contributions to this concept and related epidemiological issues in recent literature can be found in [100].

System (1.1) has been the subject of a large literature (see, e.g., [196] and the references therein). From now on we shall call it Model (E).

1.1.2 Large Time Behavior

In this section we shall report the outcomes of [67] concerning the large time behavior of System (1.1) in a nonempty bounded domain Ω of $\mathbb{R}^2$, with a sufficiently smooth boundary $\partial\Omega$.

The first equation in (1.1) is subject to a boundary condition of the form

$$\beta(x)\frac{\partial u_1}{\partial n}(x,t) + \alpha(x)\, u_1(x,t) = 0, \quad \text{for } x \in \partial\Omega,\ t > 0, \tag{1.3}$$

where $\frac{\partial}{\partial n}$ denotes the outward normal derivative on $\partial\Omega$, and α and β are sufficiently smooth nonnegative functions such that $\alpha(x) + \beta(x) > 0$, for $x \in \partial\Omega$.

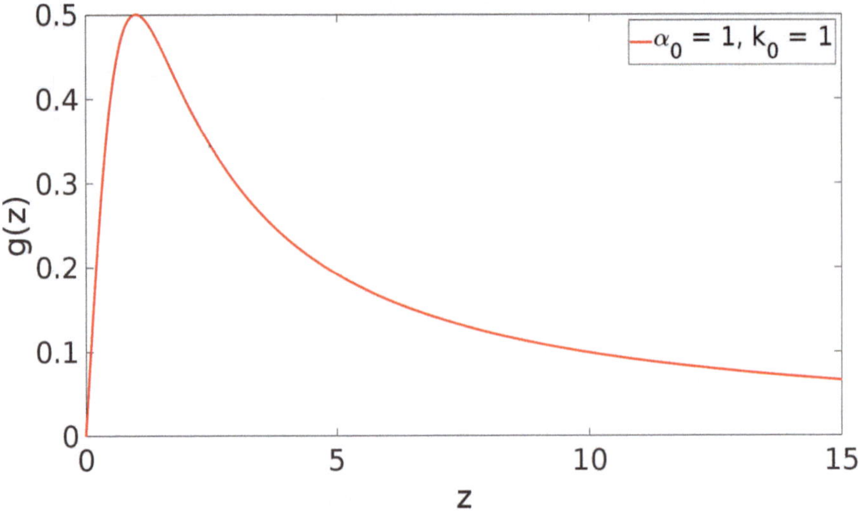

Fig. 1.2 Force of infection with behavioral response [77], according to Equation (1.2), with $k_0 = \alpha_0 = \beta_0 = 1$, $p = 1$, and $q = 2$

If $\alpha(x) = 0$ (a pure Neumann condition), (1.3) means the complete isolation of System (1.1) at $\partial\Omega$.

We shall assume that a_{11} and a_{22} are positive real constants and that $g : [0, +\infty) \longrightarrow (0, +\infty)$ satisfies the following assumptions which include most of the cases described by (1.2):

(i) If $z_1, z_2 \in [0, +\infty)$, $0 < z_1 < z_2$, then $0 < g(z_1) < g(z_2)$.
(ii) $g(0) = 0$.
(iii) g is twice continuously differentiable, and $g''(z) < 0$, for $z \in (0, +\infty)$.

Furthermore we assume that the kernel

$$k : \overline{\Omega} \times \overline{\Omega} \longrightarrow [0, +\infty)$$

is a given continuous function such that if we denote

$$a_{12} = \max_{x \in \overline{\Omega}} \int_{\Omega} k(x, x')dx',$$

then

$$\limsup_{z \to +\infty} \frac{g(z)}{z} < \frac{a_{11}\, a_{22}}{a_{12}} .$$

Let $X := C(\overline{\Omega}; \mathbb{R}^2)$ and consider the initial–boundary value problem for (1.1), supplemented by the boundary condition

$$\frac{\partial u_1}{\partial n}(x,t) = 0, \quad \text{for } x \in \partial\Omega,\ t > 0 \tag{1.4}$$

and the initial condition

$$u(\cdot, 0) = u^0 = (u_1^0, u_2^0) \in X_+, \tag{1.5}$$

where $u = (u_1, u_2)$ and

$$X_+ := \{v = (v_1, v_2) \in X : \ v_1(x), v_2(x) \geq 0, \ \text{for all } x \in \overline{\Omega}\}.$$

Definition 1.1 We say that $u = (u_1, u_2)$ is a (classical) solution to the above mentioned initial–boundary value problem (1.1), (1.4), (1.5) if

$$u \in \left(C^{2,1}(\Omega \times (0, +\infty)) \cap C^{1,0}(\overline{\Omega} \times (0, +\infty))\right)$$
$$\times \left(C^{0,1}(\Omega \times (0, +\infty)) \cap C(\overline{\Omega} \times (0, +\infty))\right)$$

satisfies (1.1), (1.4), and

$$\lim_{t \to 0+} u(x,t) = u^0(x), \quad \text{for any } x \in \Omega.$$

It can be shown the following existence result (see [67] and [112] for the case of even more general boundary conditions).

Theorem 1.2 *There exists a unique (classical) solution $u = (u_1, u_2)$ of the above problem. Both components of the solution are nonnegative.*

For a matter of simplicity, in (1.3) we have assumed that $\alpha(x) = 0$ and $\beta(x) = 1$, for $x \in \partial\Omega$. In this case the eigenvalue problem below

$$\begin{cases} -\Delta\phi(x) = \lambda\phi(x), & x \in \Omega, \\ \dfrac{\partial\phi}{\partial n}(x) = 0, & x \in \partial\Omega \end{cases}$$

admits $\lambda_1 = 0$ as a principal eigenvalue; as a corresponding eigenfunction, we may choose

$$\phi_1 \equiv 1, \quad x \in \overline{\Omega}.$$

Let us denote

$$H(u_2)(x) = \int_\Omega k(x,x')\, u_2(x')dx', \qquad u_2 \in C(\overline{\Omega}),\ x \in \overline{\Omega}.$$

For $\phi_1 \equiv 1$,

$$H(\phi_1)(x) = \int_\Omega k(x, x')\phi_1(x')dx' = \int_\Omega k(x, x')dx'.$$

We shall denote

$$\overline{H} = \max_{x\in\overline{\Omega}} \int_\Omega k(x, x')dx', \quad \underline{H} = \min_{x\in\overline{\Omega}} \int_\Omega k(x, x')dx'.$$

We may notice that if $k(x, x')$ is an isotropic symmetric probability density function, multiplied by a_{12},

$$\overline{H} = \underline{H} = a_{12}.$$

Under the above assumptions Theorem 3.1 in [67] becomes

Theorem 1.3 *If*

$$\theta_1 := \frac{g'(0)a_{12}}{a_{11}\, a_{22}} < 1,$$

then the trivial solution is globally asymptotically stable in X_+ for System (1.1).
If $\theta_1 > 1$, then the trivial solution is unstable for System (1.1).

For the case of more general boundary conditions, we refer to [67]. See later for the possible existence of travelling waves in an unbounded domain.

1.2 Regional Control

Let us consider System (1.1) in $\Omega \subset \mathbb{R}^2$, a nonempty bounded domain with a sufficiently smooth boundary $\partial\Omega$; for $t \in (0, T)$, where $a_{11} \geq 0$, $a_{22} > 0$, $d_1 > 0$, and $T > 0$ are constants.

The public health concern consists of reducing the disease in the relevant habitat, as fast as possible, at an optimal cost. The underlying issue of our presentation is that very often the entire domain of interest for the epidemic is either unknown or difficult to manage, for an affordable implementation of suitable environmental sanitation programs. Think of malaria, schistosomiasis, and alike, in Africa, Asia, etc.

This has led the authors [VC and SA] of this monograph to face the problem of an effective diminishment of an epidemic in the whole habitat Ω by implementing possible control strategies only in a suitably chosen subregion $\omega \subset \Omega$ (Figure 1.3). A satisfactory mathematical treatment of this issue has been obtained in [10] (see

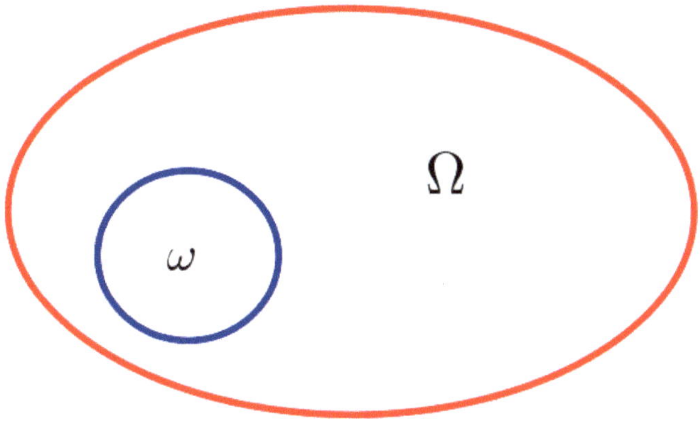

Fig. 1.3 Think globally, act locally

also [11, 14]). This practice may have an enormous importance in real cases with respect to both financial and practical affordability. Furthermore, since we propose to act on the elimination of the pollution only, this practice means an additional nontrivial social benefit on the human population, since it would not be limited in his personal and social habits.

Let $\omega \subset \Omega$ be a nonempty open subset; we denote by $\mathbb{I}_\omega$ the characteristic function of ω and use the convention

$$\mathbb{I}_\omega(x)w(x) = 0, \qquad x \in \mathbb{R}^2 \setminus \omega,$$

even if function w is not defined on the whole set $\mathbb{R}^2 \setminus \omega$.

Our goal is to study System (1.1) subject to suitable controls, as follows:

$$\begin{cases} \dfrac{\partial u_1}{\partial t}(x,t) - d_1 \Delta u_1(x,t) = -a_{11}u_1(x,t) + \displaystyle\int_\Omega k(x,x')u_2(x',t)dx' \\ \qquad -\gamma_1(x,t)\mathbb{I}_\omega(x)u_1(x,t), & (x,t) \in Q, \\ \dfrac{\partial u_2}{\partial t}(x,t) = -(a_{22} + \gamma_2(x,t)\mathbb{I}_\omega(x))u_2(x,t) \\ \qquad +(1 - \gamma_3(x,t)\mathbb{I}_\omega(x))g(u_1(x,t)), & (x,t) \in Q, \end{cases}$$

subject to suitable boundary and initial conditions. Here $Q = \Omega \times (0,T)$, $\Sigma = \partial\Omega \times (0,T)$.

The idea underlying the controls is the following:

$$\gamma_1(x,t)u_1(x,t)\mathbb{I}_\omega(x)$$

represents the sanification of the environment, say by the combined action of different chemical and/or mechanical practices, in the subregion ω; $\gamma_1(x,t)u_1(x,t)$

represents the corresponding reduction of the polluting agent at location $x \in \Omega$ and time $t > 0$.

$$\gamma_2(x,t)u_2(x,t)\mathbb{I}_\omega(x)$$

describes the medical treatment of the infected population in the subregion ω; $\gamma_2(x,t)u_2(x,t)$ is then the corresponding healed population at location $x \in \Omega$ and time $t > 0$.

$$\gamma_3(x,t)\mathbb{I}_\omega(x)$$

is the reduction of the contact rate of humans–environment by means of behavioral and/or distancing practices, implemented in the subregion ω.

Each of the above-mentioned control practices implies a cost to be taken into account in the total cost functional for optimal control problems. Below we have, respectively, denoted by $\zeta_1(\gamma_1)$ the cost for obtaining the actual control γ_1, by $\zeta_2(\gamma_2 u_2)$ the cost of the treatment per individual, and by $\zeta_3(\gamma_3)$ the cost for obtaining the actual control γ_3.

When the costs of the controls to reduce the epidemic of the damages produced by the disease and of the intervention in the subset ω in the time interval $[0, T]$ are considered, then we get the total cost function:

$$\begin{aligned} J(\gamma,\omega) &= \int_0^T\int_\Omega C(x)\zeta_1(\gamma_1(x,t))\mathbb{I}_\omega(x)dx\,dt \\ &+ \int_0^T\int_\Omega \zeta_2(\gamma_2(x,t)u_2(x,t))\gamma_2(x,t)u_2(x,t)\mathbb{I}_\omega(x)dx\,dt \\ &+ \int_0^T\int_\Omega C(x)\zeta_3(\gamma_3(x,t))\mathbb{I}_\omega(x)dx\,dt \\ &+ \int_0^T\int_\Omega l(u_2(x,t))dx\,dt + \alpha_1 \text{ area}(\omega) + \beta_1 \text{ perimeter}(\omega). \end{aligned} \tag{1.6}$$

Here ω is a subset of Ω and $\gamma = (\gamma_1, \gamma_2, \gamma_3)$ belongs to a suitable set $\mathcal{G}_T$ of feasible values of the $\gamma's$, and $C(x)$ is the spatial density of the total population. As far as the last three terms in (1.6) are concerned, the term $\int_0^T\int_\Omega l(u_2(x,t))dx\,dt$ is meant to take into account the costs, over the whole domain and the total observation/control period, deriving from loss of work hours, hospitalization, drugs, etc., while the terms α_1 area$(\omega) + \beta_1$ perimeter(ω) take into account costs of transport of intervention devices (nets, chemicals, personnel, etc.) which may depend on the geometry of the subregion of intervention ω. For our scopes, the geometry of a planar set is indeed characterized by its area and perimeter, the coefficients α_1 and β_1 take into account the specific logistic structure of the habitat, and we may assume $\alpha_1, \beta_1 \geq 0$. This aspect has been taken care in some detail for the malaria model in Chapter 2.

1.3 Eradicability

In this section we face the problem of possible eradicability of an epidemic modelled by System (1.1) by just reducing the concentration of the pollutant in a nonempty and sufficiently large subset of the spatial domain. Later, in applications, relevant optimal control problems will be presented, as mentioned in the previous section. Here we will show that, under suitable assumptions, it is indeed possible to diminish exponentially the epidemic process.

Consider again the man–environment epidemic system in $\Omega \subset \mathbb{R}^N$ $(N \geq 1)$, a nonempty bounded domain with a sufficiently smooth boundary $\partial\Omega$. Let $\omega \subset \Omega$ be a nonempty bounded domain with a C^2-boundary, such that $\overline{\omega} \subset \Omega$ and $\Omega \setminus \overline{\omega}$ a domain. Denote by $\mathbb{I}_\omega$ the characteristic function of ω. We use the convention

$$\mathbb{I}_\omega(x)h(x) = 0, \qquad x \in \mathbb{R}^N \setminus \omega,$$

even if function h is not defined on the whole set $\mathbb{R}^N \setminus \omega$.

Our goal is to study if there exists a control $v \in L^\infty_{loc}(\overline{\omega} \times [0, +\infty))$ (which implies that $supp(v(t)) \subset \overline{\omega}$, for $t \geq 0$) such that the (weak) solution (u_1, u_2) of the following system

$$\begin{cases} \dfrac{\partial u_1}{\partial t}(x,t) = d_1 \Delta u_1(x,t) - a_{11} u_1(x,t) + \displaystyle\int_\Omega k(x,x') u_2(x',t) dx' \\ \qquad + \mathbb{I}_\omega(x)\, v(x,t), & (x,t) \in \Omega \times (0,+\infty), \\ \dfrac{\partial u_2}{\partial t}(x,t) = -a_{22} u_2(x,t) + g(u_1(x,t)), & (x,t) \in \Omega \times (0,+\infty), \\ \dfrac{\partial u_1}{\partial n}(x,t) + \alpha u_1(x,t) = 0, & (x,t) \in \partial\Omega \times (0,+\infty), \\ u_1(x,0) = u_1^0(x), \quad u_2(x,0) = u_2^0(x), & x \in \Omega \end{cases} \tag{1.7}$$

satisfies

$$u_1(x,t) \geq 0, \quad u_2(x,t) \geq 0, \qquad \text{a.e. } x \in \Omega, \text{ for any } t \geq 0, \tag{1.8}$$

and

$$\lim_{t\to+\infty} \|u_1(t)\|_{L^\infty(\Omega)} = \lim_{t\to+\infty} \|u_2(t)\|_{L^\infty(\Omega)} = 0. \tag{1.9}$$

Here $d_1, a_{22} \in (0, +\infty)$ and $a_{11}, \alpha \in [0, +\infty)$ are constants.

We work under the following assumptions:

(H1) $g : \mathbb{R} \to [0, +\infty)$ is a function satisfying

$g(x) = 0, \quad$ for $x \in (-\infty, 0]$,
g is Lipschitz continuous and increasing,
$g(r) \leq a_{21} r$, for any $r \in [0, +\infty)$, where $a_{21} > 0$ is a constant.

(H2) $k \in L^{\infty}(\Omega \times \Omega),\ k(x, x') \geq 0$, a.e. in $\Omega \times \Omega$,

$$\int_{\Omega} k(x, x')dx > 0, \qquad \text{a.e. } x' \in \Omega.$$

(H3) $u_1^0,\ u_2^0 \in L^{\infty}(\Omega), \quad u_1^0(x),\ u_2^0(x) \geq 0$, a.e. in Ω.

Definition 1.4 We say that (u_1, u_2) is a (weak) solution to (1.7) if for any $T \in (0, +\infty)$,

$u_1 \in C([0, T]; L^2(\Omega)) \cap AC((0, T]; L^2(\Omega)) \cap L^2(0, T; H^1(\Omega)) \cap L^2_{loc}((0, T]; H^2(\Omega))$,

$u_2 \in L^2(\Omega \times (0, T))$ such that for almost any $x \in \Omega$, $u_2(x, \cdot) \in AC([0, T])$,

$$\begin{cases} \dfrac{\partial u_1}{\partial t}(x,t) = d_1 \Delta u_1(x,t) - a_{11}u_1(x,t) + \displaystyle\int_{\Omega} k(x,x')u_2(x',t)dx' \\ \qquad + \mathbb{I}_{\omega}(x)\, v(x,t), & \text{a.e. } (x,t) \in \Omega \times (0,+\infty), \\ \dfrac{\partial u_2}{\partial t}(x,t) = -a_{22}u_2(x,t) + g(u_1(x,t)), & \text{a.e. } (x,t) \in \Omega \times (0,+\infty), \\ \dfrac{\partial u_1}{\partial n}(x,t) + \alpha u_1(x,t) = 0, & \text{a.e. } (x,t) \in \partial\Omega \times (0,+\infty), \\ u_1(x,0) = u_1^0(x), \quad u_2(x,0) = u_2^0(x), & \text{a.e. } x \in \Omega. \end{cases}$$

Let us prove that Problem (1.7) admits a unique (weak) solution (u_1, u_2). For an arbitrary but fixed $T \in (0, +\infty)$, consider the space

$$V = C([0, T]; L^2(\Omega)) \times C([0, T]; L^2(\Omega)),$$

with the distance

$$d((\zeta_1, \zeta_2), (\zeta_3, \zeta_4)) = \sup_{t \in [0,T]} \left(e^{-\xi t} (\|\zeta_1(t) - \zeta_3(t)\|^2_{L^2(\Omega)} + \|\zeta_2(t) - \zeta_4(t)\|^2_{L^2(\Omega)}) \right)^{\frac{1}{2}}.$$

Here ξ is a positive constant to be made precise later. (V, d) is a complete metric space (actually it is even a Banach space).

For any $(y_1, y_2) \in V$, let $\mathcal{T}(y_1, y_2)$ be the weak solution to

$$\begin{cases} \dfrac{\partial u_1}{\partial t}(x,t) = d_1 \Delta u_1(x,t) - a_{11}u_1(x,t) + \displaystyle\int_{\Omega} k(x,x')y_2(x',t)dx' \\ \qquad + \mathbb{I}_{\omega}(x)v(x,t), & (x,t) \in \Omega \times (0,T), \\ \dfrac{\partial u_1}{\partial n}(x,t) + \alpha u_1(x,t) = 0, & (x,t) \in \partial\Omega \times (0,T), \\ \dfrac{\partial u_2}{\partial t}(x,t) = -a_{22}u_2(x,t) + g(y_1(x,t)), & (x,t) \in \Omega \times (0,T), \\ u_1(x,0) = u_1^0(x), \quad u_2(x,0) = u_2^0(x), & x \in \Omega. \end{cases} \tag{1.10}$$

Problem (1.10) contains a linear parabolic equation and a collection of ordinary differential equations. Notice that $\int_{\Omega} k(x, x')y_2(x', t)dx' \in L^2(\Omega \times (0, T))$ and

$g(y_1(x,t)) \in L^2(\Omega \times (0,T))$. It follows that there exists a unique weak solution to (1.10).

For any $(y_1, y_2), (\tilde{y}_1, \tilde{y}_2) \in V$, let $(u_1, u_2) = \mathcal{T}(y_1, y_2)$ be the weak solution to (1.10) and $(\tilde{u}_1, \tilde{u}_2) = \mathcal{T}(\tilde{y}_1, \tilde{y}_2)$ be the weak solution to (1.10) corresponding to $y_i := \tilde{y}_i$. We get that $(w_1, w_2) = (u_1 - \tilde{u}_1, u_2 - \tilde{u}_2)$ is the weak solution to

$$\begin{cases} \dfrac{\partial w_1}{\partial t}(x,t) = d_1 \Delta w_1(x,t) - a_{11} w_1(x,t) + \displaystyle\int_\Omega k(x,x') \ (y_2 - \tilde{y}_2)(x',t)dx', \\ \qquad\qquad\qquad\qquad\qquad\qquad\qquad\qquad (x,t) \in \Omega \times (0,T), \\ \dfrac{\partial w_1}{\partial n}(x,t) + \alpha w_1(x,t) = 0, \qquad (x,t) \in \partial\Omega \times (0,T), \\ \dfrac{\partial w_2}{\partial t}(x,t) = -a_{22} w_2(x,t) + g(y_1(x,t)) - g(\tilde{y}_1(x,t)), \ (x,t) \in \Omega \times (0,T), \\ u_1(x,0) = 0, \quad u_2(x,0) = 0, \qquad x \in \Omega. \end{cases}$$

Let us multiply the equation in w_1 by w_1 and the equation in w_2 by w_2, add them, and integrate over $\Omega \times (0,t)$ $(t \in [0,T])$. We get after an easy calculation that

$$\begin{aligned} &\frac{1}{2}\|w_1(t)\|^2_{L^2(\Omega)} + \frac{1}{2}\|w_2(t)\|^2_{L^2(\Omega)} \\ &\le \int_0^t \int_\Omega w_1(x,s) \int_\Omega k(x,x')(y_2 - \tilde{y}_2)(x',s)dx' \, dx \, ds \\ &\quad + \int_0^t \int_\Omega w_2(x,s)(g(y_1(x,s)) - g(\tilde{y}_1(x,s)))dx \, ds \\ &\le \|k\|_{L^2(\Omega\times\Omega)} \int_0^t \|w_1(s)\|_{L^2(\Omega)} \|y_2(s) - \tilde{y}_2(s)\|_{L^2(\Omega)} ds \\ &\quad + L \int_0^t \|w_2(s)\|_{L^2(\Omega)} \|y_1(s) - \tilde{y}_1(s)\|_{L^2(\Omega)} ds \end{aligned}$$

(here L is a Lipschitz constant for g). So, there exists a positive constant C such that for any $t \in [0,T]$,

$$\begin{aligned} &\|w_1(t)\|^2_{L^2(\Omega)} + \|w_2(t)\|^2_{L^2(\Omega)} \\ &\le C \int_0^t (\|y_1(s) - \tilde{y}_1(s)\|^2_{L^2(\Omega)} + \|y_2(s) - \tilde{y}_2(s)\|^2_{L^2(\Omega)}) ds \\ &\le C \int_0^t e^{\xi t} e^{-\xi s} (\|y_1(s) - \tilde{y}_1(s)\|^2_{L^2(\Omega)} + \|y_2(s) - \tilde{y}_2(s)\|^2_{L^2(\Omega)}) ds \\ &\le \frac{C}{\xi} e^{\xi t} d((y_1, y_2), (\tilde{y}_1 - \tilde{y}_2))^2. \end{aligned}$$

It follows that

$$d(\mathcal{T}(y_1, y_2), \mathcal{T}(\tilde{y}_1, \tilde{y}_2))^2 \leq \frac{C}{\xi} d((y_1, y_2), (\tilde{y}_1 - \tilde{y}_2))^2.$$

If we choose $\xi > C$, then $\mathcal{T}$ is a contraction and so by Banach's fixed point theorem it has a unique fixed point.

We may conclude that (1.7) has a unique weak solution (u_1, u_2).

Our hypothesis on v does not assure the nonnegativity of the components u_1 and u_2. However, the nonnegativity of u_1 and u_2 is a natural requirement (due to the biological significance of u_1 and u_2). The nonnegativity of the components of the weak solution follows for different choices of v via the comparison results for parabolic equations and applying Banach's fixed point result in (V_+, d), where

$$V_+ = \{(\zeta_1, \zeta_2) \in V : 0 \leq \zeta_1(x,t),\ \zeta_2(x,t),\ \text{a.e. } x \in \Omega,\ \text{for any } t \in [0,T]\}.$$

Definition 1.5 We say that the disease modelled by (1.7) is *eradicable* if, for any u_0^1 and u_0^2 satisfying (H3), there exists $v \in L^\infty_{loc}(\overline{\omega} \times [0, +\infty))$ such that (1.7) admits a unique weak solution (u_1, u_2) with nonnegative components (i.e., satisfying (1.8)) and satisfying (1.9).

Preliminary eradicability results for System (1.7) have been obtained in [10]. For eradicability (stabilizability) results related to some reaction–diffusion systems in population dynamics, we refer to [1, 4, 10–15, 17, 27], while for general distributed systems see [149].

A strong eradicability result (which provides a stabilizing control with a simple structure) has been established in [4], using a different approach from the one we will show in this section. Here we will report the results obtained in [11], with simplified proofs, providing a stabilizing feedback control with a very simple structure. This is an easily implementable feedback control.

1.3.1 An Eigenvalue Problem

For an arbitrary but fixed constant $\gamma > 0$, consider the following eigenvalue problem associated with Problem (1.7):

$$\begin{cases} -d_1 \Delta\varphi(x) + a_{11}\varphi(x) - \displaystyle\int_\Omega k(x,x')\psi(x')dx' \\ \qquad + \mathbb{I}_\omega(x)\gamma\varphi(x) = \lambda\varphi(x), & x \in \Omega, \\ \dfrac{\partial\varphi}{\partial n}(x) + \alpha\varphi(x) = 0, & x \in \partial\Omega, \\ -a_{21}\varphi(x) + a_{22}\psi(x) = 0, & x \in \Omega. \end{cases} \tag{1.11}$$

Problem (1.11) is equivalent to

$$\begin{cases} -d_1\Delta\varphi(x) + a_{11}\varphi(x) - \dfrac{a_{21}}{a_{22}}\displaystyle\int_\Omega k(x,x')\varphi(x')dx' \\ \qquad + \mathbb{I}_\omega(x)\gamma\varphi(x) = \lambda\varphi(x), & x \in \Omega, \\ \dfrac{\partial\varphi}{\partial n}(x) + \alpha\varphi(x) = 0, & x \in \partial\Omega, \\ \psi(x) = \dfrac{a_{21}}{a_{22}}\varphi(x), & x \in \Omega. \end{cases} \tag{1.12}$$

By Krein–Rutman's theorem (see Appendix A), it follows that Problem (1.12), hence (1.11), admits a real principal eigenvalue $\lambda_{1\gamma}$ and a corresponding strictly positive eigenvector in $\tilde{\varphi}$, i.e., $\tilde{\varphi} \geq \theta_0$, for $x \in \overline{\Omega}$, where θ is a positive constant (see Appendix A and [11]).

$\lambda_{1\gamma}$ is a simple eigenvalue for (1.11) and (1.12); there is no other eigenvalue with nonnegative eigenvectors. $\lambda_{1\gamma}$ is an eigenvalue (also) for

$$\begin{cases} -d_1\Delta\varphi^*(x) + a_{11}\varphi^*(x) + \mathbb{I}_\omega(x)\gamma\varphi^*(x) \\ \qquad - \dfrac{a_{21}}{a_{22}}\displaystyle\int_\Omega k(x',x)\varphi^*(x')dx' = \lambda\varphi^*(x), \ x \in \Omega, \\ \dfrac{\partial\varphi^*}{\partial n}(x) + \alpha\varphi^*(x) = 0, & x \in \partial\Omega, \end{cases} \tag{1.13}$$

and $\tilde{\varphi}^*$ is a corresponding positive eigenvector.

We may define and deduce similar results for the principal eigenvalue $\lambda_1(\omega)$ and for corresponding eigenvectors

$$\begin{cases} -d_1\Delta\varphi(x) + a_{11}\varphi(x) - \dfrac{a_{21}}{a_{22}}\displaystyle\int_{\Omega\setminus\overline{\omega}} k(x,x')\varphi(x')dx' = \lambda\varphi(x), \ x \in \Omega\setminus\overline{\omega}, \\ \varphi(x) = 0, & x \in \partial\omega, \\ \dfrac{\partial\varphi}{\partial n}(x) + \alpha\varphi(x) = 0, & x \in \partial\Omega \end{cases}$$

and for the adjoint problem

$$\begin{cases} -d_1\Delta\varphi^*(x) + a_{11}\varphi^*(x) - \dfrac{a_{21}}{a_{22}}\displaystyle\int_{\Omega\setminus\overline{\omega}} k(x',x)\varphi^*(x')dx' = \lambda\varphi^*(x), \ x \in \Omega\setminus\overline{\omega}, \\ \varphi^*(x) = 0, & x \in \partial\omega, \\ \dfrac{\partial\varphi^*}{\partial n}(x) + \alpha\varphi^*(x) = 0, & x \in \partial\Omega. \end{cases}$$

The following lemma follows via Krein–Rutman's theorem as well (see Appendix A).

Lemma 1.6 *The mapping* $\gamma \mapsto \lambda_{1\gamma}$ *is increasing on* $(0, +\infty)$ *and*

$$\lim_{\gamma \to +\infty} \lambda_{1\gamma} = \lambda_1(\omega).$$

Remark 1.7 By similar arguments we may deduce that the mapping $\gamma \mapsto \lambda_{1\gamma}$ is continuous on $(0, +\infty)$ and that

$$\lim_{\varepsilon \to 0+} \lambda_{1\gamma}(\varepsilon) = \lambda_{1\gamma},$$

where $\lambda_{1\gamma}(\varepsilon)$ is the principal eigenvalue for (1.12) and (1.13), corresponding to $a_{22} := a_{22} - \varepsilon$ (for any $0 < \varepsilon < a_{22}$).

1.3.2 The Main Result

Theorem 1.8 *If* $\lambda_1(\omega) > 0$*, then for* $\gamma > 0$ *sufficiently large, the feedback control* $v := -\gamma u_1$ *makes the disease eradicable.*

Conversely, if g *is differentiable at* 0 *and* $g'(0) = a_{21}$ *and if the disease is eradicable, then* $\lambda_1(\omega) \geq 0$.

Proof If $\lambda_1(\omega) > 0$, then by Lemma 1.6 we deduce that for $\gamma > 0$ sufficiently large, $\lambda_{1\gamma} > 0$. In addition, for $\varepsilon \in (0, a_{22})$ small enough, we have that (see Remark 1.7)

$$\lambda_{1\gamma}(\varepsilon) > 0.$$

Let us take such a γ and such an ε. Let us note first that it is possible to prove in the same manner as for (1.7) that the next problem admits a unique weak solution (u_1, u_2).

$$\begin{cases} \dfrac{\partial u_1}{\partial t}(x,t) = d_1 \Delta u_1(x,t) - a_{11}u_1(x,t) + \displaystyle\int_\Omega k(x,x')u_2(x',t)dx' & \\ \qquad - \mathbb{I}_\omega(x)\gamma u_1(x,t), & (x,t) \in \Omega \times (0,+\infty), \\ \dfrac{\partial u_1}{\partial n}(x,t) + \alpha u_1(x,t) = 0, & (x,t) \in \partial\Omega \times (0,+\infty), \\ \dfrac{\partial u_2}{\partial t}(x,t) = -a_{22}u_2(x,t) + g(u_1(x,t)), & (x,t) \in \Omega \times (0,+\infty), \\ u_1(x,0) = u_1^0(x), \quad u_2(x,0) = u_2^0(x), & x \in \Omega. \end{cases} \tag{1.14}$$

Let us prove that, actually, u_1 and u_2 are nonnegative and satisfy additional qualitative properties as well.

If we multiply the first equation by u_1 and the second one by u_2, add them, and integrate over $\Omega \times (0, t)$, we get that

$$\frac{1}{2}\int_{\Omega}|u_1(x,t)|^2dx+\frac{1}{2}\int_{\Omega}|u_2(x,t)|^2dx\le\frac{1}{2}\int_{\Omega}|u_1^0(x)|^2dx+\frac{1}{2}\int_{\Omega}|u_2^0(x)|^2dx$$
$$+\int_0^t\int_{\Omega}|u_1(x,s)|\int_{\Omega}k(x,x')|u_2(x',s)|dx'\,dx\,ds$$
$$+a_{21}\int_0^t\int_{\Omega}|u_1(x,s)||u_2(x,s)|dx\,ds$$

$$\le\frac{1}{2}\|u_1^0\|^2_{L^2(\Omega)}+\frac{1}{2}\|u_2^0\|^2_{L^2(\Omega)}+\int_0^t\|u_1(s)\|_{L^2(\Omega)}\|u_2(s)\|_{L^2(\Omega)}\|k\|_{L^2(\Omega\times\Omega)}ds$$
$$+a_{21}\int_0^t\|u_1(s)\|_{L^2(\Omega)}\|u_2(s)\|_{L^2(\Omega)}ds.$$

Let $T>0$ be arbitrary, but fixed. We get that for any $t\in[0,T]$:

$$\|u_1(t)\|^2_{L^2(\Omega)}+\|u_2(t)\|^2_{L^2(\Omega)}\le\|u_1^0\|^2_{L^2(\Omega)}+\|u_2^0\|^2_{L^2(\Omega)}$$
$$+(\|k\|_{L^2(\Omega\times\Omega)}+a_{21})\int_0^t(\|u_1(s)\|^2_{L^2(\Omega)}+\|u_2(s)\|^2_{L^2(\Omega)})ds,$$

and so we get via Gronwall's inequality that

$$\|u_1(t)\|^2_{L^2(\Omega)}+\|u_2(t)\|^2_{L^2(\Omega)}\le(\|u_1^0\|^2_{L^2(\Omega)}+\|u_2^0\|^2_{L^2(\Omega)})e^{(\|k\|_{L^2(\Omega\times\Omega)}+a_{21})T}=M_1.$$

Since $g(u_1)$ is nonnegative, we may infer that for any $t\in[0,T]$ and for almost any $x\in\Omega$,

$$u_2(x,t)=e^{-a_{22}t}u_2^0(x)+\int_0^t e^{-a_{22}(t-s)}g(u_1(x,s))ds\ge 0.$$

It follows that for any $t\in[0,T]$ and for almost any $x\in\Omega$,

$$0\le\int_{\Omega}k(x,x')u_2(x',t)dx'\le\|k\|_{L^\infty(\Omega\times\Omega)}\,\text{meas}(\Omega)^{\frac{1}{2}}M_1^{\frac{1}{2}}=M_2$$

($\text{meas}(\Omega)$ is the Lebesgue measure of Ω).

The comparison result for parabolic equations (as in Appendix A) implies that

$$0\le u_1(x,t)\le\overline{u}_1(t),\quad \text{a.e. } x\in\Omega,\ \text{for any } t\in[0,T],$$

where $\overline{u}_1$ is the solution to

$$\begin{cases}\overline{u}_1'(t)=M_2, & t\in(0,T),\\ \overline{u}_1(0)=\|u_1^0\|_{L^\infty(\Omega)}.\end{cases}$$

Hence, $u_1 \in L^\infty(\Omega \times (0,T))$,

$$0 \le u_1(x,t) \le \|u_1^0\|_{L^\infty(\Omega)} + M_2 T = M_3, \quad \text{a.e. } x \in \Omega, \text{ for any } t \in [0,T].$$

It follows that

$$0 \le u_2(x,t) \le \|u_2^0\|_{L^\infty(\Omega)} + a_{21} M_3 T = M_4, \quad \text{a.e. } x \in \Omega, \text{ for any } t \in [0,T].$$

Since $g(r) \le a_{21} r$, for all $r \ge 0$, we can get by a comparison theorem (as in Appendix A) that

$$\begin{cases} 0 \le u_1(x,t) \le y(x,t), \text{ a.e. } x \in \Omega, \text{ for any } t \ge 0, \\ 0 \le u_2(x,t) \le z(x,t), \text{ a.e. } x \in \Omega, \text{ for any } t \ge 0, \end{cases} \tag{1.15}$$

where (y,z) is the solution to

$$\begin{cases} \frac{\partial y}{\partial t}(x,t) = d_1 \Delta y(x,t) - a_{11} y(x,t) + \int_\Omega k(x,x')\, z(x',t) dx' \\ \qquad - \mathbb{I}_\omega(x) \gamma y(x,t), & (x,t) \in \Omega \times (0,+\infty), \\ \frac{\partial y}{\partial n}(x,t) + \alpha y(x,t) = 0, & (x,t) \in \partial\Omega \times (0,+\infty), \\ \frac{\partial z}{\partial t}(x,t) = -a_{22} z(x,t) + a_{21} y(x,t), & (x,t) \in \Omega \times (0,+\infty), \\ y(x,0) = u_1^0(x), \quad z(x,0) = u_2^0(x), & x \in \Omega. \end{cases} \tag{1.16}$$

Let φ_ε be a positive eigenvector to (1.12) corresponding to $\lambda := \lambda_{1\gamma}(\varepsilon)$, and $a_{22} := a_{22} - \varepsilon$, $(\varepsilon \in (0, a_{22}))$, i.e., φ_ε satisfies

$$\begin{cases} -d_1 \Delta \varphi(x) + a_{11} \varphi(x) - \frac{a_{21}}{a_{22} - \varepsilon} \int_\Omega k(x,x')\, \varphi(x') dx' \\ \qquad + \mathbb{I}_\omega(x) \gamma \varphi(x) = \lambda_{1\gamma}(\varepsilon) \varphi(x), & x \in \Omega, \\ \frac{\partial \varphi}{\partial n}(x) + \alpha \varphi(x) = 0, & x \in \partial\Omega \end{cases}$$

and is a weak solution to

$$\begin{cases} \frac{\partial \varphi}{\partial t}(x,t) = d_1 \Delta \varphi(x,t) - a_{11} \varphi(x,t) + \frac{a_{21}}{a_{22} - \varepsilon} \int_\Omega k(x,x') \varphi(x',t) dx' \\ \qquad - \mathbb{I}_\omega(x) \gamma \varphi(x,t) + \lambda_{1\gamma}(\varepsilon) \varphi(x,t), & (x,t) \in \Omega \times (0,+\infty), \\ \frac{\partial \varphi}{\partial n}(x,t) + \alpha \varphi(x,t) = 0, & (x,t) \in \partial\Omega \times (0,+\infty), \\ \varphi(x,0) = \varphi_\varepsilon(x), & x \in \Omega, \end{cases}$$

as well.

By comparison's principle we get that

$$\varphi_\varepsilon(x) \ge \overline{\varphi}(x,t), \quad \text{a.e. } x \in \Omega, \text{ for any } t \ge 0,$$

where $\overline{\varphi}$ is the weak solution to

$$\begin{cases} \dfrac{\partial \overline{\varphi}}{\partial t}(x,t) = d_1 \Delta \overline{\varphi}(x,t) \\ \qquad - (a_{11} + \gamma - \lambda_{1\gamma}(\varepsilon))\overline{\varphi}(x,t), \; (x,t) \in \Omega \times (0,+\infty), \\ \dfrac{\partial \overline{\varphi}}{\partial n}(x,t) + \alpha \overline{\varphi}(x,t) = 0, \qquad (x,t) \in \partial\Omega \times (0,+\infty), \\ \overline{\varphi}(x,0) = \varphi_\varepsilon(x), \qquad x \in \Omega. \end{cases}$$

We get that

$$\overline{\varphi}(x,t) = e^{-(a_{11}+\gamma-\lambda_{1\gamma}(\varepsilon))t} \varphi_*(x,t),$$

where φ_* is the weak solution to

$$\begin{cases} \dfrac{\partial \varphi_*}{\partial t}(x,t) = d_1 \Delta \varphi_*(x,t), \qquad (x,t) \in \Omega \times (0,+\infty), \\ \dfrac{\partial \varphi_*}{\partial n}(x,t) + \alpha \varphi_*(x,t) = 0, \quad (x,t) \in \partial\Omega \times (0,+\infty), \\ \varphi_*(x,0) = \varphi_\varepsilon(x), \qquad x \in \Omega. \end{cases}$$

A classical result (see [154]) implies that for any $t_0 > 0$, there exists $\theta_{t_0} > 0$ such that

$$\varphi_*(x,t_0) \ge \theta_{t_0}, \quad \text{for any } x \in \overline{\Omega}.$$

This implies that there exists $\theta_1^0 > 0$ such that

$$\varphi_\varepsilon(x) \ge \theta_1^0, \quad \text{for any } x \in \overline{\Omega}.$$

Let φ_ε^* be the eigenvector to (1.13) corresponding to $\lambda := \lambda_{1\gamma}(\varepsilon)$, $a_{22} := a_{22} - \varepsilon$ and satisfying

$$\varphi_\varepsilon^*(x) \ge \theta > 0 \qquad \text{in } \overline{\Omega}.$$

The existence of such θ follows in the same manner as the existence of θ_1^0 for φ_ε.

By multiplying the first equation in (1.16) by $\varphi_\varepsilon^*(x)$ and the second one by $\psi_\varepsilon^*(x) = \dfrac{1}{a_{22}-\varepsilon} \displaystyle\int_\Omega k(x',x)\varphi_\varepsilon^*(x')dx'$ and integrating over Ω, we get that

$$\frac{d}{dt}\left(\int_{\Omega}(y(x,t)\varphi_{\varepsilon}^{*}(x)+z(x,t)\psi_{\varepsilon}^{*}(x))dx\right)$$

$$+\int_{\Omega}(\lambda_{1\gamma}(\varepsilon)y(x,t)\varphi_{\varepsilon}^{*}(x)+\varepsilon z(x,t)\psi_{\varepsilon}^{*}(x))dx=0,$$

for any $t \geq 0$. It follows that

$$(y(t),z(t)) \longrightarrow (0,0)$$

in $L^1(\Omega) \times L^1(\Omega)$, as $t \to +\infty$, at an exponential rate.

Let us prove now that

$$(y(t),z(t)) \longrightarrow (0,0)$$

in $L^\infty(\Omega) \times L^\infty(\Omega)$, as $t \to +\infty$, at an exponential rate.

Let us note first that y is the weak solution to

$$\begin{cases} \dfrac{\partial y}{\partial t}(x,t)=\mathcal{A}_{\varepsilon}y(x,t)+\displaystyle\int_{\Omega}k(x,x')\left[z(x',t)-\frac{a_{21}}{a_{22}-\varepsilon}y(x',t)\right]dx', & \\ & (x,t)\in\Omega\times(0,+\infty), \\ \dfrac{\partial y}{\partial n}(x,t)+\alpha y(x,t)=0, & (x,t)\in\partial\Omega\times(0,+\infty), \\ y(x,0)=u_1^0(x), & x\in\Omega, \end{cases}$$

where

$$(\mathcal{A}_{\varepsilon}\zeta)(x)=d_1\Delta\zeta(x)-a_{11}\zeta(x)+\frac{a_{21}}{a_{22}-\varepsilon}\int_{\Omega}k(x,x')\zeta(x')dx'-\mathbb{I}_{\omega}(x)\gamma\zeta(x).$$

The mapping

$$t\mapsto\int_{\Omega}k(x,x')\Big[z(x',t)-\frac{a_{21}}{a_{22}-\varepsilon}y(x',t)\Big]dx'$$

belongs to $C([0,+\infty);L^\infty(\Omega))$ and

$$\lim_{t\to+\infty}\int_{\Omega}k(x,x')\Big[z(x',t)-\frac{a_{21}}{a_{22}-\varepsilon}y(x',t)\Big]dx'=0 \quad \text{in } L^\infty(\Omega).$$

We may infer that there exists $q \in C([0,+\infty))$ with $\lim_{t\to+\infty} q(t)=0$, and

$$\int_{\Omega}k(x,x')\Big[z(x',t)-\frac{a_{21}}{a_{22}-\varepsilon}y(x',t)\Big]dx'\leq q(t)\varphi_{\varepsilon}(x), \quad \text{a.e. } x\in\Omega,$$

for any $t \in [0,+\infty)$.

The comparison result for integro-parabolic equations (see Appendix A) implies that

$$0 \le y(x,t) \le \overline{y}(x,t), \quad \text{a.e. } x \in \Omega, \text{ for any } t \ge 0, \tag{1.17}$$

where $\overline{y}$ is the weak solution to

$$\begin{cases} \dfrac{\partial \overline{y}}{\partial t}(x,t) = \mathcal{A}_\varepsilon \overline{y}(x,t) + q(t)\varphi_\varepsilon(x), & (x,t) \in \Omega \times (0,+\infty), \\ \dfrac{\partial \overline{y}}{\partial n}(x,t) + \alpha \overline{y}(x,t) = 0, & (x,t) \in \partial\Omega \times (0,+\infty), \\ \overline{y}(x,0) = c_1 \varphi_\varepsilon(x), & x \in \Omega. \end{cases} \tag{1.18}$$

Here c_1 is a positive constant such that

$$u_1^0(x) \le c_1 \varphi_\varepsilon(x), \quad \text{a.e. } x \in \Omega.$$

It is obvious that the unique weak solution to (1.18) satisfies

$$\overline{y}(x,t) = h(t)\varphi_\varepsilon(x), \quad \text{a.e. } x \in \Omega, \text{ for any } t \ge 0, \tag{1.19}$$

where h is the solution to

$$\begin{cases} h'(t) = -\lambda_{1\gamma}(\varepsilon)h(t) + q(t), & t > 0, \\ h(0) = c_1. \end{cases}$$

We calculate and get that

$$h(t) = e^{-\lambda_{1\gamma}(\varepsilon)t}\Big[c_1 + \int_0^t e^{\lambda_{1\gamma}(\varepsilon)s} q(s)ds\Big], \quad t \ge 0.$$

By L'Hôspital's rule we get that

$$\lim_{t\to+\infty} h(t) = 0.$$

This (alongside with (1.17) and (1.19)) implies that

$$\lim_{t\to+\infty} y(t) = 0 \quad \text{in } L^\infty(\Omega),$$

and consequently

$$\lim_{t\to+\infty} u_1(t) = 0 \quad \text{in } L^\infty(\Omega).$$

On the other hand,

$$0 \le z(x,t) = e^{-(a_{22}-\varepsilon)t} u_2^0(x) + \int_0^t e^{-(a_{22}-\varepsilon)(t-s)} a_{21} y(x,s)ds$$

$$\leq e^{-(a_{22}-\varepsilon)t}\Big[\|u_2^0\|_{L^\infty(\Omega)} + a_{21}\int_0^t e^{(a_{22}-\varepsilon)s}\|y(s)\|_{L^\infty(\Omega)}ds\Big],$$

a.e. $x \in \Omega$, for any $t \geq 0$. By L'Hôspital's rule we get that

$$\lim_{t\to+\infty} z(t) = 0 \quad \text{in } L^\infty(\Omega),$$

and consequently

$$\lim_{t\to+\infty} u_2(t) = 0 \quad \text{in } L^\infty(\Omega).$$

Conversely, assume that $g'(0) = a_{21}$ and that the disease is eradicable. For any $\varepsilon \in (0, a_{21})$, there exists $\delta(\varepsilon) > 0$ such that $g(r) \geq (a_{21} - \varepsilon)r$, for all $r \in [0, \delta(\varepsilon)]$.

As in Appendix A we can prove that

$$\lim_{\varepsilon\to 0} \lambda_1(\omega, \varepsilon) = \lambda_1(\omega),$$

where $\lambda_1(\omega, \varepsilon)$ is the principal eigenvalue to

$$\begin{cases} -d_1\Delta\varphi(x) + a_{11}\varphi(x) - \dfrac{a_{21}-\varepsilon}{a_{22}}\displaystyle\int_{\Omega\setminus\overline{\omega}} k(x,x')\varphi(x')dx' = \lambda\varphi(x), & x \in \Omega\setminus\overline{\omega}, \\ \varphi(x) = 0, & x \in \partial\omega, \\ \dfrac{\partial\varphi}{\partial n}(x) + \alpha\varphi(x) = 0, & x \in \partial\Omega \end{cases}$$

(for any $\varepsilon \in (0, a_{22})$).

For any u_1^0 and u_2^0 satisfying (H3) and $u_1^0(x) > 0$, a.e. $x \in \Omega$, and for any $0 < \varepsilon < a_{22}$, there exist $T(\varepsilon) > 0$ and $v \in L^\infty_{loc}(\overline{\omega} \times [0, +\infty))$ (stabilizing control), such that the solution (u_1, u_2) to (1.7) satisfies

$$\|(u_1(t), u_2(t))\|_{L^\infty(\Omega)\times L^\infty(\Omega)} < \delta(\varepsilon), \qquad \text{for any } t \geq T(\varepsilon).$$

By the comparison result (Appendix A), we get that for any such $v \in L^\infty_{loc}(\overline{\omega} \times [0, +\infty))$, the solution (u_1, u_2) to (1.7) satisfies

$$\begin{cases} u_1(x,t) \geq \tilde{u}_1(x,t) \geq 0, & \text{a.e. } x \in \Omega\setminus\overline{\omega}, \text{ for any } t \geq T(\varepsilon), \\ u_2(x,t) \geq \tilde{u}_2(x,t) \geq 0, & \text{a.e. } x \in \Omega\setminus\overline{\omega}, \text{ for any } t \geq T(\varepsilon), \end{cases}$$

where $(\tilde{u}_1, \tilde{u}_2)$ is the solution to

$$
\begin{cases}
\dfrac{\partial \tilde{u}_1}{\partial t}(x,t) = d_1 \Delta \tilde{u}_1(x,t) - a_{11}\tilde{u}_1(x,t) \\
\qquad + \displaystyle\int_{\Omega\setminus\omega} k(x,x')\tilde{u}_2(x',t)dx', & x \in \Omega \setminus \overline{\omega},\ t > T(\varepsilon), \\
\tilde{u}_1(x,t) = 0, & x \in \partial\omega,\ t > T(\varepsilon), \\
\dfrac{\partial \tilde{u}_1}{\partial n}(x,t) + \alpha\tilde{u}_1(x,t) = 0, & x \in \partial\Omega,\ t > T(\varepsilon), \\
\dfrac{\partial \tilde{u}_2}{\partial t}(x,t) = -a_{22}\tilde{u}_2(x,t) + (a_{21}-\varepsilon)\tilde{u}_1(x,t), & x \in \Omega \setminus \overline{\omega},\ t > T(\varepsilon), \\
\tilde{u}_1(x,T(\varepsilon)) = u_1(x,T(\varepsilon)), & x \in \Omega \setminus \overline{\omega}, \\
\tilde{u}_2(x,T(\varepsilon)) = u_2(x,T(\varepsilon)), & x \in \Omega \setminus \overline{\omega}.
\end{cases}
\tag{1.20}
$$

Since $\lim_{t\to+\infty} u_1(t) = \lim_{t\to+\infty} u_2(t) = 0$ in $L^\infty(\Omega)$, we may conclude that

$$
\lim_{t\to+\infty} \tilde{u}_1(t) = \lim_{t\to+\infty} \tilde{u}_2(t) = 0 \quad \text{in } L^\infty(\Omega \setminus \overline{\omega}).
$$

By the backward uniqueness theorem for parabolic equations, we conclude that $u_1(\cdot, T(\varepsilon))$ is not identically 0, and consequently $u_2(\cdot, T(\varepsilon))$ is not identically 0.

Denote by φ^* an eigenvector corresponding to the principal eigenvalue $\lambda_1(\omega,\varepsilon)$, for the problem

$$
\begin{cases}
-d_1\Delta\varphi(x) + a_{11}\varphi(x) - \dfrac{a_{21}-\varepsilon}{a_{22}} \displaystyle\int_{\Omega\setminus\overline{\omega}} k(x',x)\varphi(x')dx' = \lambda\varphi(x), & x \in \Omega \setminus \overline{\omega}, \\
\varphi(x) = 0, & x \in \partial\omega, \\
\dfrac{\partial\varphi}{\partial n}(x) + \alpha\varphi(x) = 0, & x \in \partial\Omega,
\end{cases}
$$

and satisfying

$$
\varphi^*(x) > 0, \qquad x \in \Omega \setminus \overline{\omega}.
$$

Consider $p(t) = \int_{\Omega\setminus\overline{\omega}} \tilde{u}_1(x,t)\varphi^*(x)dx + \int_{\Omega\setminus\overline{\omega}} \tilde{u}_2(x,t)\psi^*(x)dx$, for $t \geq T(\varepsilon)$, where

$$
\psi^*(x) = \frac{1}{a_{22}} \int_{\Omega\setminus\overline{\omega}} k(x',x)\varphi^*(x')dx', \qquad x \in \Omega \setminus \overline{\omega}.
$$

By multiplying the first equation in (1.20) by φ^* and the second by ψ^* and integrating over $\Omega \setminus \overline{\omega}$, we get

$$\begin{cases} p'(t) + \lambda_1(\omega, \varepsilon) \int_{\Omega\setminus\overline{\omega}} \tilde{u}_1(x,t)\varphi^*(x)dx = 0, & t \geq T(\varepsilon), \\ p(T(\varepsilon)) = \int_{\Omega\setminus\overline{\omega}} \tilde{u}_1(x, T(\varepsilon))\varphi^*(x)dx + \int_{\Omega\setminus\overline{\omega}} \tilde{u}_2(x, T(\varepsilon))\psi^*(x)dx > 0. \end{cases}$$

Since $\lim_{t\to+\infty} \tilde{u}_1(t) = \lim_{t\to+\infty} \tilde{u}_2(t) = 0$ in $L^\infty(\Omega)$, we may infer that $\lim_{t\to+\infty} p(t) = 0$ and that $\lambda_1(\omega, \varepsilon) > 0$. It follows that $\lambda_1(\omega) \geq 0$. □

Definition 1.9 We say that a disease modelled by (1.7) is *locally eradicable* if there exists $r_0 > 0$ such that for any u_1^0 and u_2^0 satisfying (H3) and

$$\|u_1^0\|_{L^\infty(\Omega)},\ \|u_2^0\|_{L^\infty(\Omega)} \leq r_0,$$

there exists $v \in L^\infty_{loc}(\overline{\omega} \times [0, +\infty))$ such that the solution (u_1, u_2) to (1.7) satisfies

$$u_1(x,t) \geq 0,\ u_2(x,t) \geq 0, \quad \text{a.e. } x \in \Omega, \text{ for any } t \geq 0,$$

and

$$\lim_{t\to+\infty} \|u_1(t)\|_{L^\infty(\Omega)} = \lim_{t\to+\infty} \|u_2(t)\|_{L^\infty(\Omega)} = 0.$$

It is obvious that if an epidemic is eradicable, then it is also locally eradicable.

Remark 1.10 Denote by $\tilde{\lambda}_1(\omega)$ the principal eigenvalue of the problem

$$\begin{cases} -d_1 \Delta\varphi(x) + a_{11}\varphi(x) - \dfrac{g'(0)}{a_{22}} \displaystyle\int_{\Omega\setminus\overline{\omega}} k(x,x')\varphi(x')dx' = \lambda\varphi(x), & x \in \Omega \setminus \overline{\omega}, \\ \varphi(x) = 0, & x \in \partial\omega, \\ \dfrac{\partial\varphi}{\partial n}(x) + \alpha\varphi(x) = 0, & x \in \partial\Omega. \end{cases}$$

It follows in the same manner that if $\tilde{\lambda}_1(\omega) > 0$, then the disease is locally eradicable, and for $\gamma \geq 0$ sufficiently large, $v := -\gamma u_1$ is a feedback control which makes the disease eradicable.

Conversely, if the disease is locally eradicable, then $\tilde{\lambda}_1(\omega) \geq 0$.

Since $g'(0) \leq a_{21}$, it follows that

$$\lambda_1(\omega) \leq \tilde{\lambda}_1(\omega).$$

We conclude now that

1^0 If $\lambda_1(\omega) > 0$, the disease is eradicable.

2^0 If $\tilde{\lambda}_1(\omega) > 0$ and $\lambda_1(\omega) < 0$, the disease is locally eradicable but is not (globally) eradicable.

3^0 If $\tilde{\lambda}_1(\omega) < 0$, the disease is not locally eradicable and consequently is not eradicable.

1.3.3 Final Comments

If we have in mind the fact that the principal eigenvalue $\lambda_{1\gamma}$ (and consequently $\lambda_1(\omega)$) dictates the asymptotic behavior of the weak solution (u_1, u_2), then it is very important to find the position and the geometry of ω which maximize or provide a positive $\lambda_1(\omega)$. This problem is important for systems of population dynamics.

It is important to notice that in case of eradicability, the stabilizing control is a feedback control, as proposed by us, and has a very simple form. So it can be easily implemented.

1.4 A Vector Valued Approach in the Optimal Control of a Class of Epidemic Systems

In this section we report the results obtained in [43] and [44] concerning a vector valued approach for the optimal control of an epidemic of orofecal origin (such as cholera, typhoid fever, and so on), according to an extension of Model (E) (see Section 1.1; [29, 67–75], and the references therein).

Let us consider an open bounded subset Ω of $\mathbb{R}^2$ (representing a city placed on the coast of the sea), whose boundary $\partial\Omega$ is sufficiently smooth and is the union of two disjoint curves Γ_1 (representing the coast) and Γ_2. As in Section 1.1 $u_1(t, x)$ and $u_2(t, x)$, denote the density of infective agent and infected people at time t in point x, respectively.

As in [29, 72], the epidemic is enhanced by the fact that people use to eat raw seafood which are cultivated along the coast Γ_1 and are polluted by infected sewage. Such a process has been described by a boundary condition of the form

$$\frac{\partial u_1}{\partial n}(x, t) + \alpha\, u_1(x, t) = \int_\Omega k_2(x, x')\, u_2(x', t)\, dx', \quad \text{on } \Gamma_1 \times [0, +\infty)$$

(for general boundary conditions see Appendix A; see also [117]). On the other hand, it is assumed that the relevant habitat (say a coastal town) is perfectly isolated from the hinterland, i.e., it is imposed that

$$\frac{\partial u_1}{\partial n}(x, t) = 0, \quad \text{on } \Gamma_2 \times [0, +\infty).$$

The spread of the epidemic can be controlled by a sanitation program in the habitat Ω and on the coast Γ_1 (such as the treatment of sewage), which reduces the dispersal actions described above by suitable factors $\gamma_1 = \gamma_1(x,t)$ and $\gamma_2 = \gamma_2(x,t)$, with $0 \le \gamma_i \le 1$.

Obviously the sanitation program has the purpose of reducing the epidemic, i.e., of obtaining

$$\int_\Omega u_2(x,T)\,dx \le c_0, \qquad \text{for some } T > 0,$$

where $c_0 < \int_\Omega u_2^0(x)\,dx$ is a preassigned positive constant representing an "acceptable" level of infected human population. The following control problem has been faced:

(P) Look for $T > 0$, γ_1, γ_2 such that u_2 satisfies the terminal constraint

$$\int_\Omega u_2(x,T)\,dx \le c_0,$$

where u_1, u_2 is a solution of the reaction–diffusion system

$$\begin{aligned}
&\frac{\partial u_1}{\partial t}(x,t) - \Delta u_1(x,t) + a_1 u_1(x,t)\\
&\qquad = \gamma_1(x,t)\int_\Omega k_1(x,x')\,u_2(x',t)\,dx', \qquad (x,t) \in \Omega \times (0,T),\\
&\frac{\partial u_2}{\partial t}(x,t) + a_2 u_2(x,t) = g(u_1(x,t)), \qquad (x,t) \in \Omega \times (0,T),
\end{aligned}$$

subject to the following initial and boundary conditions:

$$\begin{aligned}
&u_1(x,0) = u_1^0(x), \quad u_2(x,0) = u_2^0(x), \quad x \in \Omega,\\
&\frac{\partial u_1}{\partial n}(x,t) + \alpha u_1(x,t)\\
&\qquad = \gamma_2(x,t)\int_\Omega k_2(x,x')\,u_2(x',t)\,dx', \quad (x,t) \in \Gamma_1 \times (0,T),\\
&\frac{\partial u_1}{\partial n}(x,t) = 0, \qquad (x,t) \in \Gamma_2 \times (0,T).
\end{aligned}$$

The epidemic has several costs. First we have to consider the costs of the sanitation activities which depend on their reductive effects γ_1, γ_2, i.e.,

$$J_1(T,\gamma_1,\gamma_2) = \int_0^T \int_\Omega h_1(\gamma_1(x,t))\,dx\,dt,$$

$$J_2(T,\gamma_1,\gamma_2) = \int_0^T \int_{\Gamma_1} h_2(\gamma_2(x,t))\,d\sigma\,dt.$$

Moreover there is the cost of the epidemic itself which depends on the number of infected people and therefore has the form

$$J_3(T, \gamma_1, \gamma_2) = \int_0^T \int_\Omega f(u_2(x,t))\, dx\, dt.$$

Finally we may take into account the psychological and economic effects of the duration of the epidemic by introducing an additional cost $J_4(T, \gamma_1, \gamma_2)$ with an appropriate structure.

The purpose of public authorities is to choose a strategy T, γ_1, γ_2, i.e., a sanitary strategy during the time interval $[0, T]$ producing the reductive effects γ_1, γ_2, which leads to a minimization of the costs. In general there is not a single strategy which minimizes all costs $J_1 - J_4$; this may correspond to an "utopia." On the other hand, by minimizing only one cost we might have too high values of the other ones. A compromise is required. For example, one might look for strategies which minimize a linear combination of $J_1 - J_4$; this has been analyzed in [29] which minimizes the sum of the cost functionals (see also [43]).

In [44] a so-called Pareto control has been proposed: at first "efficient" strategies are identified which reduce each cost but cannot improve all costs at the same time; among them a decision maker may choose the best one according to their own utility function. Such "efficient" strategies are known as Pareto optimal solutions. Necessary conditions are provided for a Pareto optimal solution, derived by an extension of the Pontryagin maximum principle (see [144, Chapter 17] and the references therein; see also [43]).

1.5 Travelling Waves and Their Control

For the integrodifferential epidemic system (1.1), the problem of possible existence of travelling waves has been faced in [196] when $\Omega = \mathbb{R}^n$, $n = 1, 2, 3$.

As kernel of the integral term, an isotropic one has been assumed, i.e.,

$$k(x, y) = \tilde{k}(x - y) \quad \text{with} \quad \tilde{k}(x) = \tilde{k}(y) \quad \text{if} \quad |x| = |y|,\ x, y \in \mathbb{R}^n,$$

where $|\cdot|$ denotes the usual norm in $\mathbb{R}^n$.

By a suitable scaling of the variables x, t, and u_2, Model (E), i.e., System (1.1), can be rewritten in the following form:

$$\begin{cases} \dfrac{\partial u_1}{\partial t}(x,t) = \Delta u_1(x,t) - u_1(x,t) + \displaystyle\int_\Omega K(x-y)u_2(y,t)dy, \\ \dfrac{\partial u_2}{\partial t}(x,t) = -\beta u_2(x,t) + g(u_1(x,t)), \end{cases} \tag{1.21}$$

where $\beta = a_{22}/a_{11}$, g has been divided by a_{11}^2, and $K(x) = \left(\sqrt{\frac{d}{a_{11}}}\right)^{n/2} \tilde{k}\left(\sqrt{\frac{d}{a_{11}}}x\right)$.

System (1.21) is supplemented by initial conditions

$$u_1(x,0) = \phi_1(x) \geq 0, \quad u_2(x,0) = \phi_2(x) \geq 0, \quad x \in \mathbb{R}^n. \tag{1.22}$$

It can be shown (see, e.g., [155]) that System (1.21), subject to bounded, uniformly continuous and nonnegative initial conditions (ϕ_1, ϕ_2), admits a unique nonnegative classical solution (u_1, u_2) for $t > 0$.

Let $\Gamma(t,x)$ and $\Gamma_1(t,x)$ be the Green functions associated with the parabolic equations $\frac{\partial u}{\partial t} = \Delta u$ and $\frac{\partial u}{\partial t} = \Delta u - u$, respectively. Then $\Gamma_1(t,x) = \Gamma(t,x)\,e^{-t}$. The solution to (1.21), subject to the initial conditions (1.22), can be expressed in the following form:

$$\begin{aligned} u_1(x,t) = &\int_{\mathbb{R}^n} \Gamma_1(t, x-y)\,\phi_1(y)\,dy \\ &+ \int_0^t ds \int_{\mathbb{R}^n} \Gamma_1(t-s, x-y) \int_{\mathbb{R}^n} K(y-z)\,u_2(z,s)\,dz\,dy, \\ u_2(x,t) = &\ e^{-\beta t}\phi_2(x) + \int_0^t e^{-\beta(t-r)} g(u_1(x,r))\,dr. \end{aligned}$$

By some manipulations (see [188, 196]), it can be shown that u_1 is the solution of the following integral equation:

$$u_1(x,t) = u_0(x,t) + \int_0^t ds \int_{\mathbb{R}^n} g(u_1(x-y, t-s))\,k_2(s,y)\,dy,$$

where

$$k_2(t,x) = \int_0^t \int_{\mathbb{R}^n} \Gamma_1(t-s, x-y)\,K(y)\,e^{-\beta s}\,dy\,ds.$$

The following assumptions are made on System (1.21), with

$$\mathcal{K}(\lambda) = \int_{\mathbb{R}^n} e^{-\lambda y_1} K(y)\,dy,$$

where y_1 is the first coordinate of y.

(H1)' $K : \mathbb{R}^n \to [0, +\infty)$ is continuous and isotropic, i.e., $K(x) = K(y)$ if $|x| = |y|$.

(H2)' $\mathcal{K}(0) > 0$, and there exists some $\lambda_0 > 0$ such that $\mathcal{K}(\lambda_0) = \infty$, and $\mathcal{K}(\lambda) < \infty$ for all $\lambda \in [0, \lambda_0)$, where λ_0 may be infinity.

(H3)' $g : [0, +\infty) \to [0, +\infty)$ is Lipschitz continuous with $g(0) = 0$, differentiable at 0, and satisfies $g'(0)\mathcal{K}(0) > \beta$, $0 < g(u) \leq g'(0)u$, for all $u > 0$.

(H4)' $\lim_{u\to\infty} \frac{g(u)}{u} = 0$, and there exists $u^* > 0$ such that g is increasing on $[0, u^*]$, $g(u)\mathcal{K}(0) > \beta u$ for $u \in (0, u^*)$, and $g(u)\mathcal{K}(0) < \beta u$ for $u > u^*$.

(H5)' $\limsup_{u\to\infty} \frac{g(u)}{u} < \frac{\beta}{\mathcal{K}(0)}$, $\frac{g(u)}{u}$ is strictly decreasing, and $u\, g(u)$ is strictly increasing for $u > 0$.

It is not difficult to show that the first condition of either (H4)' or (H5)' implies the boundedness of solutions of System (1.21), subject to bounded initial conditions.

Let (c^*, λ^*) denote the unique positive solution of the system

$$(\beta + \lambda c)(1 + \lambda c - c^2) = g'(0)\mathcal{K}(\lambda),$$
$$\frac{2\lambda - c}{1 + \lambda c - c^2} - \frac{c}{\beta + \lambda c} = \frac{1}{\mathcal{K}(\lambda)} \int_{\mathbb{R}^n} y_1 e^{-\lambda y_1} K(y)\, dy.$$

The following theorem holds [188].

Theorem 1.11 *Let (H1)'–(H3)' hold and c^* be defined as above. Denote by $u(x, t, \phi)$ the unique solution of System (1.21)–(1.22). Then the following statements hold:*

(i) For any continuous function $\phi = (\phi_1, \phi_2) : \mathbb{R}^n \to [0, +\infty) \times [0, +\infty)$, with the property that for every $\kappa_1 > 0$, there exists $\kappa_2 > 0$ such that $\phi_1(y) + \phi_2(y) \leq \kappa_2 e^{-\kappa_1 |y|}$, $y \in \mathbb{R}^n$, we have

$$\lim_{t\to\infty, |x|\geq ct} u(x, t, \phi) = (0, 0), \quad \text{for all } c > c^*.$$

(ii) Assume in addition that either (H4)' or (H5)' holds, and let u^* denote the unique solution of

$$g(u)\mathcal{K}(0) = \beta u,$$

and $v^* = \frac{g(u^*)}{\beta}$.

Then for any bounded and uniformly continuous function $\phi = (\phi_1, \phi_2) : \mathbb{R}^n \to [0, +\infty) \times [0, +\infty)$, with $\phi_1(\cdot) + \phi_2(\cdot) \not\equiv 0$,

$$\lim_{t\to\infty, |x|\leq ct} u(t, x, \phi) = (u^*, v^*), \quad \text{for all } c \in (0, c^*).$$

Within the above framework, c^* is known as the *asymptotic speed* of spread for solutions of (1.21) with initial functions having compact supports.

Example 1.12 A simple example for $n = 1$ is the following one.
The kernel K is the standard normal density function, i.e.,

$$K(x) = \frac{1}{\sqrt{2\pi}} \exp(-x^2/2),$$

$\beta = 1$, and

$$g(z) = \frac{2z}{1+z}.$$

It is not difficult to see that in this case System (1.21) satisfies the assumptions $(H1)'$–$(H4)'$, with

$$u^* = v^* = 1.$$

By Theorem 1.11, for any continuous initial conditions ϕ_1 and ϕ_2, with compact supports, we have

$$\lim_{t\to\infty, |x|\geq ct} u(x,t) = (0,0), \quad \text{for all } c > c^*,$$

and

$$\lim_{t\to\infty, |x|\leq ct} u(x,t) = (1,1), \quad \text{for all } c \in (0, c^*).$$

1.5.1 *Travelling Wave Solution*

Consider the case $n = 1$. A solution of System (1.21) is known as a *travelling wave solution* if it is of the form

$$u(x,t) = U(x+ct), \quad x \in \mathbb{R},\ t > 0,$$

for some $c > 0$.

The parameter c is called the *wave speed* and the function U is called the *wave profile.*

In [188], under the following conditions:

$$U : \mathbb{R} \to [0, +\infty), \text{ is positive and bounded; } \lim_{\xi\to-\infty} U(\xi) = 0,$$

the following theorem is shown to hold.

Theorem 1.13 *Let Assumptions* $(H1)'$–$(H3)'$ *hold, and let* c^*, u^*, *and* v^* *be as in Theorem 1.11. Then*

(i) There is no travelling wave solution of System (1.21) with wave speed $c \in (0, c^*)$.
(ii) Assume in addition $(H4)'$, *and that*

$$|g(z) - g(z')| \leq g'(0)|z - z'|, \quad \text{for } z, z' \in [0, u^*],$$

$$g''(0) \in \mathbb{R} \quad \text{exists.}$$

Then System (1.21) admits a monotone travelling wave connecting $(0, 0)$ *and* (u^*, v^*) *with speed* $c \geq c^*$. *Moreover the monotone travelling wave with speed* $c > c^*$ *is unique up to translation.*

For a possible extension of the above results to the time periodic case, the reader may refer to [36]. Other interesting results can be found in [192] and [54].

1.5.2 A Controlled Example

Below we shall present a series of computational experiments concerning the case presented in Example 1.12, i.e., System (1.21), in which the parameter β is spatially dependent, so to include possible regional controls. It would be an interesting open problem for an interested reader to obtain the emerged behaviors by analytical methods.

Consider the following system of PDEs, for $(x, t) \in \Omega \times (0, T)$:

$$\begin{cases} \dfrac{\partial u_1}{\partial t}(x,t) = \dfrac{\partial^2 u_1}{\partial x^2}(x,t) - u_1(x,t) + \displaystyle\int_a^b k(x,x')\, u_2(x',t)\, dx', \\ \dfrac{\partial u_2}{\partial t}(x,t) = -\beta(x) u_2(x,t) + g(u_1(x,t)), \end{cases} \tag{1.23}$$

where $\beta(x) > 0$, for $x \in \Omega$,

$$k(x, x') = \frac{1}{\sqrt{2\pi}} e^{-(x-x')^2/2} \quad \text{and} \quad g(u) = \frac{2u}{1+u}.$$

System (1.23) is subject to suitable initial conditions $u_1(x, 0) = u_1^0(x) > 0$, $u_2(x, 0) = u_2^0(x) > 0$.

For computational purposes, below we shall take $\Omega = (a, b)$, a sufficiently large but bounded interval; in this case homogeneous Neumann boundary conditions will be adopted.

1.5.3 Computational Issues

The numerical strategy adopted to approximate system (1.23) consists of the finite element method for space discretization and the finite difference method for time discretization. See Appendix B for an introduction to this kind of numerical methods for PDEs.

Space Discretization We first apply a standard Galerkin procedure to the weak formulation of system (1.23). Suppose that the computational domain $\Omega = (a, b)$ is discretized by a uniform grid of $N - 1$ linear finite elements ($\mathbf{P^1}$), yielding a total number of N discretization nodes.

In the following, we describe in detail this procedure applied to system (1.23). Let us first denote by V the Sobolev space $H^1(\Omega)$, by $(\cdot, \cdot)$ the L^2 inner product, and by $a(\cdot, \cdot) : V \times V \to \mathbb{R}$ the bilinear form

$$a(u, v) := \int_\Omega \frac{\partial u}{\partial x} \frac{\partial v}{\partial x} dx, \quad \text{for any } u, v \in V.$$

Then, integrating by part, the variational formulation of problem (1.23) reads as follows:

find u_1 and u_2 such that

$$\begin{cases} \dfrac{\partial}{\partial t}(u_1(t), \phi) + a(u_1(t), \phi) = (R_1(u_1(t), u_2(t)), \phi), & \text{for all } \phi \in V, \\ \dfrac{\partial}{\partial t}(u_2(t), \phi) = (R_2(u_1(t), u_2(t)), \phi), & \text{for all } \phi \in V, \end{cases} \tag{1.24}$$

subject to the initial conditions $u_1(x, 0) = u_1^0(x)$ and $u_2(x, 0) = u_2^0(x)$ and where R_1 and R_2 are the reaction terms in (1.23).

Let us now denote by V^h the $\mathbf{P}^1$ finite element space associated with a discretization of the interval domain Ω into $N - 1$ sub-intervals of equal size $h = (b - a)/(N - 1)$. For a detailed description of this space, see Appendix B.

The space semi-discretization of problem (1.24) then reads as follows:

find $u_{1,h}(\cdot, t), u_{2,h}(\cdot, t) : (0, T] \to V^h$ such that

$$\begin{cases} \dfrac{\partial}{\partial t}(u_{1,h}(t), \phi) + a(u_{1,h}(t), \phi) = (R_1(u_{1,h}(t), u_{2,h}(t)), \phi), & \text{for any } \phi \in V^h, \\ \dfrac{\partial}{\partial t}(u_{2,h}(t), \phi) = (R_2(u_{1,h}(t), u_{2,h}(t)), \phi), & \text{for any } \phi \in V^h, \end{cases}$$

which is equivalent to

$$\begin{cases} \dfrac{\partial}{\partial t}(u_{1,h}(t), \phi_i) + a(u_{1,h}(t), \phi_i) = (R_1(u_{1,h}(t), u_{2,h}(t)), \phi_i), & i = 1, \ldots, N, \\ \dfrac{\partial}{\partial t}(u_{2,h}(t), \phi_i) = (R_2(u_{1,h}(t), u_{2,h}(t)), \phi_i), & i = 1, \ldots, N, \end{cases} \tag{1.25}$$

where ϕ_i, $i = 1, \dots, N$, are the *hat basis functions* of the $\mathbf{P}^1$ finite element space V^h.

Expanding the unknowns $u_{1,h}(\cdot, t)$ and $u_{2,h}(\cdot, t)$ with respect to the basis $\{\phi_i\}_{i=1}^N$, we have

$$u_{1,h}(x,t) = \sum_{j=1}^{N} U_{1,j}(t)\phi_j(x), \quad u_{2,h}(x,t) = \sum_{j=1}^{N} U_{2,j}(t)\phi_j(x), \tag{1.26}$$

where $U_{1,j},\ U_{2,j} : [0, T] \to \mathbb{R}$ are unknown scalar functions to be determined.

In the next steps, we follow closely the approximation of the nonlinear problem presented in Appendix B. Hence, denoting by A the stiffness matrix associated with the bilinear form $a(\cdot, \cdot)$ and by M the lumped mass matrix (see the appendix for the construction of such matrices) and substituting the expressions (1.26) into system (1.25), the semi-discrete problem then reads as follows:

find $\mathbf{U}_1(t), \mathbf{U}_2(t) : (0, T] \to \mathbb{R}^N$ such that

$$\begin{cases} \mathrm{M}\dfrac{d\mathbf{U}_1}{dt}(t) + \mathbf{A}\mathbf{U}_1(t) = \mathbf{M}\mathbf{R}_1(\mathbf{U}_1(t), \mathbf{U}_2(t)), \\ \mathrm{M}\dfrac{d\mathbf{U}_2}{dt}(t) = \mathbf{M}\mathbf{R}_2(\mathbf{U}_1(t), \mathbf{U}_2(t)), \end{cases}$$

where

$$\begin{aligned} \mathbf{U}_1(t) &= [\dots, U_{1,j}(t), \dots]^T, \quad j = 1, \dots, N, \\ \mathbf{U}_2(t) &= [\dots, U_{2,j}(t), \dots]^T, \quad j = 1, \dots, N, \\ \mathbf{R}_1(\mathbf{U}_1(t), \mathbf{U}_2(t)) &= [\dots, R_1(U_{1,j}(t), U_{2,j}(t)), \dots]^T, \quad j = 1, \dots, N, \\ \mathbf{R}_2(\mathbf{U}_1(t), \mathbf{U}_2(t)) &= [\dots, R_2(U_{1,j}(t), U_{2,j}(t)), \dots]^T, \quad j = 1, \dots, N. \end{aligned}$$

Denoting by x_j the generic discretization node, the nonlocal component of the generic element $R_1(U_{1,j}(t), U_{2,j}(t))$ of the vector $\mathbf{R}_1(\mathbf{U}_1(t), \mathbf{U}_2(t))$ is computed as follows:

$$\begin{aligned} \int_\Omega k(x_j, x')u_{2,h}(x', t)dx' &= \int_\Omega k(x_j, x') \sum_{j=1}^{N} U_{2,j}(t)\phi_j(x')dx' \\ &\approx \int_\Omega \sum_{i=1}^{N} k(x_j, x_i)\phi_i(x') \sum_{j=1}^{N} U_{2,j}(t)\phi_j(x')dx' \\ &= \mathbf{K}(x_j)^T \mathbf{M}\mathbf{U}_2(t), \end{aligned}$$

where

$$\mathbf{K}(x_j) = [k(x_j, x_1), \dots, k(x_j, x_i), \dots, k(x_j, x_N)]^T.$$

Time Discretization After the spatial discretization, we obtain a semi-discrete problem that consists of a system of ordinary differential equations (ODEs). We solve these ODEs by employing a first order semi-implicit finite difference scheme, where the linear diffusion terms are approximated by Backward Euler, whereas the nonlinear reaction terms are approximated by Forward Euler.

In formulae, at the generic time step n,
given $\mathbf{U}_1^n$ and $\mathbf{U}_2^n$, we have to find $\mathbf{U}_1^{n+1}$ and $\mathbf{U}_2^{n+1}$ such that

$$\begin{cases} \mathrm{M}\dfrac{\mathbf{U}_1^{n+1} - \mathbf{U}_1^n}{\Delta t} + \mathrm{A}\mathbf{U}_1^{n+1} = \mathrm{M}\mathbf{R}_1(\mathbf{U}_1^n, \mathbf{U}_2^n), \\ \mathrm{M}\dfrac{\mathbf{U}_2^{n+1} - \mathbf{U}_2^n}{\Delta t} = \mathrm{M}\mathbf{R}_2(\mathbf{U}_1^n, \mathbf{U}_2^n). \end{cases}$$

As a result, at each time step it is required the solution of the linear system of algebraic equations

$$\begin{bmatrix} \dfrac{1}{\Delta t}\mathrm{M} + \mathrm{A} & 0 \\ 0 & \dfrac{1}{\Delta t}\mathrm{M} \end{bmatrix} \begin{bmatrix} \mathbf{U}_1^{n+1} \\ \mathbf{U}_2^{n+1} \end{bmatrix} = \begin{bmatrix} \dfrac{1}{\Delta t}\mathrm{M}\mathbf{U}_1^n + \mathrm{M}\mathbf{R}_1(\mathbf{U}_1^n, \mathbf{U}_2^n) \\ \dfrac{1}{\Delta t}\mathrm{M}\mathbf{U}_2^n + \mathrm{M}\mathbf{R}_2(\mathbf{U}_1^n, \mathbf{U}_2^n) \end{bmatrix},$$

which has the dimension $2 \times N$ degrees of freedom. In the implementation, the linear system is solved by Gaussian elimination with the built-in MATLAB function. For further details on the numerical discretization of parabolic problems, we refer the reader to [168].

1.5.3.1 Parameter Calibration

As a computational domain $\Omega = (a, b)$, we have taken the interval $(-60, 2)$, discretized by a uniform grid of 2000 $\mathbf{P}^1$ finite elements, yielding a total number of $N = 2001$ discretization nodes. The total simulation time is $T = 200$. The time step size is $\Delta t = 0.01$.

The initial distributions, used in all next numerical simulations, are $u_1^0(x) = u_2^0(x) = \Phi(x)$, where

$$\Phi(x) = \begin{cases} 0, & x \leq -1, \\ \dfrac{1}{2}(1 + x), & -1 < x < 1, \\ 1, & x \geq 1. \end{cases}$$

We consider three different scenarios, depending on the calibration of the parameter function $\beta(x)$:

- In **experiment 1**, we do not apply any control, and thus we set $\beta(x) = 1$ everywhere.

- In **experiment 2**, we set

$$\beta(x) = \begin{cases} 2.1, & x \leq -20, \\ 1, & x > -20. \end{cases}$$

We denote the interval $(-60, -20)$ as the control region.

- In **experiment 3**, we set

$$\beta(x) = \begin{cases} 1, & x < -30, \\ 3, & -30 \leq x \leq -10, \\ 1, & x > -10. \end{cases}$$

We denote the interval $(-30, -10)$ as the control region.

1.5.3.2 Numerical Simulations

We report here, in two graphical representations, the numerical results related to the three experiments, obtained with the MATLAB code

```
run_trav_wave.m,
```

available on the Birkhäuser website provided in the Preface and also at the link

```
http://www.mat.unimi.it/users/scacchi/ACS_codes/
run_trav_wave.m
```

In **experiment 1**, a travelling wave propagates from the right border ($x = 2$) to the left border ($x = -60$) of the domain, see Figures 1.4 and 1.5.

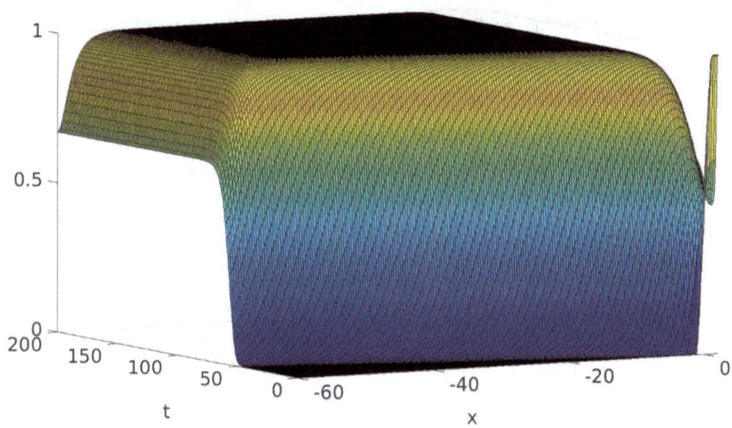

Fig. 1.4 Experiment 1. We solve System (1.23) on the space interval $(-60, 2)$ during the time span $(0, 200)$, with discretization and control parameters as detailed in the Subsection **Parameter calibration** of Section 1.5.3. Here we do not apply any control, thus $\beta(x) = 1$ everywhere. Space-time evolution of u_1 (3D representation)

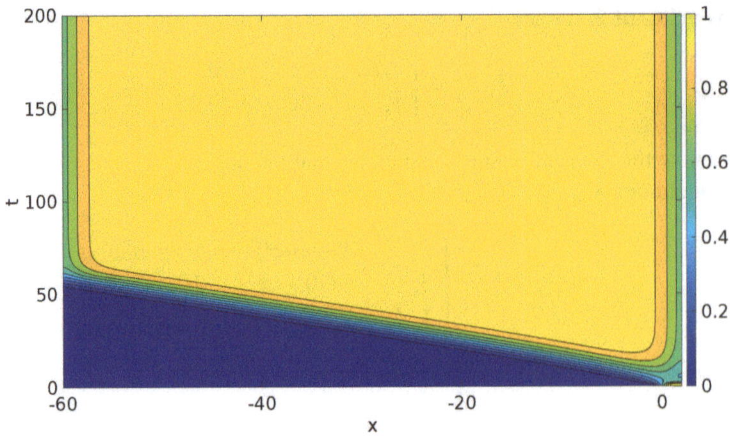

Fig. 1.5 **Experiment 1**. Same parameters as in Figure 1.4. Space-time evolution of u_1 (2D representation)

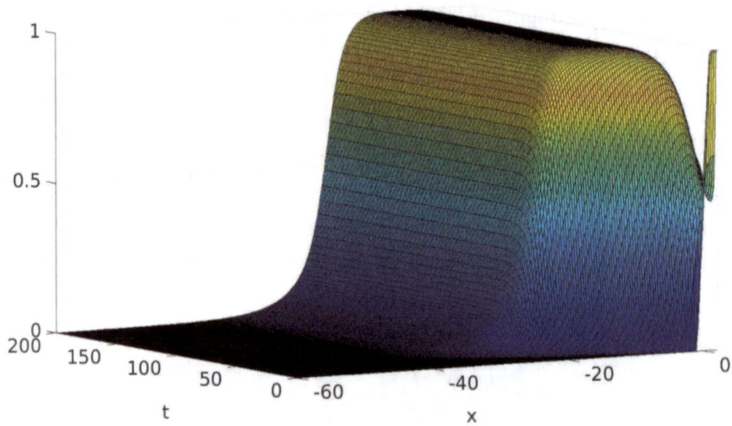

Fig. 1.6 **Experiment 2**. We solve System (1.23) on the space interval $(-60, 2)$ during the time span $(0, 200)$, with discretization and control parameters as detailed in the Subsection **Parameter calibration** of Section 1.5.3. Here the control region is the interval $(-60, -20)$, where we set $\beta(x) = 2.1$, whereas $\beta(x) = 1$ elsewhere. Space-time evolution of u_1 (3D representation)

In **experiment 2**, a travelling wave propagates from the right border of the domain ($x = 2$) to the right border of the control region ($x = -20$), where it remains blocked, see Figures 1.6 and 1.7.

In **experiment 3**, a travelling wave propagates from the right border of the domain ($x = 2$) to the right border of the control region ($x = -10$), where it stops. However, after a delay, a travelling wave originates from the left border of the control region ($x = -30$) and propagates to the left border of the domain ($x = -60$), see Figures 1.8 and 1.9.

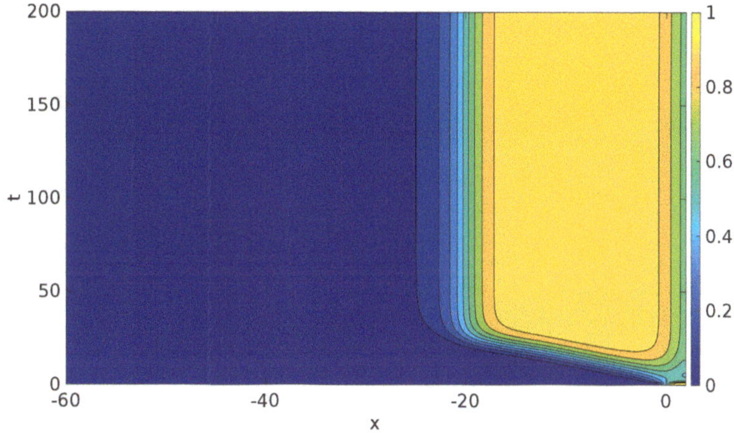

Fig. 1.7 **Experiment 2**. Same parameters as in Figure 1.6. Space-time evolution of u_1 (2D representation)

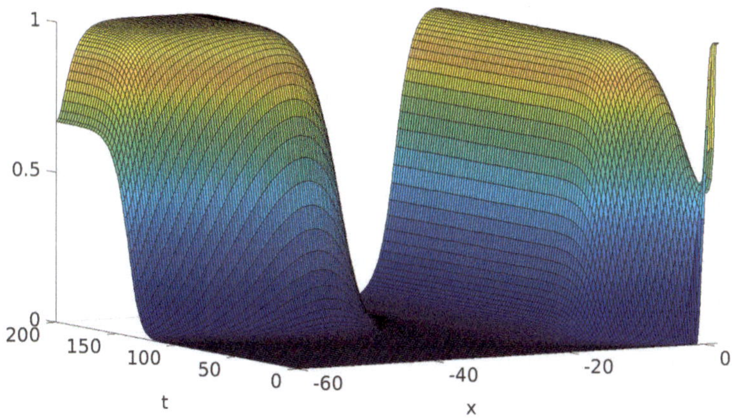

Fig. 1.8 **Experiment 3**. We solve System (1.23) on the space interval $(-60, 2)$ during the time span $(0, 200)$, with discretization and control parameters as detailed in the Subsection **Parameter calibration** of Section 1.5.3. Here the control region is the interval $(-30, -10)$, where we set $\beta(x) = 3$, whereas $\beta(x) = 1$ elsewhere. Space-time evolution of u_1 (3D representation)

1.6 Nonlocal Evolution Equations

In the first equation of System (1.21) the Laplace operator models the random dispersal of the pollutant; to model other possible mechanisms of dispersion a more general nonlocal operator might be used (see, e.g., [107, 108, 172]), which can be written in the form

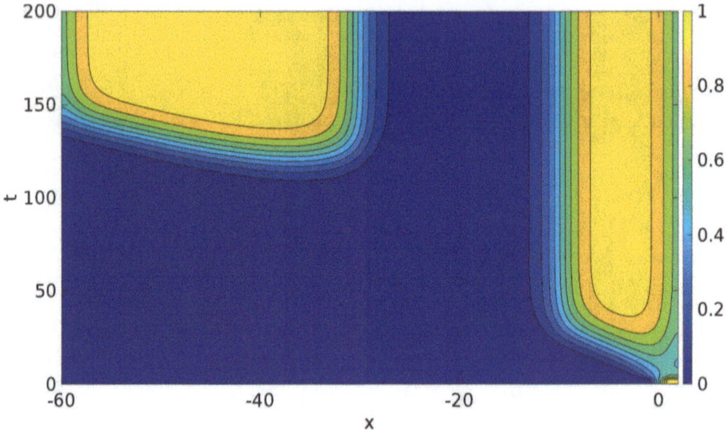

Fig. 1.9 Experiment 3. Same parameters as in Figure 1.8. Space-time evolution of u_1 (2D representation)

$$
\begin{aligned}
J * u(x) - u(x) &= \int_{\mathbb{R}^n} J(x-y)u(y)dy - u(x)\int_{\mathbb{R}^n} J(x-y)dy \\
&= \int_{\mathbb{R}^n} J(x-y)u(y)dy - u(x), \quad x \in \mathbb{R}^n, \qquad (1.27)
\end{aligned}
$$

where $u : \mathbb{R}^n \to [0, +\infty)$ is a given function and $J : \mathbb{R}^n \to [0, +\infty)$ is a so-called kernel.

Suppose $u(x)$ represents the concentration of a population at location $x \in \mathbb{R}^n$, and J is a probability density function; $J(x-y)$ may represent the probability density of jumping of the population from another point $y \in \mathbb{R}^n$ to point x.

Then (see, e.g., [83, 108])

$$J * u(x) = \int_{\mathbb{R}^n} J(x-y)u(y)dy, \quad x \in \mathbb{R}^n,$$

represents the rate at which the population available at all points $y \in \mathbb{R}^n$ is transferred, by any possible physical mechanism, to point $x \in \mathbb{R}^n$. Vice versa the term

$$-u(x) = -u(x)\int_{\mathbb{R}^n} J(x-y)dy$$

is the rate at which the population available at location $x \in \mathbb{R}^n$ is transferred anywhere else in $\mathbb{R}^n$, by the same mechanism.

In terms of nonlocal diffusion (of the pollutant), System (1.21) can be rewritten as

$$\begin{cases} \dfrac{\partial u_1}{\partial t}(x,t) = [J * u_1(\cdot,t)](x) - u_1(x,t) + [K * u_2(\cdot,t)](x), \\ \dfrac{\partial u_2}{\partial t}(x,t) = -\beta u_2(x,t) + g(u_1(x,t)). \end{cases}$$

In [85] and [84] general reaction diffusion systems with nonlocal operators of the form

$$\frac{\partial u}{\partial t}(x,t) = D\Delta u(x,t) + f(t, u(x,t), [J_1 * s_1(u(\cdot,t)](x), \ldots, [J_n * s_n(u(\cdot,t)](x))$$

have been analyzed, with respect to the existence of travelling waves.

1.6.1 Diffusion Approximation of Nonlocal Operators

Let us consider the one-dimensional case $n = 1$; assume that the kernel J is symmetric with respect to zero, with compact support, i.e., it satisfies the following assumptions:

(H1)" $J(-x) = J(x), \quad x \in \mathbb{R},$
$J(x) = 0$, for $|x| > \delta$,

for a suitable $\delta > 0$.

Consider now a sufficiently smooth function $v : \mathbb{R} \to [0, +\infty)$, so that we may assume

$$\frac{d^3 v}{dx^3}(x) = O(x), \qquad \text{as } x \to 0.$$

By a Taylor expansion, we have

$$v(x - x') = v(x) - x'\frac{dv}{dx}(x) + \frac{1}{2}(x')^2\frac{d^2 v}{dx^2}(x) - \frac{1}{6}(x')^3\frac{d^3 v}{dx^3}(\bar{x}),$$

for any $x' \in \mathbb{R}$ and a suitable $\bar{x} \in (x - x', x)$.

As a consequence, for any $x \in \mathbb{R}$, we have

$$\begin{aligned} J * v(x) &= \int_{\mathbb{R}} J(x')v(x - x')dx' \\ &= v(x)\int_{\mathbb{R}} J(x')dx' - \frac{dv}{dx}(x)\int_{\mathbb{R}} x'J(x')dx' \\ &\quad + \frac{1}{2}\frac{d^2 v}{dx^2}(x)\int_{\mathbb{R}} (x')^2 J(x')dx' \end{aligned}$$

$$-\frac{1}{6}\int_{\mathbb{R}}(x')^3 J(x')\frac{d^3 v}{dx^3}(\bar{x})dx'.$$

Now, in the second integral the integrand is odd, and hence it is zero. The fourth integral is of the order $(\delta)^3$, and hence it can be neglected. Finally we may write the approximation

$$\begin{aligned} J * v(x) &= \int_{\mathbb{R}} J(x')v(x-x')dx' \simeq v(x)\int_{\mathbb{R}} J(x')dx' \\ &+\frac{1}{2}\frac{d^2 v}{dx^2}(x)\int_{\mathbb{R}}(x')^2 J(x')dx' \\ &= \bar{h}v(x) + \bar{\bar{h}}\frac{d^2 v}{dx^2}(x), \end{aligned} \tag{1.28}$$

where

$$\bar{h} = \int_{\mathbb{R}} J(x')dx'$$

and

$$\bar{\bar{h}} = \frac{1}{2}\int_{\mathbb{R}}(x')^2 J(x')dx'.$$

Clearly (1.28) represents the so-called diffusion approximation of $J * v$.

Epidemic models with nonlocal diffusion have been proposed in [137]; in [33] one may find the existence of travelling fronts for Kendall's model with nonlocal diffusion. The existence of travelling waves for the diffusion approximation of Kendall's model can be found in [128].

Chapter 2
Controlling the Spread of a Vector Borne Epidemic: The Case of Malaria

2.1 A Spatially Structured Malaria Model

Malaria is found throughout the tropical and subtropical regions of the world and causes more than 300 million acute illnesses and at least one million deaths annually [194], especially in Africa. Human malaria is caused by one or a combination of four species of plasmodia: *Plasmodium falciparum, P. vivax, P. malariae, and P. ovale*. The parasites are transmitted through the bite of infected female mosquitoes of the genus *Anopheles*. Mosquitoes can become infected by feeding on the blood of infected people, and the parasites then undergo another phase of reproduction in the infected mosquitoes.

The earliest attempt to provide a quantitative understanding of the dynamics of malaria transmission was that of Ross [171]. Ross' models consisted of a few differential equations describing changes in densities of susceptible and infected people and susceptible and infected mosquitoes. Macdonald [152] extended Ross' basic model, analyzed several factors contributing to malaria transmission, and concluded that "the least influence is the size of the mosquito population, upon which the traditional attack has always been made."

The work of Macdonald had a very beneficial impact on the collection, analysis, and interpretation of epidemic data on malaria infection [158] and guided the enormous global malaria eradication campaign of his era.

The classical Ross–Macdonald model is highly simplified. Subsequent contributions have been made to extend the Ross–Macdonald malaria models considering a variety of epidemiological features of malaria [32, 97] (see also [35, 86, 87, 98, 138, 173, 182]).

Here we will consider an oversimplified model, concentrating on a problem of diminishment of a spatially structured malaria epidemic, by acting on the segregation of the human and the mosquito populations via, e.g., treated bed nets (see, e.g., [183]).

S. Aniţa et al., *Mathematical Modeling and Control in Life and Environmental Sciences*, Modeling and Simulation in Science, Engineering and Technology,
https://doi.org/10.1007/978-3-031-49971-5_2

By following recent papers by Ruan et al. [173] and Chamchod and Britton [79], we divide the human population into two classes, susceptible and infectious, whereas the mosquito population is divided into three classes, susceptible, infectious, and removed because of death; i.e., we suppose that the infection in the humans does not result in death or isolation. For the transmission of the pathogen, it is assumed that a susceptible human may receive the infection only by contacting with infective mosquitoes, and a susceptible mosquito may receive the infection only from infectious humans.

For simplicity, assume the total populations of both humans and mosquitoes are constants, denoted by H and M, respectively. Let $X(t)$ and $Y(t)$ denote the numbers of infected humans and mosquitoes at time t, respectively. Let a be the rate of biting on humans by a single mosquito (the number of bites per unit time). Then the number of bites on humans, per unit time per human, is $\frac{a}{H}$. If b is the proportion of infected bites on humans that produce an infection, the interaction between the infected mosquitoes $Y(t)$ and the uninfected humans $H - X(t)$ will produce new infected humans at a rate of $\frac{a}{H}b(H - X(t))Y(t)$. Malaria on humans is an SIS system for which human infectives after recovery go back in the susceptible state; we will denote by r the per capita rate of recovery in humans so that $1/r$ is the duration of the disease in humans [69].

Therefore, the equation for the rate of change of the number of infected humans is

$$\frac{dX}{dt} = -rX(t) + \frac{a}{H}b(H - X(t))Y(t).$$

Similarly, if μ is the per capita rate of mortality in mosquitoes, so that $1/\mu$ is the life expectancy of mosquitoes, and c is the transmission efficiency from humans to mosquito, then we have the equation for the rate of change in the number of infected mosquitoes:

$$\frac{dY}{dt} = -\mu Y(t) + \frac{a}{H}cX(t)(M - Y(t)).$$

Here we wish to consider a spatially structured system, so that we will refer to spatial densities of infected humans and infected mosquitoes. Specifically, let $u_1(x,t)$ denote the spatial density of the population of infected mosquitoes at a spatial location $x \in \overline{\Omega}$ and time $t \geq 0$, while $u_2(x,t)$ denotes the spatial distribution of the human infective population. In accordance with the above, the spatial density $C(x)$ of the total human population will be assumed constant in time, so that $C(x) - u_2(x,t)$ will provide the spatial distribution of susceptible humans, at a spatial location $x \in \overline{\Omega}$ and time $t \geq 0$.

Hence the "local incidence rate" for humans, at point $x \in \overline{\Omega}$ and time $t \geq 0$, is taken of the form [151]

$$(i.r.)_H(x,t) = g(x, u_1(x,t), u_2(x,t)) = (C(x) - u_2(x,t))h\left(\frac{u_1(x,t)}{C(x)}\right),$$

depending upon the local densities of both populations via a suitable functional response h.

Here we have taken into account that, in accordance with the Ross–Macdonald model [151], the function h, describing the force of infection of humans by mosquitoes, depends upon the relative concentration of the total mosquito population with respect to the total human population because of the specific biting habits of humans by mosquitoes.

On the other hand, we have here extended the (linear) response of the Ross–Macdonald model, by using a possibly nonlinear functional response $h\left(\frac{u_1(x,t)}{C(x)}\right)$. This choice may allow possible saturation effects, σ-type responses, etc. (see [69, 77]). For example, behavioral changes can be taken into account; for a very large density of the infective population the force of infection represented by $h\left(\frac{u_1(x,t)}{C(x)}\right)$ may tend to reduce itself because of reduction of open exposure of the human population. This aspect had been already proposed in [77], which has motivated a large recent literature (see [99] and the references therein).

As pointed out in [158], seasonality of the aggressivity to humans by the mosquito population might also be considered in the functional response h; this has been a topic of the authors' research program in [74], and [12, 13].

Concerning the "local incidence" for mosquitoes, at point $x \in \overline{\Omega}$and time $t \geq 0$, as in the previous models [67], we assume that it is due to contagious bites to human infectives at any point $x' \in \Omega$ of the habitat, within a spatial neighborhood of x represented by a suitable probability kernel $k(x, x')$, depending on the specific structure of the local ecosystem (see also [179]); as a trivial simplification one may assume $k(x, \cdot)$ as a Gaussian density centered at x; hence the "local incidence" for mosquitoes, at point $x \in \Omega$ and time $t \geq 0$, is taken as

$$(i.r.)_M(x,t) = \int_\Omega k(x, x')u_2(x', t)dx'.$$

As proposed in [35], we have included spatial diffusion of the infective mosquito population (with a constant diffusion coefficient to avoid purely technical complications), but we assume that the human population does not diffuse. More refined models may include some kind of mobility of the human population, due to migration mechanisms (see Section 1.6), but this would require further nontrivial complications that we leave to further investigations.

Finally, from now on, we will denote by $-a_{11}u_1(x,t)$ the natural decay of the infected mosquito population, while $-a_{22}u_2(x,t)$ will denote the recovery rate (back to susceptibility) of the human infective population.

All the above leads to the following oversimplified model for the spatial spread of malaria epidemics, in which we have ignored the possible acquired immunity of humans after exposure to the contagion, the possible differentiation of the mosquito population, etc. (see, e.g., [138]).

$$\begin{cases} \dfrac{\partial u_1}{\partial t}(x,t) = d\Delta u_1(x,t) - a_{11}u_1(x,t) + \displaystyle\int_\Omega k(x,x')u_2(x',t)dx', \\ \dfrac{\partial u_2}{\partial t}(x,t) = -a_{22}u_2(x,t) + g(x,u_1(x,t),u_2(x,t)) \end{cases} \tag{2.1}$$

in $\Omega \times (0,+\infty)$, where $a_{11} > 0$, $a_{22} > 0$, and $d > 0$ are constants.

We may realize that Model (2.1) is a possible extension of Model **E** presented in System (1.1), with a major modification in the dependence of the local human incidence rate upon both u_1 and u_2.

As anticipated in Section 1.2, under realistic assumptions concerning the costs of all controls, we shall study the following controlled system:

$$\begin{cases} \dfrac{\partial u_1}{\partial t}(x,t) - d\Delta u_1(x,t) = \eta(x)u_1(x,t) + \displaystyle\int_\Omega k(x,x')u_2(x',t)dx' \\ \qquad -\gamma_1(x,t)\mathbb{I}_\omega(x)u_1(x,t), & (x,t) \in Q, \\ \dfrac{\partial u_2}{\partial t}(x,t)(x,t) = -(a_{22} + \gamma_2(x,t)\mathbb{I}_\omega(x))u_2(x,t) \\ \qquad +(1-\gamma_3(x,t)\mathbb{I}_\omega(x))g(u_1(x,t)), & (x,t) \in Q, \\ \dfrac{\partial u_1}{\partial n}(x,t) = 0, & (x,t) \in \Sigma, \\ u_1(x,0) = u_1^0(x),\ u_2(x,0) = u_2^0(x), & x \in \Omega \end{cases} \tag{2.2}$$

(which is a slight extension of (2.1)).

Here $Q = \Omega \times (0,+\infty)$ and $\Sigma = \partial\Omega \times (0,+\infty)$, where Ω is a bounded domain in $\mathbb{R}^2$, with a sufficiently smooth boundary $\partial\Omega$, ω is an open subset of Ω, and $\mathbb{I}_\omega$ is its characteristic function; $\eta \in L^\infty(\Omega)$.

2.2 Preliminary Results

We shall work under the following hypotheses:

(H1) $\eta \in L^\infty(\Omega), k \in L^\infty(\Omega \times \Omega),\ k(x,x') \geq 0$, a.e. in $\Omega \times \Omega$,

$$\int_\Omega k(x,x')dx > 0, \qquad \text{a.e. } x' \in \Omega.$$

(H2) $C \in L^\infty(\Omega)$, $C(x) \geq \tau > 0$, a.e. in Ω, where τ is a positive constant.
(H3) $h : \mathbb{R} \to [0, +\infty)$.
$h|_{[0,+\infty)}$ is continuously differentiable and increasing,
$h(x) = 0, \quad$ for $x \in (-\infty, 0]$,
$h(x) \leq a_{21}x$, for any $x \in [0, +\infty)$, where a_{21} is a positive constant.
(H4) $u_1^0,\ u_2^0 \in L^\infty(\Omega), \quad 0 \leq u_1^0(x),\ 0 \leq u_2^0(x) \leq C(x), \quad$ a.e. $x \in \Omega$.

Based on epidemiological reasoning (see later in Section 2.4), we shall assume that the space of parameters is the following one:

$$\mathcal{G} = \{\gamma = (\gamma_1, \gamma_2, \gamma_3) \in L^\infty(Q) \times L^\infty_{loc}(\overline{Q}) \times L^\infty(Q) \ :\ 0 \leq \gamma_1(x,t) < \Gamma_1,$$

$$0 \leq \gamma_2(x,t),\ 0 \leq \gamma_3(x,t) < \Gamma_3,\ \text{a.e. } (x,t) \in Q\},$$

where $\Gamma_1 > 0$ and $1 \geq \Gamma_3 > 0$ are suitable parameters, or

$$\mathcal{G}_T = \{\gamma = (\gamma_1, \gamma_2, \gamma_3) \in L^\infty(Q_T)^3 \ :\ 0 \leq \gamma_1(x,t) < \Gamma_1, 0 \leq \gamma_2(x,t),$$

$$0 \leq \gamma_3(x,t) < \Gamma_3,\ \text{a.e. } (x,t) \in Q_T\},$$

where $Q_T = \Omega \times (0, T)$, for a suitable $T > 0$.

Notice that the controls act only in a subdomain $\omega \subset \Omega$. For the sake of simplicity, we prefer however to take $\gamma \in \mathcal{G}$ or $\gamma \in \mathcal{G}_T$ and to consider that the action (the control) on the two populations is done by the vector control $\gamma \mathbb{I}_\omega$ (in Sections 2.3, 2.4, and 2.5). On the other hand, considering $\gamma \in \mathcal{G}_T$ will be very useful when we investigate a certain approximating optimal control problem (and where the vector control on the two populations actually acts in the whole habitat Ω).

By usual arguments (see, e.g., [12–14]), it is possible to prove the following existence and uniqueness theorem.

Theorem 2.1 *Under the assumptions **(H1)–(H4)**, for any $T > 0$ and $\gamma \in \mathcal{G}_T$, Problem (2.2) (where instead of Q we take $Q_T = \Omega \times (0, T)$, and instead of Σ we take $\Sigma_T = \partial\Omega \times (0, T)$) has a unique solution (u_1^γ, u_2^γ).*

The solution satisfies

$$0 \leq u_1^\gamma(x,t), \quad 0 \leq u_2^\gamma(x,t) \leq C(x), \quad a.e.\ (x,t) \in Q_T.$$

This implies the following corollary:

Corollary 2.2 *For any $\gamma \in \mathcal{G}$, Problem (2.2) has a unique solution (u_1^γ, u_2^γ) (it is a solution to (2.2) corresponding to $Q := Q_T$ and $\Sigma := \Sigma_T$ for any $T > 0$). The solution satifies*

$$0 \leq u_1^\gamma(x,t), \quad 0 \leq u_2^\gamma(x,t) \leq C(x), \quad a.e.\ (x,t) \in Q.$$

Thanks to the comparison result (established as in Appendix A) for the following problem (where $\gamma = (\gamma_1, \gamma_2, \gamma_3) \in \mathcal{G}$):

$$\begin{cases} \dfrac{\partial u_1}{\partial t}(x,t) - d\Delta u_1(x,t) = \eta(x)u_1(x,t) + \displaystyle\int_\Omega k(x,x')u_2(x',t)dx' \\ \qquad -\gamma_1(x,t)\mathbb{I}_\omega(x)u_1(x,t) + f_1(x,t), & (x,t) \in Q, \\ \dfrac{\partial u_2}{\partial t}(x,t) = -(a_{22} + \gamma_2(x,t)\mathbb{I}_\omega(x))u_2(x,t) \\ \qquad +(1 - \gamma_3(x,t)\mathbb{I}_\omega(x))g(x, u_1(x,t), u_2(x,t)) + f_2(x,t), & (x,t) \in Q, \\ \dfrac{\partial u_1}{\partial n}(x,t) = 0, & (x,t) \in \Sigma, \\ u_1(x,0) = u_1^0(x), \ u_2(x,0) = u_2^0(x), & x \in \Omega, \end{cases} \tag{2.3}$$

we may state that the following holds.

Lemma 2.3 *Let $u_1^0, u_2^0, u_3^0, u_4^0 \in L^\infty(\Omega)$, $f_1, f_2, f_3, f_4 \in L^\infty_{loc}(\overline{Q})$ such that*

$$\begin{array}{lll} 0 \le u_1^0(x) \le u_3^0(x), & 0 \le u_2^0(x) \le u_4^0(x) \le C(x), & a.e.\ x \in \Omega, \\ 0 \le f_1(x,t) \le f_3(x,t), & 0 \le f_2(x,t) \le f_4(x,t), & a.e.\ (x,t) \in Q \end{array}$$

and that

$$0 \le u_2(x,t) \le C(x), \qquad a.e.\ (x,t) \in Q,$$

where (u_1, u_2) is the solution to (2.3), and (u_3, u_4) is the solution to (2.3), for $f_1 := f_3$, and $f_2 := f_4$. Then

$$0 \le u_1(x,t) \le u_3(x,t), \quad 0 \le u_2(x,t) \le u_4(x,t), \quad a.e.\ x \in \Omega, \ for\ all\ t \ge 0.$$

Notice that the solutions (u_1, u_2) and (u_3, u_4) exist and are unique.

As a consequence we obtain in particular the strict positivity of the solution of Problem (2.2).

2.3 Eradicability

Assume that ω is a fixed open subset of Ω and $\Omega \setminus \overline{\omega} \neq \emptyset$. Concerning eradicability (the costs to be paid for the controls γ_1, γ_2, and γ_3 are not taken into account), the arguments are parallel to the ones adopted in Section 1.2, where it was shown that eradicability is deeply related to the magnitude of the principal eigenvalue of an appropriate operator.

Under the assumptions **(H1)**–**(H4)**, the following results hold true (for details see [16]).

Proposition 2.4 *If $u_2^0(x) > 0$, a.e. $x \in \Omega \setminus \overline{\omega}$, then for any $\gamma = (\gamma_1, \gamma_2, \gamma_3) \in \mathcal{G}$, $u_2^\gamma(\cdot, t)$ cannot tend to 0 (in $L^\infty(\Omega)$), as $t \to +\infty$, faster than $\exp(-a_{22}t)$, where (u_1^γ, u_2^γ) is the solution to (2.2).*

The proof follows immediately via the comparison result for ODEs.

The next result shows that, under the assumptions **(H1)–(H4)** and for any $\varepsilon > 0$ sufficiently small, there exists at least a constant control $\gamma = (\gamma_1, \gamma_2, \gamma_2) \in [0, \Gamma_1) \times [0, +\infty) \times [0, \Gamma_3)$ such that for ω appropriately chosen we have that $u_1^\gamma(\cdot, t) \to 0$ and $u_2^\gamma(\cdot, t) \to 0$ in $L^\infty(\Omega)$ (as $t \to +\infty$) faster than or as fast as $\exp(-(a_{22} - \varepsilon)t)$.

For $\varepsilon \in (0, a_{22})$, consider the following eigenvalue problem:

$$\begin{cases} -d\Delta\psi_1(x) - (a_{22} - \varepsilon)\psi_1(x) - \eta(x)\psi_1(x) & \\ \qquad - \displaystyle\int_\Omega k(x, x')\psi_2(x')dx' + \gamma_1 \mathbb{I}_\omega(x)\psi_1(x) = \lambda\psi_1(x), & x \in \Omega, \\ \dfrac{\partial\psi_1}{\partial n}(x) = 0, & x \in \partial\Omega, \\ (\varepsilon + \gamma_2 \mathbb{I}_\omega(x))\psi_2(x) - (1 - \gamma_3 \mathbb{I}_\omega(x))a_{21}\psi_1(x) = 0, & x \in \Omega. \end{cases}$$

It is equivalent to

$$\begin{cases} -d\Delta\psi_1(x) - (a_{22} - \varepsilon)\psi_1(x) - \eta(x)\psi_1(x) & \\ \qquad - \displaystyle\int_\Omega k(x, x')\frac{(1 - \gamma_3 \mathbb{I}_\omega(x'))a_{21}}{\varepsilon + \gamma_2 \mathbb{I}_\omega(x')}\psi_1(x')dx' & \\ \qquad + \gamma_1 \mathbb{I}_\omega(x)\psi_1(x) = \lambda\psi_1(x), & x \in \Omega, \\ \dfrac{\partial\psi_1}{\partial n}(x) = 0, & x \in \partial\Omega, \\ \psi_2(x) = \dfrac{(1 - \gamma_3 \mathbb{I}_\omega(x))a_{21}}{\varepsilon + \gamma_2 \mathbb{I}_\omega(x)}\psi_1(x), & x \in \Omega. \end{cases} \tag{2.4}$$

By applying Krein–Rutman's theorem (see Appendix A),we may conclude that there exists one eigenvalue $\lambda_{1\varepsilon}^\omega$ which is real and simple and admits at least one eigenvector ψ_1 which is positive. Actually, it follows that there exist ψ_1^0 and ψ_2^0 positive constants such that

$$\psi_1(x) \geq \psi_1^0, \qquad \psi_2(x) \geq \psi_2^0,$$

for $x \in \overline{\Omega}$, where ψ_2 is the corresponding function from (2.4).

By arguing as in Chapter 1, we get the following theorem:

Theorem 2.5 *If $\varepsilon \in (0, a_{22})$, γ is a constant control in $\mathcal{G}$ and ω is an open subset of Ω such that $\lambda_{1\varepsilon}^\omega \geq 0$, then*

$$\lim_{t\to+\infty} u_1^\gamma(\cdot,t) = \lim_{t\to+\infty} u_2^\gamma(\cdot,t) = 0,$$

in $L^\infty(\Omega)$ *faster than or as fast as* $\exp(-(a_{22}-\varepsilon)t)$.

A natural question to be addressed here is whether, for a given $\varepsilon \in (0, a_{22})$, there exist an open subset ω of Ω and a constant control $\gamma \in \mathcal{G}$, such that $\lambda_{1\varepsilon}^{\omega} \geq 0$. This matter has been addressed by the following lemma (see [16]).

Lemma 2.6 *Assume that* $\gamma \in \mathcal{G}$ *is a constant control.*

(i) The mapping $\varepsilon \mapsto \lambda_{1\varepsilon}^{\omega}$ *is continuous and strictly increasing on* $(0, a_{22})$.
(ii) If $\{\omega_n\}_{n\in\mathbf{N}^*}$ *is an increasing sequence of open subsets of* Ω *such that* $\cup_{n=1}^{\infty}\omega_n = \Omega$, *then*

$$\lim_{n\to+\infty} \lambda_{1\varepsilon}^{\omega_n} = \lambda_{1\varepsilon}^{\Omega}.$$

Remark 2.7 If $\varepsilon \in (0, a_{22})$ is fixed, then $\lambda_{1\varepsilon}^{\Omega}$ is a function depending on $\gamma_1 \in [0, \Gamma_1)$, $\gamma_2 \in [0, +\infty)$, and $\gamma_3 \in [0, \Gamma_3)$, which is increasing with respect to each of them.

Moreover, if $\gamma_1 \to \Gamma_1$ and $\gamma_2 \to +\infty$ (their superior bounds), then

$$\lambda_{1\varepsilon}^{\Omega} \nearrow \lambda_{1\varepsilon},$$

where $\lambda_{1\varepsilon}$ is the principal eigenvalue to

$$\begin{cases} -d\Delta\psi(x) - (a_{22}-\varepsilon)\psi(x) - \eta(x)\psi(x) + \Gamma_1\psi(x) = \lambda\psi(x), & x \in \Omega, \\ \dfrac{\partial\psi_1}{\partial n}(x) = 0, & x \in \partial\Omega. \end{cases}$$

By Rayleigh's formula it follows that

$$\lambda_{1\varepsilon} \geq -(a_{22}-\varepsilon) - \|\eta\|_\infty + \Gamma_1.$$

Hence, if we put all together, we may infer that:

If $\Gamma_1 > a_{22} + \|\eta\|_\infty$, then for $\varepsilon \in (0, a_{22})$ sufficiently small, γ_1 sufficiently close to Γ_1, γ_2 sufficiently large, and ω sufficiently large, we have that $\lambda_{1\varepsilon}^{\omega} \geq 0$, and consequently the constant control $\gamma = (\gamma_1, \gamma_2, \gamma_3)$ (γ_3 is arbitrary but fixed in $[0, \Gamma_3)$) is such that the solution (u_1^γ, u_2^γ) to (2.2) tends to $(0, 0)$ in $L^\infty(\Omega)\times L^\infty(\Omega)$ as $t \to +\infty$, faster than or as fast as $\exp(-(a_{22}-\varepsilon)t)$.

Remark 2.8 If $\Gamma_1 = +\infty$, then it follows in the same manner that if $\gamma_1 \to +\infty$ and $\gamma_2 \to +\infty$, then

$$\lambda_{1\varepsilon}^{\Omega} \nearrow +\infty.$$

We conclude that for $\varepsilon \in (0, a_{22})$ sufficiently small, γ_1 sufficiently close to Γ_1, γ_2 sufficiently large, and ω sufficiently large, we have that $\lambda_{1\varepsilon}^{\omega} \geq 0$, and consequently the constant control $\gamma = (\gamma_1, \gamma_2, \gamma_3)$ (γ_3 is arbitrary but fixed in $[0, \Gamma_3)$) is such that t he solution (u_1^γ, u_2^γ) to (2.2) tends to $(0, 0)$ in $L^\infty(\Omega) \times L^\infty(\Omega)$ as $t \to +\infty$, faster than or as fast as $\exp(-(a_{22} - \varepsilon)t)$. This means that, for a suitably chosen control effort, eradicability of the epidemic in the whole habitat Ω is feasible, even though the region of intervention ω is strictly contained in Ω.

2.4 The Optimal Control Problem

Concerning the costs for the controls, we wish to emphasize that our explicit modelling of the costs of the three controls aims to taking into account realistic situations concerning malaria.

If we denote by $\widetilde{\gamma}_1(x, t)$ the total cost of the insecticides at location $x \in \Omega$ and time $t > 0$ and if $\tilde{\Gamma}_1 > 0$ is the cost per unit, then the quantity of insecticide per human being at location $x \in \Omega$ and time $t > 0$ is $\dfrac{\tilde{\gamma}_1(x, t)}{\tilde{\Gamma}_1 C(x)}$. Correspondingly we may assume that the harvesting rate $\gamma_1(x, t)$ is a suitable function f of $\dfrac{\widetilde{\gamma}_1(x, t)}{C(x)}$. We have an interaction of prey–predator type between the infected mosquitoes population (predator) and the insecticide (prey) and consequently a reaction term of the form

$$f\left(\frac{\tilde{\gamma}_1(x, t)}{C(x)}\right) u_1(x, t).$$

Obviously $f : [0, +\infty) \to [0, +\infty)$, $f(0) = 0$, and f is continuous and strictly increasing. This implies that $f([0, +\infty))$ is a bounded or unbounded interval of the form $[f(0), f(+\infty)) = [0, f(+\infty)) = [0, \Gamma_1)$ (where $\Gamma_1 = f(+\infty) = \lim_{r\to+\infty} f(r) \in (0, +\infty]$). A possible choice for such a function is the following:

$$f(z) = \frac{\Gamma_1 b_1 z^\theta}{1 + b_1 z^\theta},$$

with $\Gamma_1, b_1 > 0$, and $0 < \theta \leq 1$. The function f is saturating for large values of z to Γ_1.

The assumption $0 < \theta \leq 1$, instead of $1 < \theta$, assures an initial faster growth of the harvesting effort. This also describes the fact that, for small amounts of the insecticide per human being, we get a superlinear value of the harvesting effort. For either cases we have a saturation level Γ_1.

As a consequence, the cost corresponding to the time interval $[0, T]$ ($T \in (0, +\infty)$) and to the subset ω should appear as follows:

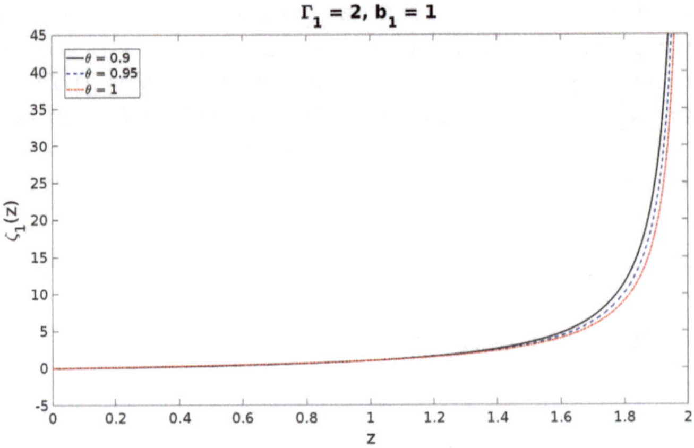

Fig. 2.1 Variation of the function ζ_1, for different values of the parameter θ

$$\int_0^T \int_\Omega \tilde{\gamma}_1(x,t)\mathbb{I}_\omega(x)dx\,dt = \int_0^T \int_\Omega C(x)f^{-1}(\gamma_1(x,t))\mathbb{I}_\omega(x)dx\,dt.$$

We will denote $\zeta_1 := f^{-1}$. Therefore $\zeta_1(z) = \left[\frac{z}{b_1(\Gamma_1 - z)}\right]^{\frac{1}{\theta}}$. It is obvious that $\tilde{\gamma}_1(x,t) = C(x)\zeta_1(\gamma_1(x,t))$ for any $x \in \Omega$, and $t > 0$, where $\zeta_1 : [0, \Gamma_1) \to [0, +\infty)$ is a bijective, strictly increasing, and continuous function (see Figure 2.1).

Let Γ_2 be the cost of the treatment to cure one person when $\gamma_2 u_2 \to +\infty$ (and a protocol treatment is used). Actually, when $\gamma_2 u_2$ is not large, the cost to cure one person is bigger. The cost is a function ζ_2 of $\gamma_2(x,t)u_2(x,t)$. The total cost paid for the treatment at location $x \in \Omega$ and time $t > 0$ is

$$\zeta_2(\gamma_2(x,t)u_2(x,t))\gamma_2(x,t)u_2(x,t).$$

Obviously $\zeta_2 : [0, +\infty) \to [0, +\infty)$ is a continuous and strictly decreasing function satisfying $\lim_{r\to+\infty} \zeta_2(r) = \Gamma_2$. A possible choice of ζ_2 is the following:

$$\zeta_2(z) = \frac{\Gamma_2 a_2 z + c_2}{1 + a_2 z},$$

Where $a_2 > 0$ and $c_2 > \Gamma_2$ (see Figure 2.2).

As a consequence, the cost corresponding to the time interval $[0, T]$ and to the subset ω should appear as

$$\int_0^T \int_\Omega \zeta_2(\gamma_2(x,t)u_2(x,t))\gamma_2(x,t)u_2(x,t)\mathbb{I}_\omega(x)dx\,dt.$$

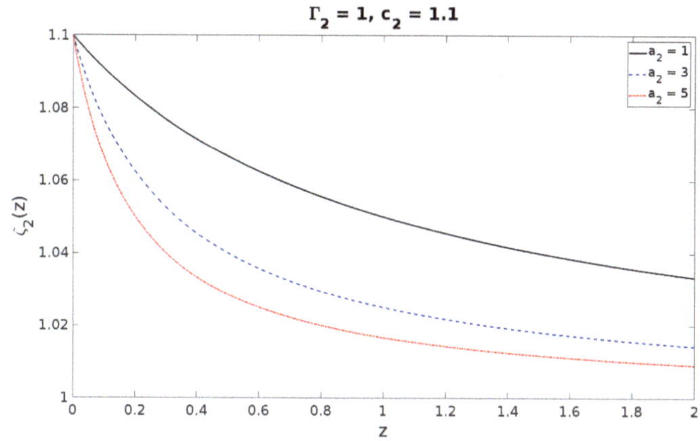

Fig. 2.2 Variation of the function ζ_2, for different values of the parameter a_2

Concerning the possible reduction of the contact rate of mosquitoes–humans, this can be obtained by means of physical devices such as treated bed nets or because of behavioral changes in the human population. The use of treated nets requires a cost proportional to the total human population here represented by the spatial density $C(x)$. We may proceed as for γ_1; if we let $\tilde{\gamma}_3(x,t)$ represent the total cost for the treated bed nets at location $x \in \Omega$ and time $t > 0$ and if $\tilde{\Gamma}_3$ is the cost per unit, then the number of the treated bed nets at location $x \in \Omega$ and time $t > 0$ is $\dfrac{\tilde{\gamma}_3(x,t)}{\tilde{\Gamma}_3}$. The reduction of the contact rate by using the treated bed nets is a function f_3 of $\dfrac{\tilde{\gamma}_3(x,t)}{C(x)}$, where $f_3 : [0,+\infty) \to [0,1)$ is a continuous and strictly increasing function, with $f_3(0) = 0$. This shows that $f_3([0,+\infty)) = [f_3(0), f_3(+\infty)) = [0, \Gamma_3)$ (where $\Gamma_3 = f_3(+\infty) \in (0,1]$). A total segregation rate would mean $\Gamma_3 = 1$. Actually, such a situation does not occur in practice and the value of Γ_3 is less than 1.

As a consequence, we may claim that the cost functional corresponding to the time interval $[0,T]$ and to the subset ω should appear as

$$\int_0^T\int_\Omega \tilde{\gamma}_3(x,t)\mathbb{I}_\omega(x)dx\,dt = \int_0^T\int_\Omega C(x)f_3^{-1}(\gamma_3(x,t))\mathbb{I}_\omega(x)dx\,dt$$
$$= \int_0^T\int_\Omega C(x)\zeta_3(\gamma_3(x,t))\mathbb{I}_\omega(x)dx\,dt, \qquad (2.5)$$

where $\zeta_3 := f_3^{-1}$. We have of course that $\tilde{\gamma}_3(x,t) = C(x)\zeta_3(\gamma_3(x,t))$ for any $x \in \Omega$ and $t > 0$.

Function $\zeta_3 : [0, \Gamma_3) \to [0,+\infty)$ is bijective, strictly increasing, and continuous. The graph of ζ_3 has the shape of the graph of ζ_1 in Figure 2.1 (except the fact that $\Gamma_3 < 1$).

As a consequence of the above, we shall assume that

$$0 \le \gamma_1(t) < \Gamma_1, \quad 0 \le \gamma_2(t), \quad 0 \le \gamma_3(t) < \Gamma_3, \quad \text{a.e. } t \in (0, T),$$

where $\Gamma_1 \in \mathbb{R}$, $\Gamma_3 \in (0, 1]$.

For the sake of technical simplification, we consider the case in which the controls are spatially homogeneous.

Based on the above, we may assume that the specific cost functions ζ_1, ζ_2, and ζ_3 satisfy the following assumptions:

(H5) $\zeta_1 : [0, \Gamma_1) \to [0, +\infty)$ is a continuously differentiable function, bijective, strictly increasing and ζ_1' is strictly increasing,
$\zeta_3 : [0, \Gamma_3) \to [0, +\infty)$ is a continuously differentiable function, bijective, strictly increasing and ζ_3' is strictly increasing.

(H6) $\zeta_2 : [0, +\infty) \to (\Gamma_2, \tilde{\Gamma}_2]$ is a continuous differentiable function, bijective, and strictly decreasing, and
$\tilde{\zeta}_2 : [0, +\infty) \to [0, +\infty)$, given by $\tilde{\zeta}_2(s) = \zeta_2(s)s$, $s \ge 0$, has a strictly decreasing derivative $\tilde{\zeta}_2'$.

When the costs of the controls to reduce the epidemic of the damages produced by the disease and of the intervention in the subset ω in the time interval $[0, T]$ are taken into account, for the total cost function, we get

$$\begin{aligned} J(\gamma, \omega) = & \int_0^T \int_\Omega C(x)\zeta_1(\gamma_1(t))\mathbb{I}_\omega(x)dx\, dt \\ & + \int_0^T \int_\Omega \zeta_2(\gamma_2(t)u_2(x,t))\gamma_2(t)u_2(x,t)\mathbb{I}_\omega(x)dx\, dt \\ & + \int_0^T \int_\Omega C(x)\zeta_3(\gamma_3(t))\mathbb{I}_\omega(x)dx\, dt \\ & + \int_0^T \int_\Omega l(u_2(x,t))dx\, dt + \alpha \text{ area}(\omega) + \beta \text{ perimeter}(\omega). \end{aligned} \tag{2.6}$$

As far as the last three terms in (2.6) are concerned, the term $\int_0^T \int_\Omega l(u_2(x,t))dx\, dt$ is meant to take into account the costs, over the whole domain and the total observation/control period, deriving from loss of work hours, hospitalization, drugs, etc., while the terms α area(ω) + β perimeter(ω) take into account costs of the logistics for providing intervention devices (nets, chemicals, personnel, etc.), which may depend on the geometry of the subregion of intervention ω.

We wish to attract the attention of the reader on the fact that, as in the optimal harvesting problem treated in Chapter 3, we have to consider not only the optimization with respect to the intensity of the harvesting effort but also with respect to the optimal choice of the subregion ω where it is implemented.

As evidenced in Chapter 3, for our purposes the geometry of a set ω in 2D may be characterized by its area and perimeter. We shall once again handle these quantities by means of the so-called implicit interface representation: a subset ω of a domain $\Omega \subset \mathbb{R}^2$ may be represented in terms of a function $\varphi : \overline{\Omega} \to \mathbb{R}$, twice continuously differentiable, called its implicit function, as follows:

$$\omega = \{x \in \Omega \, : \, \varphi(x) > 0\} \quad \text{and} \quad \partial\omega = \{x \in \Omega \, : \, \varphi(x) = 0\}.$$

We may then carry out integration over the interior of the domain ω by means of the Heaviside function $H : \mathbb{R} \to \{0, 1\}$, such that

$$H(z) = \begin{cases} 1, & \text{if} \quad z \geq 0, \\ 0, & \text{if} \quad z < 0. \end{cases}$$

Hence, in terms of the implicit representation of ω, the total cost function J defined above can be rewritten in the following form:

$$\begin{aligned} \tilde{J}(\gamma, \varphi) = & \int_0^T \int_\Omega C(x)\zeta_1(\gamma_1(t))H(\varphi(x))dx\,dt \\ & + \int_0^T \int_\Omega \zeta_2(\gamma_2(t)u_2(x,t))\gamma_2(t)u_2(x,t)H(\varphi(x))dx\,dt \\ & + \int_0^T \int_\Omega C(x)\zeta_3(\gamma_3(t))H(\varphi(x))dx\,dt \\ & + \int_0^T \int_\Omega l(u_2(x,t))dx\,dt + \alpha \int_\Omega H(\varphi(x))dx \\ & + \beta\,\text{perimeter}(\omega), \end{aligned} \tag{2.7}$$

where $\gamma \in \mathcal{G}_T$ and (u_1, u_2) is the solution to

$$\begin{cases} \dfrac{\partial u_1}{\partial t}(x,t) - d\Delta u_1(x,t) = \eta(x)u_1(x,t) + \displaystyle\int_\Omega k(x,x')u_2(x',t)dx' & \\ \qquad -\gamma_1(t)H(\varphi(x))u_1(x,t), & (x,t) \in Q_T, \\ \dfrac{\partial u_2}{\partial t}(x,t) = -(a_{22} + \gamma_2(t)H(\varphi(x)))u_2(x,t) & \\ \qquad +(1 - \gamma_3(t)H(\varphi(x)))g(x, u_1(x,t), u_2(x,t)), & (x,t) \in Q_T, \\ \dfrac{\partial u_1}{\partial n}(x,t) = 0, & (x,t) \in \Sigma_T, \\ u_1(x,0) = u_1^0(x), \quad u_2(x,0) = u_2^0(x), & x \in \Omega. \end{cases}$$

For computational purposes, the Heaviside function is approximated by its mollified version $H_\sigma(s) = \dfrac{1}{2}\left(1 + \dfrac{2}{\pi}\arctan\dfrac{s}{\sigma}\right)$ ($\sigma > 0$ is a sufficiently small number), and using the Modica–Mortola formula (see [157]) to "approximate" the perimeter of $\omega = \{x \, : \, \varphi(x) > 0\}$, we get the following approximation of J:

$$
\begin{aligned}
\Phi(\gamma, \varphi) = & \int_0^T \int_\Omega C(x)\zeta_1(\gamma_1(t)) H_\sigma(\varphi(x)) dx\, dt \\
& + \int_0^T \int_\Omega \tilde{\zeta}_2(\gamma_2(t) u_2^{\gamma,\varphi}(x,t)) H_\sigma(\varphi(x)) dx\, dt \\
& + \int_0^T \int_\Omega C(x)\zeta_3(\gamma_3(t)) H_\sigma(\varphi(x)) dx\, dt \\
& + \int_0^T \int_\Omega l(u_2^{\gamma,\varphi}(x,t)) dx\, dt + \alpha \int_\Omega H_\sigma(\varphi(x)) dx \\
& + \beta[\varepsilon \int_\Omega |\nabla H_\sigma(\varphi(x))|^2 dx \\
& + \frac{1}{\varepsilon} \int_\Omega (H_\sigma(\varphi(x)))^2 (1 - H_\sigma(\varphi(x)))^2 dx]
\end{aligned}
$$

($\varepsilon > 0$ is a small constant).

Here $(u_1^{\gamma,\varphi}, u_2^{\gamma,\varphi})$ is the solution to

$$
\begin{cases}
\dfrac{\partial u_1}{\partial t}(x,t) - d\Delta u_1(x,t) = \eta(x) u_1(x,t) + \displaystyle\int_\Omega k(x,x') u_2(x',t) dx' & \\
\qquad -\gamma_1(t) H_\sigma(\varphi(x)) u_1(x,t), & (x,t) \in Q_T, \\
\dfrac{\partial u_2}{\partial t}(x,t) = -(a_{22} + \gamma_2(t) H_\sigma(\varphi(x))) u_2(x,t) & \\
\qquad +(1 - \gamma_3(t) H_\sigma(\varphi(x))) g(x, u_1(x,t), u_2(x,t)), & (x,t) \in Q_T, \\
\dfrac{\partial u_1}{\partial n}(x,t) = 0, & (x,t) \in \Sigma_T, \\
u_1(x,0) = u_1^0(x), \quad u_2(x,0) = u_2^0(x), & x \in \Omega.
\end{cases}
\tag{2.8}
$$

A rigorous proof of the existence of a global optimum of the above cost functional is a very hard task. A usual search algorithm is based on the gradient method. The following theorem provides the gradient with respect to both γ and ϕ; its proof can be found in [16].

Theorem 2.9 *For any $\gamma = (\gamma_1, \gamma_2, \gamma_3) \in \mathcal{G}_T$ and $w \in L^\infty(0,T) \times L^\infty(0,T) \times L^\infty(0,T)$ $(w = (w_1, w_2, w_3))$, such that $\gamma + \rho w \in \mathcal{G}_T$, for any $\rho > 0$ sufficiently small, and for any smooth functions $\varphi, \psi : \overline{\Omega} \to \mathbb{R}$, we have that*

$$
\begin{aligned}
d\Phi(\gamma, \varphi)(w, \psi) = & \int_0^T w_1(t)\mathcal{F}_1(t) dt + \int_0^T w_2(t)\mathcal{F}_2(t) dt + \int_0^T w_3(t)\mathcal{F}_3(t) dt \\
& + \int_\Omega \psi(x)\mathcal{F}(x) H'_\sigma(\varphi(x)) dx + \int_{\partial\Omega} \psi(x)\mathcal{G}(x) H'_\sigma(\varphi(x)) d\tilde{\sigma},
\end{aligned}
$$

where

$$\mathcal{F}_1(t) = \int_\Omega H_\sigma(\varphi(x))[u_1^{\gamma,\varphi}(x,t)q_1^{\gamma,\varphi}(x,t) + C(x)\zeta_1'(\gamma_1(t))]dx,$$

$$\mathcal{F}_2(t) = \int_\Omega H_\sigma(\varphi(x))[u_2^{\gamma,\varphi}(x,t)q_2^{\gamma,\varphi}(x,t) + u_2^{\gamma,\varphi}(x,t)\tilde{\zeta}_2'(\gamma_2(t)u_2^{\gamma,\varphi}(x,t))]dx,$$

$$\mathcal{F}_3(t) = \int_\Omega H_\sigma(\varphi(x))\left[(C(x) - u_2^{\gamma,\varphi}(x,t))h\left(\frac{u_1^{\gamma,\varphi}(x,t)}{C(x)}\right)q_2^{\gamma,\varphi}(x,t) + C(x)\zeta_3'(\gamma_3(t))\right]dx,$$

$$\begin{aligned}
\mathcal{F}(x) = \int_0^T &\Big[C(x)\zeta_1(\gamma_1(t)) + \tilde{\zeta}_2(\gamma_2(t)u_2^{\gamma,\varphi}(x,t)) + C(x)\zeta_3(\gamma_3(t)) \\
&+\gamma_1(t)u_1^{\gamma,\varphi}(x,t)q_1^{\gamma,\varphi}(x,t) + \gamma_2(t)u_2^{\gamma,\varphi}(x,t)q_2^{\gamma,\varphi}(x,t) \\
&+\gamma_3(t)(C(x) - u_2^{\gamma,\varphi}(x,t))h\left(\frac{u_1^{\gamma,\varphi}(x,t)}{C(x)}\right)q_2^{\gamma,\varphi}(x,t) + \alpha\Big]dt \\
&-2\beta\varepsilon\Delta[H_\sigma(\varphi(x))] - \frac{\beta}{\varepsilon}[4(H_\sigma(\varphi(x)))^3 - 6(H_\sigma(\varphi(x)))^2 + 2H_\sigma(\varphi(x))],
\end{aligned}$$

and

$$\mathcal{G}(x) = \frac{2\beta}{\varepsilon}\frac{\partial H_\sigma}{\partial n}(\varphi(x)).$$

Here $(q_1^{\gamma,\varphi}, q_2^{\gamma,\varphi})$ *is the solution to*

$$\begin{cases}
\dfrac{\partial q_1}{\partial t}(x,t) + d\Delta q_1(x,t) = -\eta(x)q_1(x,t) + \gamma_1(t)H_\sigma'(\varphi(x))q_1(x,t) \\
\quad -[1-\gamma_3(t)H_\sigma(\varphi(x))]\dfrac{C(x) - u_2^{\gamma,\varphi}(x,t)}{C(x)}h'\left(\dfrac{u_1^{\gamma,\varphi}(x,t)}{C(x)}\right)q_2(x,t), \ (x,t) \in Q_T, \\
\dfrac{\partial q_2}{\partial t}(x,t) = -\displaystyle\int_\Omega k(x',x)q_1(x',t)dx' \\
\qquad +[a_{22} + \gamma_2(t)H_\sigma(\varphi(x))]q_2(x,t) \\
\qquad +[1 - \gamma_3(t)H_\sigma(\varphi(x))]h\left(\dfrac{u_1^{\gamma,\varphi}(x,t)}{C(x)}\right)q_2(x,t) \\
\qquad +\tilde{\zeta}_2'(\gamma_2(x,t)u_2^{\gamma,\varphi}(x,t))\gamma_2(t)H_\sigma(\varphi(x)) + l'(u_2^{\gamma,\varphi}(x,t)), & (x,t) \in Q_T, \\
\dfrac{\partial q_1}{\partial n}(x,t) = 0, & (x,t) \in \Sigma_T, \\
q_1(x,T) = q_2(x,T) = 0, & x \in \Omega.
\end{cases} \tag{2.9}$$

Based on the above theorem, the descent towards an optimal γ can be obtained by considering $\gamma_{new} = \gamma_{old} + sw$, where

$$w_j(t) = -\mathcal{F}_j(t), \quad j = 1, 2, 3,$$

and truncating if necessary.

The descent towards an optimal φ can be obtained by introducing an artificial time ρ. We introduce a function $\tilde{\varphi}(x, \rho)$ such that, given an initial guess $\varphi_{old}(x)$, we take $\tilde{\varphi}(x, 0) = \varphi_{old}(x)$. At each iteration as $\varphi_{new}(x)$, we take

$$\varphi_{new}(x) = \tilde{\varphi}(x, s),$$

the solution to the following PDE

$$\frac{\partial \tilde{\varphi}}{\partial \rho}(x, \rho) = -\mathcal{F}, \quad x \in \Omega, \rho \in (0, s),$$

subject to the boundary condition

$$\frac{\partial H_\sigma}{\partial n}(\tilde{\varphi}(x, \rho)) = 0, \quad x \in \partial\Omega, \rho \in (0, s),$$

and the initial condition

$$\tilde{\varphi}(x, 0) = \varphi_{old}(x).$$

Since

$$\frac{\partial H_\sigma}{\partial n}(\tilde{\varphi}(x, \rho)) = H_\sigma'(\tilde{\varphi}(x, \rho))\frac{\partial \tilde{\varphi}}{\partial n}(x, \rho),$$

we may take, as boundary condition, instead of (2.4),

$$\frac{\partial \tilde{\varphi}}{\partial n}(x, \rho) = 0, \quad x \in \partial\Omega, \rho \in (0, s).$$

2.5 Computational Issues

In the first paragraph, we describe the numerical methods adopted for the solution of systems (2.8) and (2.9). In the second paragraph, we provide the values of all parameters used in the numerical tests. In the last paragraph, we report and discuss the computational results. The optimal control procedure implemented here is based on the gradient method presented in Section 2.4 and the consequent conceptual algorithm described in [16]. The reported numerical simulations have been performed with an in-house MATLAB code [24].

We wish to stress here that the search of the optimal subregion ω is a very hard computational problem; mathematically the search should explore a very highly dimensional space of shapes. This is a well-known issue in the realm of statistical shape analysis [101]. Such an intrinsic complexity has been reduced via the choice of the initial function ϕ within a parameterized family of shapes (e.g., ellipses), hence in a finite dimensional space; then the search goes on by looking for the parameters of the chosen shape that optimize the relevant cost functional. This may explain why in the reported numerical simulations the final optimal subset ω keeps the same shape as the one given via the initial function ϕ. From a purely mathematical point of view the search is supposed to move among all possible shape classes, which is clearly unaffordable from a computational point of view.

2.5.1 *Numerical Methods*

Systems (2.8) and (2.9) are discretized by finite elements in space and finite differences in time.

2.5.1.1 Space Discretization

We first apply a standard Galerkin procedure to the weak formulations of systems (2.8) and (2.9). The unit square domain Ω is discretized by a uniform grid of 80×80 bilinear finite elements ($\mathbf{Q}^1$), yielding a total amount of $N = 6561$ discretization nodes. The stiffness matrix is computed exactly, whereas the mass matrix is obtained by applying the mass lumping technique.

In the following, we describe in detail this procedure applied to system (2.8), being the treatment of system (2.9) analogous.

Let us first denote by V the Sobolev space $H^1(\Omega)$, by $(\cdot, \cdot)$ the L^2 inner product, and by $a(\cdot, \cdot) : V \times V \to \mathbb{R}$ the bilinear form

$$a(u, v) := \int_\Omega d\nabla u \cdot \nabla v dx, \quad \text{for all } u, v \in V.$$

Then, using the Green formula, the variational formulation of problem (2.8) reads as follows:
find u_1 and u_2 such that

$$\begin{cases} \dfrac{\partial}{\partial t}(u_1(t), \phi) + a(u_1(t), \phi) = (R_1(u_1(t)), \phi) + (G(u_2(t)), \phi), \text{for all } \phi \in V, \\ \dfrac{\partial}{\partial t}(u_2(t), \psi) = (R_2(u_1(t), u_2(t)), \psi), \qquad \text{for all } \psi \in V, \end{cases} \tag{2.10}$$

subject to the initial conditions $u_1(x,0) = u_1^0(x)$ and $u_2(x,0) = u_2^0(x)$, and where

$$R_1(u_1(x,t)) = \eta(x)u_1(x,t) - \gamma_1(t)\mathbb{I}_\omega(x)u_1(x,t),$$

$$G(u_2(\cdot,t),x) = \int_\Omega k(x,x')u_2(x',t)dx',$$

$$\begin{aligned} R_2(u_1(x,t),u_2(x,t)) = &-(a_{22}+\gamma_2(t)\mathbb{I}_\omega(x))u_2(x,t) \\ &+(1-\gamma_3(t)\mathbb{I}_\omega(x))g(x,u_1(x,t),u_2(x,t)). \end{aligned}$$

Let us now introduce the $\mathbf{Q}^1$ finite element space V^h, associated with a Cartesian grid discretizing the unit square domain Ω. For a detailed description of this space, see Appendix B.

The space semi-discretization of problem (2.10) then reads as follows: find $u_{1,h}(\cdot,t) : (0,T] \longrightarrow V^h$ and $u_{2,h}(\cdot,t) : (0,T] \longrightarrow V^h$ such that

$$\begin{cases} \dfrac{\partial}{\partial t}(u_{1,h}(t),\phi) + a(u_{1,h}(t),\phi) = (R_1(u_{1,h}(t)),\phi) + (G(u_{2,h}(t)),\phi), \\ \qquad \text{for all } \phi \in V^h, \\ \dfrac{\partial}{\partial t}(u_{2,h}(t),\psi) = (R_2(u_{1,h}(t),u_{2,h}(t)),\psi), \\ \qquad \text{for all } \psi \in V^h, \end{cases}$$

which is equivalent to

$$\begin{cases} \dfrac{\partial}{\partial t}(u_{1,h}(t),\phi_i)+a(u_{1,h}(t),\phi_i)=(R_1(u_{1,h}(t)),\phi_i)+(G(u_{2,h}(t)),\phi_i), \\ \dfrac{\partial}{\partial t}(u_{2,h}(t),\phi_i) = (R_2(u_{1,h}(t),u_{2,h}(t)),\phi_i), \end{cases} \tag{2.11}$$

for $i = 1,\ldots,N$, where ϕ_i, $i = 1,\ldots,N$, are the *hat basis functions* of the $\mathbf{Q}^1$ finite element space V^h (see Appendix B for the definition of this basis).

Expanding the unknowns $u_{1,h}(x,t)$ and $u_{2,h}(x,t)$ with respect to the basis $\{\phi_i\}_{i=1}^N$, we have

$$u_{1,h}(x,t) = \sum_{j=1}^N U_{1,j}(t)\phi_j(x), \quad u_{2,h}(x,t) = \sum_{j=1}^N U_{2,j}(t)\phi_j(x), \tag{2.12}$$

where $U_{1,j}$, $U_{2,j} : [0,T] \longrightarrow \mathbb{R}$ are unknown scalar functions to be determined.

In the next steps, we follow closely the approximation of the nonlinear problem presented in Appendix B. Hence, denoting by A the stiffness matrix associated with the bilinear form $a(\cdot,\cdot)$ and by M the lumped mass matrix (see the appendix for the

construction of such matrices) and substituting the expressions (2.12) into system (2.11), the semi-discrete problem then reads as follows:
find $\mathbf{U}_1(t) : (0, T] \longrightarrow \mathbb{R}^N$ and $\mathbf{U}_2(t) : (0, T] \longrightarrow \mathbb{R}^N$ such that

$$\begin{cases} \mathrm{M}\dfrac{d\mathbf{U}_1}{dt}(t) + \mathrm{A}\mathbf{U}_1(t) = \mathbf{M}\mathbf{R}_1(\mathbf{U}_1(t)) + \mathbf{M}\mathbf{G}(\mathbf{U}_2(t)), \\ \mathrm{M}\dfrac{d\mathbf{U}_2}{dt}(t) = \mathbf{M}\mathbf{R}_2(\mathbf{U}_1(t), \mathbf{U}_2(t)), \end{cases}$$

where

$$\begin{aligned}
\mathbf{U}_1(t) &= [U_{1,1}(t), \dots, U_{1,j}(t), \dots, U_{1,N}(t)]^T, \\
\mathbf{U}_2(t) &= [U_{2,1}(t), \dots, U_{2,j}(t), \dots, U_{2,N}(t)]^T, \\
\mathbf{R}_1(\mathbf{U}_1(t)) &= [R_1(U_{1,1}(t)), \dots, R_1(U_{1,j}(t)), \dots, R_1(U_{1,N}(t))]^T, \\
\mathbf{G}(\mathbf{U}_2(t)) &= [G_1(\mathbf{U}_2(t)), \dots, G_j(\mathbf{U}_2(t)), \dots, G_N(\mathbf{U}_2(t))]^T, \\
\mathbf{R}_2(\mathbf{U}_1(t), \mathbf{U}_2(t)) &= [R_2(U_{1,1}(t), U_{2,1}(t)), \dots, R_2(U_{1,j}(t), U_{2,j}(t)), \\
&\quad \dots, R_2(U_{1,N}(t), U_{2,N}(t))]^T.
\end{aligned}$$

Denoting by x_j the generic discretization node, the element $G_j(\mathbf{U}_2(t))$ of the vector $\mathbf{G}(\mathbf{U}_2(t))$ is obtained as the following approximation of $G(u_{2,h}(t), x_j)$:

$$\begin{aligned}
G(u_{2,h}(t), x_j) &= \int_\Omega k(x_j, x')u_{2,h}(x', t)dx' \\
&= \int_\Omega k(x_j, x') \sum_{j=1}^N U_{2,j}(t)\phi_j(x')dx' \\
&\approx \int_\Omega \sum_{i=1}^N k(x_j, x_i)\phi_i(x') \sum_{j=1}^N U_{2,j}(t)\phi_j(x')dx' \\
&= \mathbf{K}(x_j)^T\mathbf{M}\mathbf{U}_2(t) \\
&=: G_j(\mathbf{U}_2(t)),
\end{aligned}$$

where

$$\mathbf{K}(x_j) = [k(x_j, x_1), \dots, k(x_j, x_i), \dots, k(x_j, x_N)]^T.$$

2.5.1.2 Time Discretization

After space discretization by finite elements, we obtain the semi-discrete problem that consists of a system of ordinary differential equations (ODEs). We solve these ODEs by employing a first order semi-implicit finite difference scheme, where the linear diffusion and reaction terms are approximated by Backward Euler, whereas the nonlinear reaction terms are approximated by Forward Euler.

In formulae, at the generic time step n,
given $\mathbf{U}_1^n$ and $\mathbf{U}_2^n$, we have to find $\mathbf{U}_1^{n+1}$ and $\mathbf{U}_2^{n+1}$ such that

$$\begin{cases} \mathrm{M}\dfrac{\mathbf{U}_1^{n+1}-\mathbf{U}_1^n}{\Delta t}+\mathrm{A}\mathbf{U}_1^{n+1}=\mathrm{M}\mathbf{R}_1(\mathbf{U}_1^n)+\mathrm{M}\mathbf{G}(\mathbf{U}_2^n), \\ \mathrm{M}\dfrac{\mathbf{U}_2^{n+1}-\mathbf{U}_2^n}{\Delta t}=\mathrm{M}\mathbf{R}_2(\mathbf{U}_1^n,\mathbf{U}_2^n). \end{cases}$$

As a result, at each time step it is required the solution of the linear system of algebraic equations

$$\begin{bmatrix} \dfrac{1}{\Delta t}\mathrm{M}+\mathrm{A} & 0 \\ 0 & \dfrac{1}{\Delta t}\mathrm{M} \end{bmatrix} \begin{bmatrix} \mathbf{U}_1^{n+1} \\ \mathbf{U}_2^{n+1} \end{bmatrix} = \begin{bmatrix} \dfrac{1}{\Delta t}\mathrm{M}\mathbf{U}_1^n+\mathrm{M}\mathbf{R}_1(\mathbf{U}_1^n)+\mathrm{M}\mathbf{G}(\mathbf{U}_2^n) \\ \dfrac{1}{\Delta t}\mathrm{M}\mathbf{U}_2^n+\mathrm{M}\mathbf{R}_2(\mathbf{U}_1^n,\mathbf{U}_2^n) \end{bmatrix},$$

which in our specific case has the dimension 2×6561 degrees of freedom. The linear system is solved by the Gaussian elimination with the built-in function of MATLAB. The time step size is $\Delta t = 0.05$.

For further details on the numerical approximation of parabolic equations, we refer, e.g., to [168].

2.5.2 *Parameter Calibration*

Except otherwise stated, the values of the parameters are given as detailed here below (see Tables 2.1, 2.2, 2.3, 2.4, and 2.5).

Functions $\eta(x_1, x_2)$ and $C(x_1, x_2)$, representing the growth rate of the infected mosquitoes and the spatial density of the total human population, respectively, are given by

$$\eta(x_1, x_2) = -10,$$

and

$$C(x_1, x_2) = 10^4 \exp\left(-5\frac{(x_1-0.45)^2}{30} - 5\frac{(x_2-0.55)^2}{30}\right).$$

Functions ζ_1, ζ_2, and ζ_3 are given by (see also Figure 2.4)

$$\zeta_1(z) = \alpha_1 \frac{z}{b_1(\Gamma_1 - z)},$$

with $\alpha_1 = 0.1$, $\Gamma_1 = 500$, and $b_1 = 1$;

$$\zeta_2(z) = \alpha_2 \frac{\Gamma_2 a_2 z + c_2}{1 + a_2 z},$$

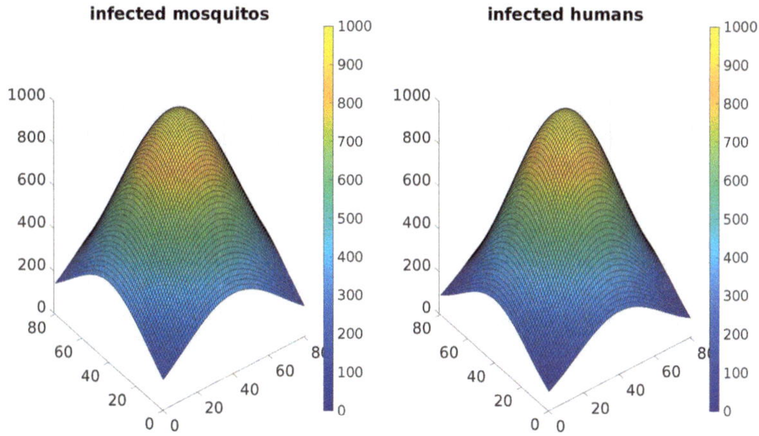

Fig. 2.3 Initial distribution of infected mosquitoes and humans, used in all numerical simulations

with $\alpha_2 = 0.01$, $\Gamma_2 = 500$, $a_2 = 0.02$, and $c_2 = 600$; and

$$\zeta_3(z) = \alpha_3 \frac{z}{b_3(\Gamma_3 - z)},$$

with $\alpha_3 = 0.01$, $\Gamma_3 = 0.85$, and $b_3 = 1$.

Functions h and l are taken as $h(r) = 4r$ and $l(r) = r$, whereas constants α, β, d, and a_{22} are set to $\alpha = 100$, $\beta = 0.01$, $d = 1e-4$, and $a_{22} = 0.5$. We recall that α is the coefficient in front of the cost functional term $\int_\Omega H(\varphi)$, which penalizes large values of the area of the control region. Regarding the diffusion coefficient d, its small value $(1e - 4)$ yields a reaction–diffusion problem with a dominant reaction. However, using higher diffusion coefficients values, we believe that non-realistic diffusion effects might occur, with a too fast spread of the outbreak.

The initial distributions of infected mosquitoes and humans, used in all next numerical simulations, are plotted in Figure 2.3 and are set as follows:

$$\begin{cases} u_1^0(x_1, x_2) = 10^3 \exp\left(-4(x_1 - 0.5)^2 - 4(x_2 - 0.5)^2\right), \\ u_2^0(x_1, x_2) = 10^3 \exp\left(-5(x_1 - 0.5)^2 - 5(x_2 - 0.5)^2\right). \end{cases}$$

2.5.3 Numerical Simulations

We report in this section the results obtained with the MATLAB code

```
run_opt_malaria.m,
```

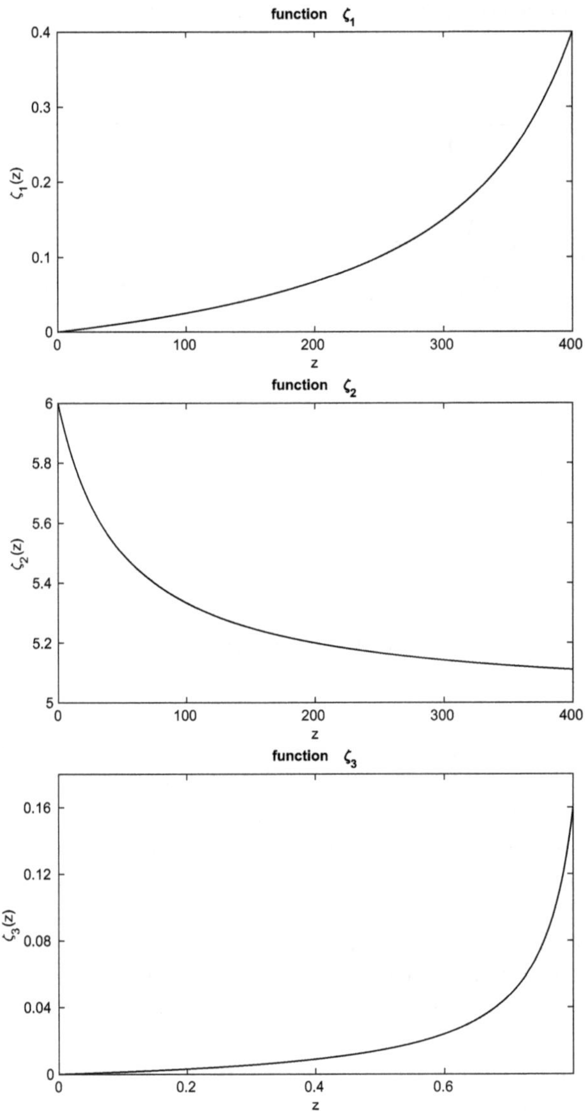

Fig. 2.4 Plots of functions ζ_1, ζ_2, and ζ_3

available on the Birkhäuser website provided in the Preface and also at the link

```
http://www.mat.unimi.it/users/scacchi/ACS_codes/
run_opt_malaria.m
```

2.5.3.1 Test 1: Reduction of the Infected Mosquitoes/Humans Distributions

In this first test, we keep fixed the control region ω with

$$\varphi(x_1, x_2) = \exp\left(-4(x-0.5)^2 - 4(y-0.5)^2\right) - 0.75,$$

and we investigate the influence of the control parameters γ_1, γ_2, and γ_3 on the cost functional and on the evolution of the infected mosquitoes and humans distributions. The kernel function $k(x, x')$ is the Dirac function centered in x, thus

$$k(x, x') = \delta(x' - x).$$

The results reported in Table 2.1 and Figures 2.5 and 2.6 show that the most sensitive parameter is γ_2, whose increase yields a significant decay of the number of infected mosquitoes and humans, with respect to non-controlled system.

Table 2.1 Test 1, reduction of the infected mosquitoes/humans distributions. Number of infected mosquitoes (# mosq) and humans (# hum) at final simulation time $T = 1$ as a function of control parameters γ_1, γ_2, and γ_3

γ_1	γ_2	γ_3	cost	# mosq	# hum
0	0	0	6899	68	671
100	0	0	6045	34	534
200	0	0	5965	28	508
300	0	0	6043	24	491
0	50	0	7092	22	199
0	100	0	8088	14	120
0	200	0	10,185	7	52
0	0	0.2	6691	66	642
0	0	0.4	6493	63	614
0	0	0.6	6321	60	588
100	50	0.2	6483	18	185
200	100	0.4	7425	10	105
300	200	0.6	9650	4	43

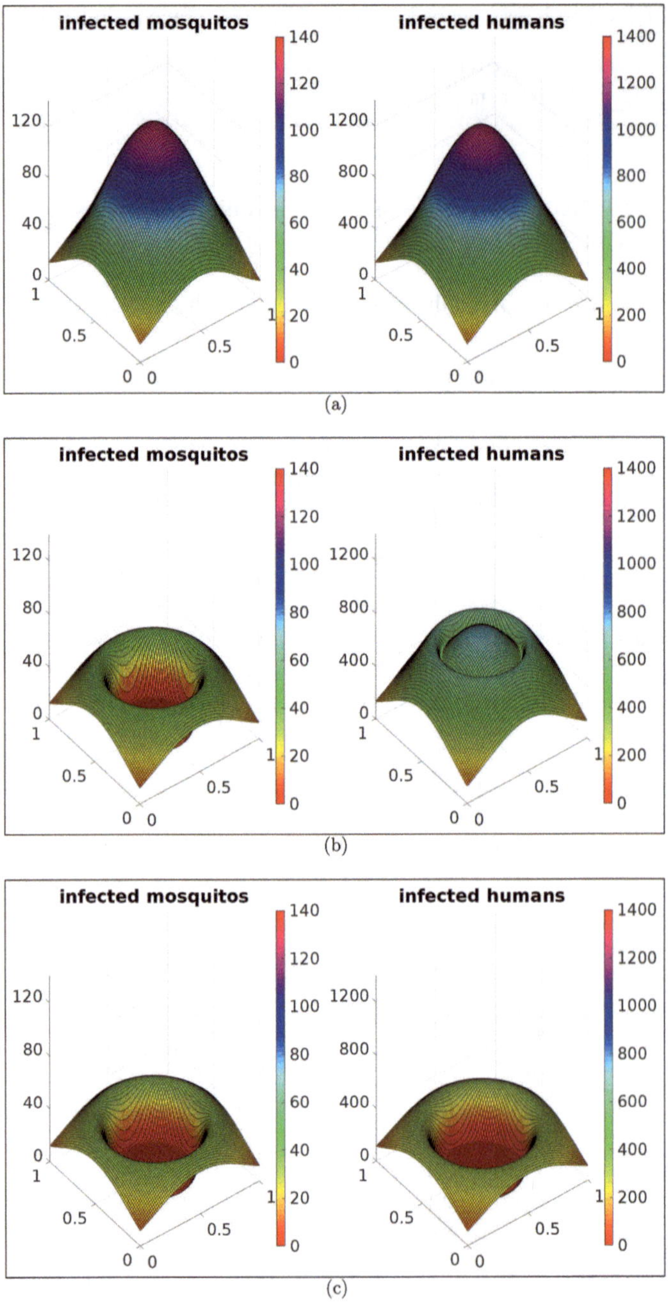

Fig. 2.5 Test 1, we solve System (2.8) with the discretization parameters detailed in Section 2.5. Distributions of infected mosquitoes and humans at final simulation time $T = 1$ for different choices of control parameters $(\gamma_1, \gamma_2, \gamma_3)$. Panel (**a**): (0,0,0), i.e., no control, panel (**b**): (200,0,0), and panel (**c**): (0,50,0). (**a**) $\gamma_1 = 0, \gamma_2 = 0, \gamma_3 = 0$ (no control). (**b**) $\gamma_1 = 200, \gamma_2 = 0, \gamma_3 = 0$. (**c**) $\gamma_1 = 0, \gamma_2 = 50, \gamma_3 = 0$

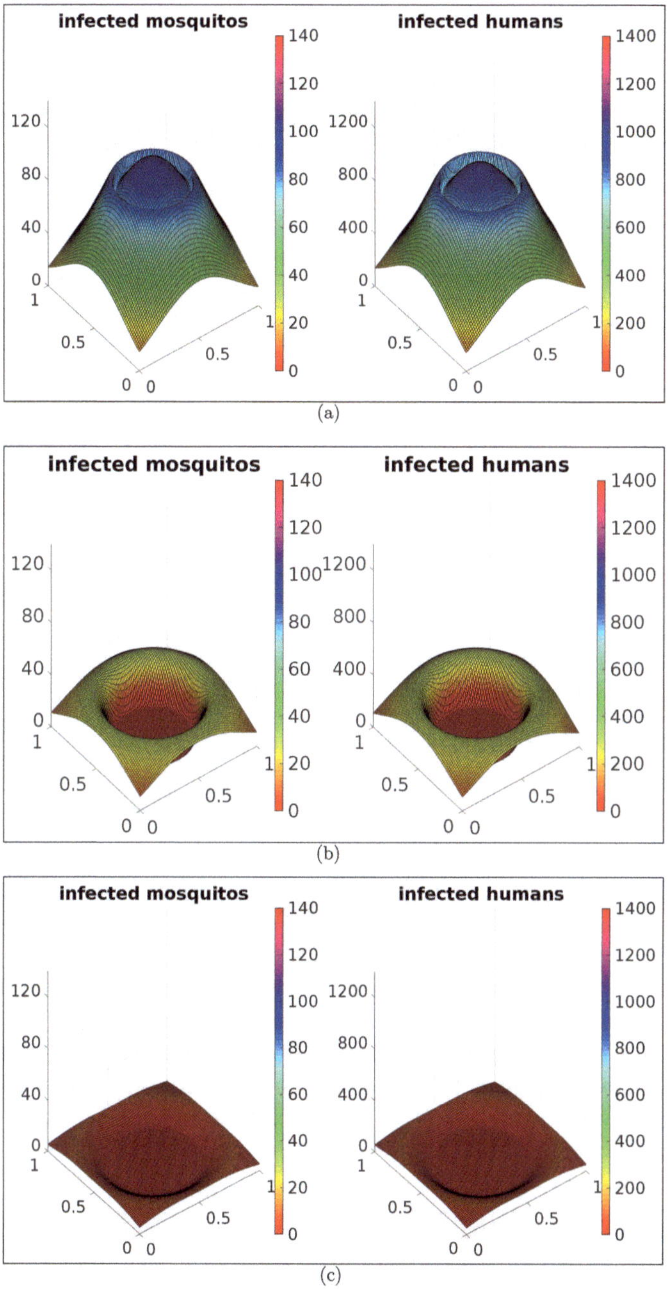

Fig. 2.6 Test 1, we solve System (2.8) with the discretization parameters detailed in Section 2.5. Distributions of infected mosquitoes and humans at final simulation time $T = 1$ for different choices of control parameters $(\gamma_1, \gamma_2, \gamma_3)$. Panel (**a**): (0,0,0.4), panel (**b**): (100,50,0.2), and panel (**c**): (300,200,0.6). (**a**) $\gamma_1 = 0$, $\gamma_2 = 0$, $\gamma_3 = 0.4$. (**b**) $\gamma_1 = 100$, $\gamma_2 = 50$, $\gamma_3 = 0.2$. (**c**) $\gamma_1 = 300$, $\gamma_2 = 200$, $\gamma_3 = 0.6$

2.5.3.2 Test 2: Optimal Control

In this test, we apply the optimal control procedure in four different scenarios (Cases 1–4), detailed here below, depending on the choice of the kernel function $k(x, x')$ and the initial shape of function φ.

- Case 1:

$$\begin{cases} k(x, x') = \exp\left(-\dfrac{(x_1 - x_1')^2 + (x_2 - x_2')^2}{0.1}\right), \\ \varphi(x_1, x_2) = \exp\left(-4(x - 0.5)^2 - 4(y - 0.5)^2\right) - 0.75. \end{cases}$$

- Case 2:

$$\begin{cases} k(x, x') = \exp\left(-\dfrac{(x_1 - x_1')^2 + (x_2 - x_2')^2}{0.1}\right), \\ \varphi(x_1, x_2) = \sin(\pi x_1)\sin(\pi x_2) - 0.4. \end{cases}$$

- Case 3:

$$\begin{cases} k(x, x') = \delta(x' - x), \\ \varphi(x_1, x_2) = \sin(\pi(x_1 + 0.05))\sin(\pi(x_2 + 0.09)) - 0.38(1 - 0.89x_2). \end{cases}$$

- Case 4: $\varphi(x_1, x_2) = 0$ has the quasi-elliptical shape reported in Figure 2.10 (the first row and second columns) and

$$k(x, x') = \delta(x' - x).$$

For each of the four cases, we perform two optimal control procedures: one minimizing the cost functional only with respect to the parameters γ_1, γ_2, and γ_3 (called in the following $(\gamma_1, \gamma_2, \gamma_3)$ control strategy); and one minimizing the cost functional both with respect to the parameters γ_1, γ_2, and γ_3 and to the function φ (called in the following $(\gamma_1, \gamma_2, \gamma_3, \varphi)$ control strategy).

The results reported in Tables 2.2, 2.3, 2.4, and 2.5 and Figures 2.7, 2.8, 2.9, 2.10, and 2.11 show that:

- Except case 1, the minimal cost is achieved by performing the $(\gamma_1, \gamma_2, \gamma_3, \varphi)$ control strategy.
- In all four cases, the area of the optimal control region is significantly reduced with respect to the initial control region.

Table 2.2 Test 2, optimal control, case 1. Evolution of cost functional, control parameters γ_1, γ_2, and γ_3 and the number of infected mosquitoes (# mosq) and humans (# hum) at final simulation time $T = 1$ as a function of the iterations (iter) of the optimal control procedure

Case 1: $(\gamma_1, \gamma_2, \gamma_3)$ control						
Iter.	γ_1	γ_2	γ_3	cost	# mosq	# hum
0	25.4	182.7	0.4	9059	1	42
1	400.0	23.6	0.8	6452	2	185
2	369.3	20.2	0.8	6136	2	196
3	328.1	15.2	0.8	5864	2	214
4	265.2	6.1	0.8	5660	3	256
5	264.0	5.8	0.6	5415	3	258
6	262.7	5.5	0.5	5414	3	261
Case 1: $(\gamma_1, \gamma_2, \gamma_3, \varphi)$ control						
Iter.	γ_1	γ_2	γ_3	cost	# mosq	# hum
0	25.4	182.7	0.4	9059	1	42
1	400.0	23.6	0.8	6531	8	421
2	200.9	11.9	0.8	6068	9	449
3	181.6	10.2	0.8	6068	10	458

Table 2.3 Test 2, optimal control, case 2. Evolution of cost functional, control parameters γ_1, γ_2, and γ_3 and the number of infected mosquitoes (# mosq) and humans (# hum) at final simulation time $T = 1$ as a function of the iterations (iter) of the optimal control procedure

Case 2: $(\gamma_1, \gamma_2, \gamma_3)$ control						
Iter.	γ_1	γ_2	γ_3	cost	# mosq	# hum
0	0.0	76.0	0.1	7950	1	56
1	268.7	66.6	0.8	7735	0	56
2	272.5	66.5	0.7	7284	0	56
3	278.5	66.4	0.6	7242	0	56
4	278.5	66.4	0.6	7242	0	56
Case 2: $(\gamma_1, \gamma_2, \gamma_3, \varphi)$ control						
Iter.	γ_1	γ_2	γ_3	cost	# mosq	# hum
0	158.4	191.9	0.39	11,853	0	16
1	400.0	143.3	0.8	10,772	4	269
2	400.0	121.0	0.8	9450	5	309
3	400.0	92.9	0.8	8354	6	339
4	400.0	55.8	0.8	7370	6	359
5	191.4	49.7	0.8	6583	7	368
6	170.1	49.0	0.2	6372	7	368
7	170.1	49.0	0.2	6372	7	368

Table 2.4 Test 2, optimal control, case 3. Evolution of cost functional, control parameters γ_1, γ_2, and γ_3 and the number of infected mosquitoes (# mosq) and humans (# hum) at final simulation time $T = 1$ as a function of the iterations (iter) of the optimal control procedure

Case 3: $(\gamma_1, \gamma_2, \gamma_3)$ control						
Iter.	γ_1	γ_2	γ_3	cost	# mosq	# hum
0	0.0	76.0	0.01	8586	3	27
1	69.8	75.7	0.1	7719	3	25
2	91.8	75.8	0.2	7696	2	24
3	108.8	75.8	0.2	7690	2	24
Case 3: $(\gamma_1, \gamma_2, \gamma_3, \varphi)$ control						
Iter.	γ_1	γ_2	γ_3	cost	# mosq	# hum
0	0.0	76.0	0.01	8586	3	27
1	69.8	75.7	0.1	7971	13	133
2	181.3	75.4	0.5	7931	19	190
3	400.0	12.4	0.8	7517	26	254
4	186.5	8.5	0.8	5968	28	278
5	163.0	7.9	0.4	5478	29	283
6	163.0	7.9	0.4	5478	29	283

Table 2.5 Test 2, optimal control, case 4. Evolution of cost functional, control parameters γ_1, γ_2, and γ_3 and the number of infected mosquitoes (# mosq) and humans (# hum) at final simulation time $T = 1$ as a function of the iterations (iter) of the optimal control procedure

Case 4: $(\gamma_1, \gamma_2, \gamma_3)$ control						
Iter.	γ_1	γ_2	γ_3	cost	# mosq	# hum
0	0.0	182.7	0.0	11,314	14	114
1	400.0	61.0	0.8	8442	17	202
2	141.1	47.9	0.8	6984	23	237
3	118.2	47.0	0.1	6636	24	241
4	118.2	47.0	0.1	6636	24	241
Case 4: $(\gamma_1, \gamma_2, \gamma_3, \varphi)$ control						
Iter.	γ_1	γ_2	γ_3	cost	# mosq	# hum
0	0.0	182.7	0.0	11,314	14	114
1	400.0	61.0	0.8	8441	38	377
2	400.0	47.5	0.8	7671	47	460
3	400.0	28.9	0.8	7128	54	529
4	121.1	26.0	0.8	6693	59	576
5	92.7	25.7	0.6	6621	60	588
6	92.4	25.7	0.6	6621	60	588

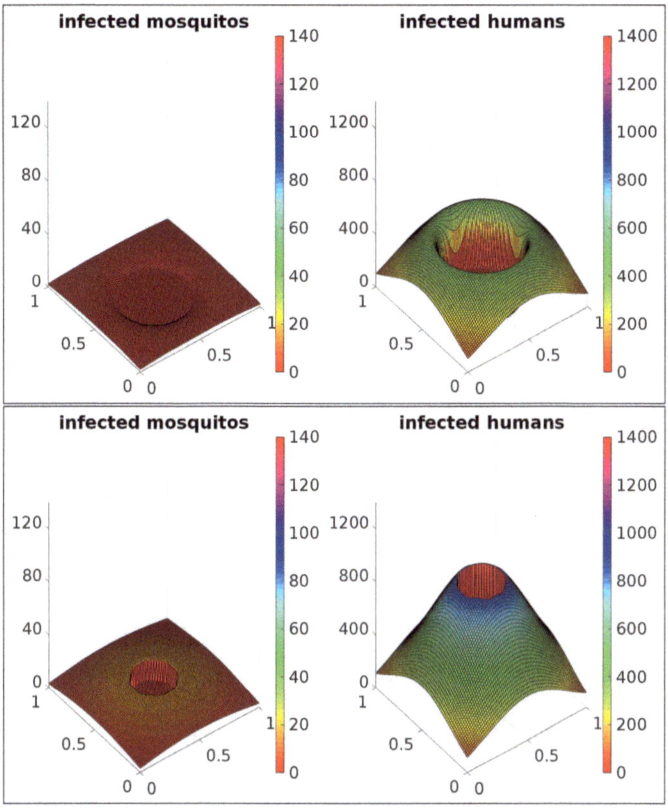

Fig. 2.7 Test 2. We solve the optimal control problem detailed in Section 2.4, with the discretization parameters given in Section 2.5. Case 1 (see parameter description reported in Section 2.5.3.2), optimal distributions of infected mosquitoes and humans at final simulation time $T = 1$ for (γ_1, γ_2, γ_3) control (first row) and (γ_1, γ_2, γ_3, φ) control (second row)

We observe that the optimization process is highly sensitive with respect to the area of the control region, which is always significantly reduced in the optimal configuration with respect to the initial one, whereas the shape and location of the optimal region remain similar to the initial one. Thus, the simulations and the optimal control procedures show that acting only on a subregion, with a significantly smaller area than the entire domain, might control successfully the epidemics.

Moreover, the results show that when the control region is located at the center of the domain (cases 1–2–4), where the epidemics is stronger, the optimal configuration has a much smaller area than in case 3, where the control region is shifted with respect to the center of the epidemics.

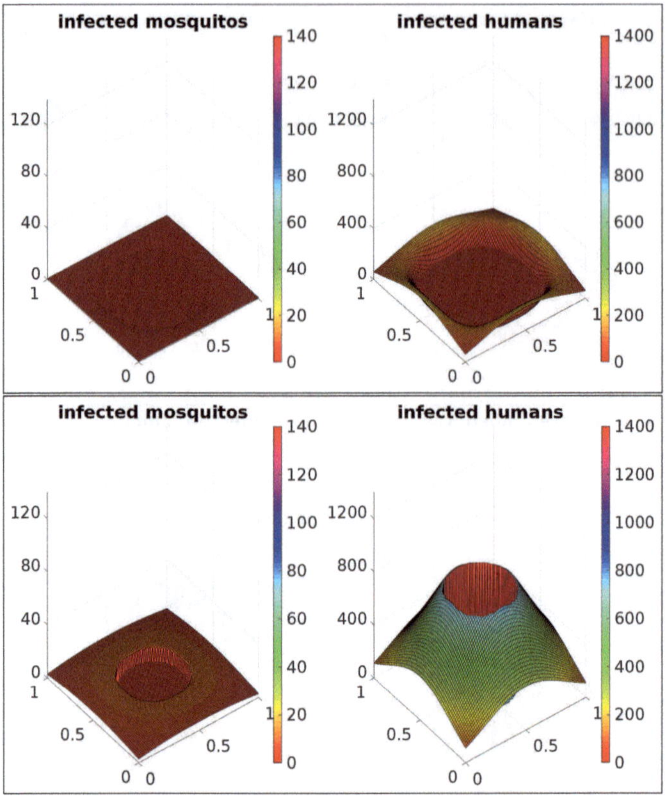

Fig. 2.8 Test 2. We solve the optimal control problem detailed in Section 2.4, with the discretization parameters given in Section 2.5. Case 2 (see parameter description reported in Section 2.5.3.2), optimal distributions of infected mosquitoes and humans at final simulation time $T = 1$ for (γ_1, γ_2, γ_3) control (first row) and (γ_1, γ_2, γ_3, φ) control (second row)

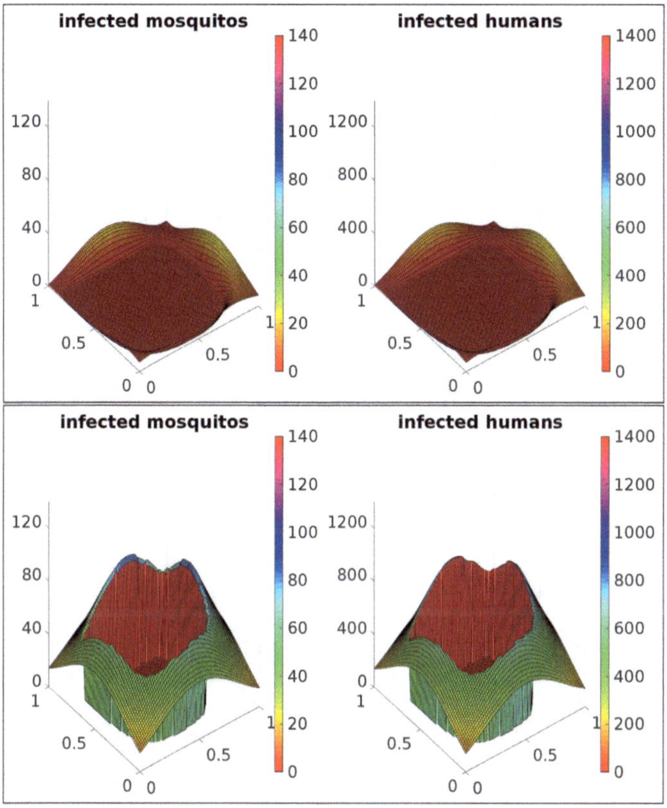

Fig. 2.9 Test 2. We solve the optimal control problem detailed in Section 2.4, with the discretization parameters given in Section 2.5. Case 3 (see parameter description reported in Section 2.5.3.2), optimal distributions of infected mosquitoes and humans at final simulation time $T = 1$ for $(\gamma_1, \gamma_2, \gamma_3)$ control (first row) and $(\gamma_1, \gamma_2, \gamma_3, \varphi)$ control (second row)

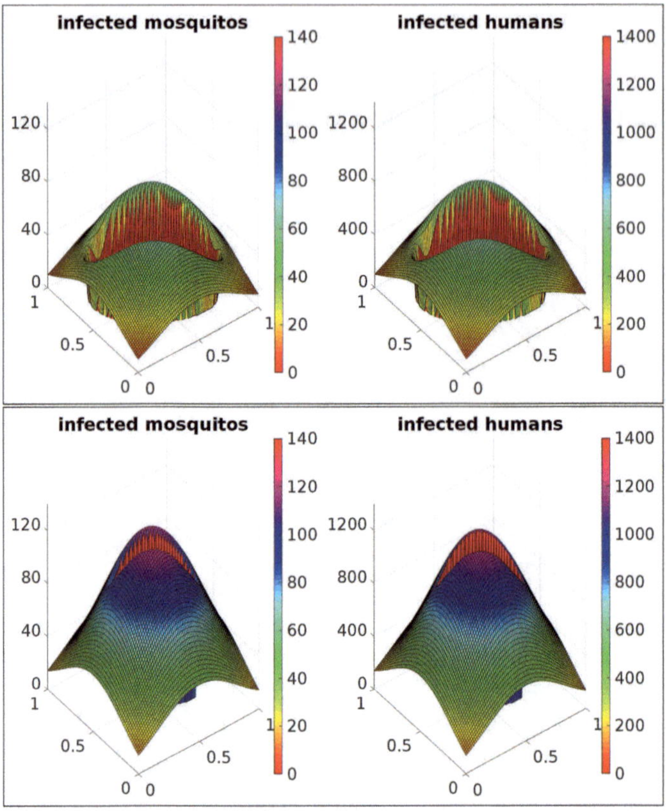

Fig. 2.10 Test 2. We solve the optimal control problem detailed in Section 2.4, with the discretization parameters given in Section 2.5. Case 4 (see parameter description reported in Section 2.5.3.2), optimal distributions of infected mosquitoes and humans at final simulation time $T = 1$ for $(\gamma_1, \gamma_2, \gamma_3)$ control (first row) and $(\gamma_1, \gamma_2, \gamma_3, \varphi)$ control (second row)

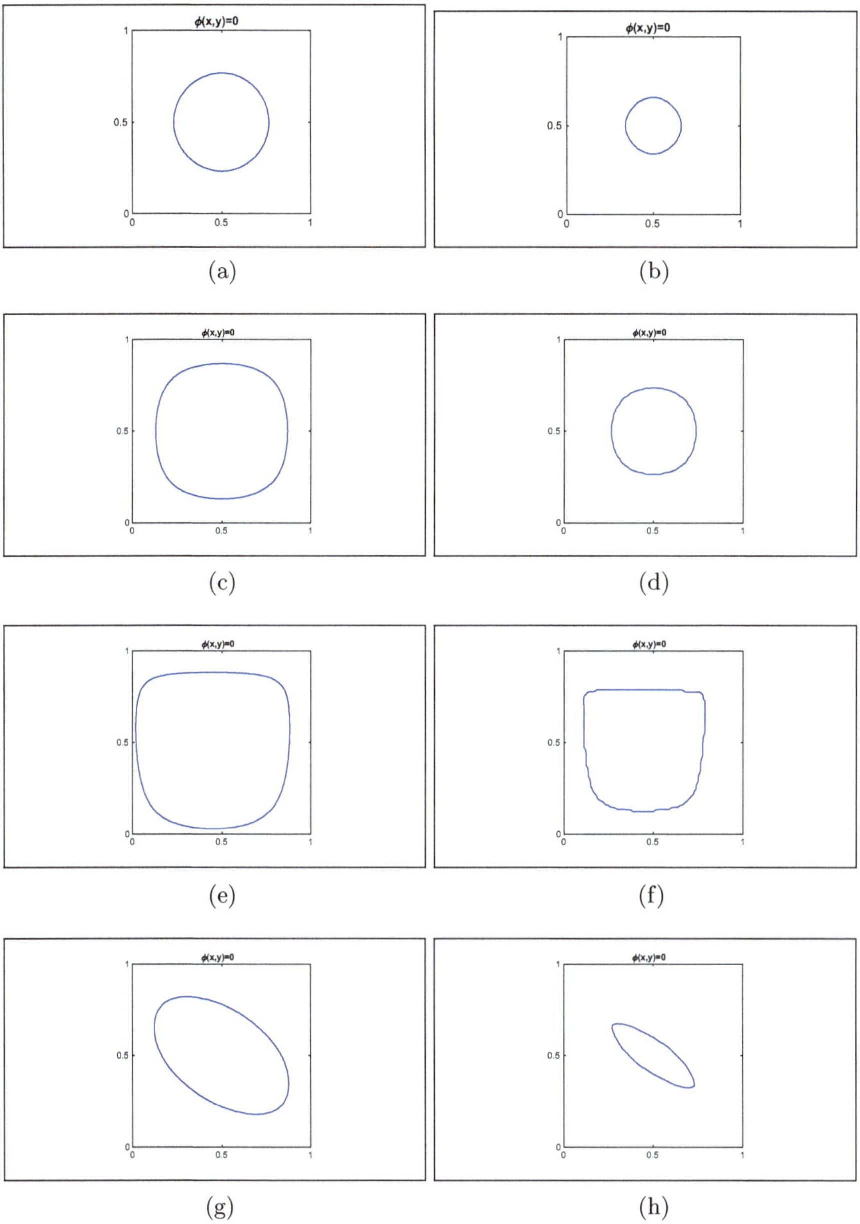

Fig. 2.11 Test 2. We solve the optimal control problem detailed in Section 2.4, with the discretization parameters given in Section 2.5. Cases 1–4 (see parameter description reported in Section 2.5.3.2), optimal control regions. (**a**) Case 1, (γ_1, γ_2, γ_3) control. (**b**) Case 1, (γ_1, γ_2, γ_3, φ) control. (**c**) Case 2, (γ_1, γ_2, γ_3) control. (**d**) Case 2, (γ_1, γ_2, γ_3, φ) control. (**e**) Case 3, (γ_1, γ_2, γ_3) control. (**f**) Case 3, (γ_1, γ_2, γ_3, φ) control. (**g**) Case 4, (γ_1, γ_2, γ_3) control. (**h**) Case 4, (γ_1, γ_2, γ_3, φ) control

2.6 Concluding Remarks

As in our previous literature, first of all we have shown, supported by numerical simulations in a variety of parameter scenarios, that it is indeed possible to eventually eradicate a man–environment epidemic, by acting only on the environment, while anyhow medically treating the human infected population. As stressed in [16], we do not claim that the simplified model presented here as a working example contains all details which might be required to act on a real epidemic system; still our hope is that our simulations may convince the public health authorities that it is possible, and it can be affordable, to control the full system by implementing the control strategies only in a suitably chosen subregion of the relevant habitat.

Part II
Optimal Harvesting

Chapter 3
Optimal Harvesting: Space Dependence

3.1 Introduction and Setting of the Problem

Optimal harvesting problems have been intensively studied for structured population dynamics by several authors. We wish further to evidence that possible space (and time) heterogeneities of the domain Ω can be taken in due account via the parameters of the relevant dynamical system.

Spatially structured harvesting problems have usually been treated by considering an effort applied to the whole domain Ω of interest for the relevant population (see, e.g., [8]). Actually in real systems, because of financial and logistic constraints, the effort is usually localized in a suitably chosen subregion ω of Ω (see [11]).

Hence the point we wish to raise in this chapter regards not only the optimization with respect to the intensity (magnitude) of the harvesting effort but also with respect to the choice of the subregion ω where it is implemented.

We wish to emphasize that such problems may be treated as shape optimization problems. Hence, to begin with, we remind that from a measure theoretical point of view the geometry of a set ω can be characterized in terms of its Minkowski functionals or intrinsic volumes. In 2D there are three such functionals, and these are proportional to more commonly known quantities such as the area, the perimeter, and the Euler–Poincaré characteristic (see, e.g., [123], [139, p.30]). Intrinsic volumes have been successfully used in material science to characterize and discriminate morphology of various media (see, e.g., [31, 162]).

Here we shall limit ourselves to refer to two of such functionals: the area and the perimeter. On the other hand, it is of course easily understandable, from the point of view of our optimal control problems, that in order to optimize the harvesting region, we have to take into account both its area and its perimeter.

Consider a domain Ω (we may remind that by domain $\Omega \subset \mathbb{R}^2$ we mean an open connected subset of $\mathbb{R}^2$; see e.g., [200, p.423]). A convenient tool for handling the shape of an open subset $\omega \subset \Omega$ in $\mathbb{R}^2$ is the so-called implicit interface

S. Aniţa et al., *Mathematical Modeling and Control in Life and Environmental Sciences*, Modeling and Simulation in Science, Engineering and Technology,
https://doi.org/10.1007/978-3-031-49971-5_3

representation, according to which the boundary of a domain is defined as the isocontour of some function φ (see, e.g., [163, p.3], [96, p.75]).

In accordance with such representation of ω, it may be expressed in terms of a function $\varphi : \Omega \to \mathbb{R}$, called its implicit function, as $\omega = \{x \in \Omega;\ \varphi(x) > 0\}$ (or $\omega = \{x \in \Omega;\ \varphi(x) < 0\}$), while its boundary is defined as the zero level set of φ: $\partial\omega = \{x \in \Omega;\ \varphi(x) = 0\}$.

A relevant tool to carry out integration over the interior of the domain ω is then the Heaviside function $H : \mathbb{R} \to \{0, 1\}$, such that

$$H(z) = \begin{cases} 1, & \text{if} \quad z \geq 0, \\ 0, & \text{if} \quad z < 0. \end{cases}$$

If φ is an implicit function of ω, in order to integrate over ω a function f defined over the whole Ω we may write $\int_\Omega f(x)H(\varphi(x))dx$. If we need to integrate over the boundary of ω, we may then refer to the Dirac Delta function as the generalized derivative of the Heaviside function H. In fact, if φ is sufficiently smooth, the directional derivative of the Heaviside function in the normal direction at a point $x \in \partial\omega$ is given by $\widehat{\delta}(x) = H'(\varphi(x))|\nabla\varphi(x)|$, and, by using the usual Dirac Delta δ on $\mathbb{R}$, we have $\widehat{\delta}(x) = \delta(\varphi(x))|\nabla\varphi(x)|$. Hence the boundary integral over $\partial\omega$ of a function f defined over the whole Ω may be formally expressed as $\int_\Omega f(x)\delta(\varphi(x))|\nabla\varphi(x)|dx$ if φ is a distance function as well.

This is actually a curve integral along $\partial\omega$. Provided φ is a distance function, the length of $\partial\omega$ is obtained and given by $\int_\Omega \delta(\varphi(x))|\nabla\varphi(x)|dx$.

In the above it has been taken into account that, in terms of the implicit function φ, the unit (outward) normal to the boundary at point $x \in \partial\omega$ can be expressed as $n(x) = -\dfrac{\nabla\varphi(x)}{|\nabla\varphi(x)|}$ (or $n(x) = \dfrac{\nabla\varphi(x)}{|\nabla\varphi(x)|}$).

We wish to mention that, for a similar regional control problem, in [53, 88, 89], a representation of the harvesting region has been expressed in terms of suitable measures.

Consider the density $y^u(x, t)$, at position $x \in \Omega$ and time $t \geq 0$, of a population that lives in a habitat $\Omega \subset \mathbb{R}^2$ and is subject to a harvesting effort $u(x, t)$ (localized in a subregion $\omega \subset \Omega$, during $t \in [0, T]$). Our main purpose is to investigate the following optimal harvesting problem:

(OH) $\qquad \underset{\omega}{Maximize}\ \underset{u \in \mathcal{U}_\omega}{Maximize}\ \Phi_\omega(u),$

where

$$\Phi_\omega(u) = \int_0^T \int_\omega [u(x,t)y^u(x,t) - cu(x,t)]dx\,dt - c_1\text{length}(\partial\omega) - c_2\text{area}(\omega),$$

and

$$\mathcal{U}_\omega = \{v \in L^\infty(\omega \times (0, T)) : 0 \leq v(x,t) \leq L, \ \text{ a.e. in } \omega \times (0, T)\}.$$

T, L, c_1, and c_2 are positive constants and $c \in [0, +\infty)$ (c represents the unitary cost of the harvesting effort, $\int_0^T \int_\omega cu(x,t)dx\,dt$ is the cost of the total harvesting effort, and $c_1 \text{length}(\partial\omega) + c_2 \text{area}(\omega)$ is the cost paid for harvesting in the specific subregion ω). It means that we wish to maximize (with respect to both the harvesting effort and the subregion, ω) the profit, which is the difference between the gain deriving from harvested population and the cost of the harvesting effort plus the cost paid to harvest in the subregion, ω.

In terms of the implicit interface representation, the problem can be rewritten as

$$\underset{\varphi}{Maximize}\ \underset{u \in \mathcal{U}_\omega}{Maximize}\ \Psi_\varphi(u),$$

where

$$\Psi_\varphi(u) = \int_0^T \int_\omega (uy^u - cu)dx\,dt - c_1 \int_\Omega \delta(\varphi(x))|\nabla\varphi(x)|dx - c_2 \int_\Omega H(\varphi(x))dx,$$

and $\omega = \{x \in \Omega;\ \varphi(x) > 0\}$.

In order to reduce the computational complexity of the control problem, it has been treated in two steps: firstly, for a fixed ω (or φ) we derive first order optimality conditions and go further using them to get the structure of the optimal harvesting effort (control). Next we investigate the second problem by adapting some shape optimization techniques.

Note that some particular harvesting problems related to structured population dynamics have been studied by several authors; see [6–9, 38, 46, 47, 60, 61, 109, 121, 122, 124, 129, 150, 159, 198, 199]. For optimal control related to biological models, see [145]. For basic results and methods in shape optimization, we refer to [63, 96, 126, 163, 178, 184].

3.2 A Spatially Structured Population with Nonlocal Logistic Term

Consider a habitat $\Omega \subset \mathbb{R}^N$, $N \in \{2, 3\}$, a bounded domain with a sufficiently smooth boundary $\partial\Omega$. In [23] the following model has been analyzed, describing the evolution of a single population species:

$$\begin{cases} \dfrac{\partial y}{\partial t}(x,t) - d\Delta y(x,t) = \eta(x)y(x,t) & \\ \qquad -y(x,t)\displaystyle\int_\Omega k(x,x')y(x',t)dx' - \mathbb{I}_\omega(x)u(x,t)y(x,t), & (x,t) \in Q, \\ \dfrac{\partial y}{\partial n}(x,t) = 0, & (x,t) \in \Sigma, \\ y(x,0) = y_0(x), & x \in \Omega, \end{cases} \tag{3.1}$$

where $T \in (0, +\infty)$, $Q = \Omega \times (0, T)$, and $\Sigma = \partial\Omega \times (0, T)$. Here $y(x, t)$ is the population density at position $x \in \overline{\Omega}$ and moment $t \in [0, T]$; $d \in (0, +\infty)$ is a diffusion coefficient, and $\eta(x)$ is the growth rate at position x (the difference between the fertility rate and the mortality rate).

The logistic term

$$y(x,t)\int_{\Omega} k(x,x')y(x',t)dx'$$

is nonlocal; it represents the impact on the growth rate due to the competition for resources with all other individuals within the habitat. For related biological models with diffusion and nonlocal terms, we refer to [9, 115].

The homogeneous Neumann boundary condition means that there is no population flow through the boundary of the habitat; $y_0(x)$ is the initial population density at position $x \in \Omega$.

In System (3.1), $u(x, t)$ represents a harvesting effort (control) that acts in the subregion $\omega \subset \Omega$ ($\mathbb{I}_\omega$ is the characteristic function of ω, i.e., $\mathbb{I}_\omega(x) = 1$, for $x \in \omega$, and $\mathbb{I}_\omega(x) = 0$, for $x \notin \omega$). For any v defined on ω, we shall denote

$$\mathbb{I}_\omega(x)v(x) = \begin{cases} v(x), & \text{if } x \in \omega, \\ 0, & \text{if } x \in \Omega \setminus \omega. \end{cases}$$

The term $\mathbb{I}_\omega(x)u(x, t)y(x, t)$ gives the density of the harvested population at position x and time t.

The first issue to be investigated is the following optimal harvesting problem:

$$\textbf{(OH1)} \qquad \underset{u\in\mathcal{U}_\omega}{Maximize}\ \gamma\int_0^T\int_\omega [u(x,t)y^u(x,t) - cu(x,t)]dx\,dt,$$

where $L, \gamma \in (0, +\infty)$ and $c \in [0, +\infty)$ are constants and y^u is the (weak) solution to (3.1).

Actually it means that we are interested to maximize the profit (i.e., the difference between the gain deriving from the harvested population and the cost of the harvesting effort). The constant γ is a dimensional constant referring to the unitary cost of the harvested population; the unitary cost of the harvesting effort is $\gamma c \in [0, +\infty)$). We may assume that $\gamma = 1$.

Here are the hypotheses we are going to use throughout this section.

(H1) $\eta, y_0 \in L^\infty(\Omega)$, $y_0(x) \geq 0$, a.e. in Ω, and $y_0 \neq 0_{L^\infty(\Omega)}$.
(H2) $k \in L^\infty(\Omega \times \Omega)$, $k(x, x') \geq 0$, a.e. in $\Omega \times \Omega$.

Let us specify first what we mean by a solution to the following initial/boundary value problem (see Appendix A):

$$\begin{cases} \dfrac{\partial w}{\partial t}(x,t) - d\Delta w(x,t) = f(x,t), \ (x,t) \in Q, \\ \dfrac{\partial w}{\partial n}(x,t) = 0, \qquad\qquad\qquad\qquad (x,t) \in \Sigma, \\ w(x,0) = w_0(x), \qquad\qquad\qquad\quad x \in \Omega. \end{cases} \tag{3.2}$$

We say that $w \in C([0,T]; L^2(\Omega)) \cap AC((0,T]; L^2(\Omega)) \cap L^2(0,T; H^1(\Omega)) \cap L^2_{loc}((0,T]; H^2(\Omega))$ is a (weak) solution to (3.2) if it satisfies

$$\begin{cases} \dfrac{\partial w}{\partial t}(x,t) - d\Delta w(x,t) = f(x,t), \text{ a.e. in } Q, \\ \dfrac{\partial w}{\partial n}(x,t) = 0, \qquad\qquad\qquad\qquad \text{a.e. in } \Sigma, \\ w(x,0) = w_0(x), \qquad\qquad\qquad\quad \text{a.e. in } \Omega. \end{cases}$$

It is known that for any $w_0 \in L^2(\Omega)$ and $f \in L^2(Q)$, Problem (3.2) admits a unique (weak) solution (see [8, p.187] and [39, p.227]). Moreover, we have that, for any $t \in (0,T]$,

$$\frac{1}{2}\int_\Omega |w(x,t)|^2 dx + d\int_0^t \int_\Omega |\nabla w(x,s)|^2 dx\, ds \\ \leq C\left[\int_\Omega |w_0(x)|^2 dx + \int_0^t \int_\Omega |f(x,s)|^2 dx\, ds\right],$$

where $C \in (0,+\infty)$ is a constant.

In a similar way, for the following problem

$$\begin{cases} \dfrac{\partial w}{\partial t}(x,t) - d\Delta w(x,t) = F(w)(x,t), \ (x,t) \in Q, \\ \dfrac{\partial w}{\partial n}(x,t) = 0, \qquad\qquad\qquad\qquad (x,t) \in \Sigma, \\ w(x,0) = w_0(x), \qquad\qquad\qquad\quad x \in \Omega, \end{cases} \tag{3.3}$$

we say that w is a (weak) solution to (3.3) if $w \in C([0,T]; L^2(\Omega)) \cap AC((0,T]; L^2(\Omega)) \cap L^2(0,T; H^1(\Omega)) \cap L^2_{loc}((0,T]; H^2(\Omega))$ and satisfies

$$\begin{cases} \dfrac{\partial w}{\partial t}(x,t) - d\Delta w(x,t) = F(w)(x,t), \text{ a.e. in } Q, \\ \dfrac{\partial w}{\partial n}(x,t) = 0, \qquad\qquad\qquad\qquad \text{a.e. in } \Sigma, \\ w(x,0) = w_0(x), \qquad\qquad\qquad\quad \text{a.e. in } \Omega. \end{cases}$$

Problem (3.1) is a particular case of (3.3).

Let us prove now that for any $u \in \mathcal{U}_\omega$, (3.1) admits a unique solution.
Notice first that if y is a solution to (3.1) (where $u \in \mathcal{U}_\omega$), then the mapping $x \mapsto \int_\Omega k(x,x')y(x',t)dx'$ belongs to $L^\infty(Q)$. It follows that the first equation in (3.1) has the form

$$\frac{\partial y}{\partial t}(x,t)(x,t) = d\Delta y(x,t) + \rho(x,t)y(x,t), \quad (x,t) \in Q,$$

where $\rho \in L^\infty(Q)$. We get now via a comparison result for parabolic equations (see Appendix A [113, 165], and [11] for parabolic equations with a nonlocal term) and using the assumptions on y_0 that

$$0 < y(x,t) \le \tilde{y}(x,t), \quad \text{a.e. } x \in \Omega, \text{ for any } t \in (0,T],$$

where $\tilde{y}$ is the solution to (3.1) corresponding to $u \equiv 0$ and $k \equiv 0$. Actually, via the comparison result we get that $\tilde{y} \in C([0,T];L^2(\Omega)) \cap L^\infty(Q)$ ($\tilde{y}(x,t) \le \|y_0\|_\infty \exp(t\|\eta\|_\infty)$, a.e. $x \in \Omega$, for any $t \in [0,T]$).

Let $X = \{\zeta \in C([0,T];L^2(\Omega)) : 0 \le \zeta(x,t) \le \tilde{y}(x,t)$, a.e. $x \in \Omega$, for any $t \in [0,T]\}$. X is a closed subset of the Banach space $C([0,T];L^2(\Omega))$ endowed with the norm $\|\zeta\| = \max_{t\in[0,T]} \{e^{-\lambda t}\|\zeta(t)\|_{L^2(\Omega)}\}$, where λ is a positive constant to be made precise later.

$\|\cdot\|_p$ or $\|\cdot\|_{L^p(\Omega)}$ shall denote the usual norm on $L^p(\Omega)$, for $p \ge 1$. $\|\cdot\|_\infty$ or $\|\cdot\|_{L^\infty(\Omega)}$ shall denote the usual norm on $L^\infty(\Omega)$.

So, $(X, dist)$ is a complete metric space, where $dist$ is the distance induced by the above-mentioned norm.

For any $\zeta \in X$ we denote by $g(\zeta)$ the solution to

$$\begin{cases} \dfrac{\partial y}{\partial t}(x,t) - d\Delta y(x,t) = \eta(x)y(x,t) \\ \qquad -y(x,t)\displaystyle\int_\Omega k(x,x')\zeta(x',t)dx' - \mathbb{I}_\omega(x)u(x,t)y(x,t), & (x,t) \in Q, \\ \dfrac{\partial y}{\partial n}(x,t) = 0, & (x,t) \in \Sigma, \\ y(x,0) = y_0(x), & x \in \Omega. \end{cases} \tag{3.4}$$

Since for any $\zeta \in X$, the mapping $x \mapsto \int_\Omega k(x,x')\zeta(x',t)dx'$ belongs to $L^\infty(Q)$, it follows that for such ζ, Problem (3.4) admits a unique solution $g(\zeta)$. By the comparison result for parabolic equations, we get that $g(\zeta) \in X$.

After an easy calculation we get that there exist two positive constants C_1 and C_2 such that for any $\zeta_1, \zeta_2 \in X$ and $t \in [0,T]$:

$$\|g(\zeta_1)(t) - g(\zeta_2)(t)\|^2_{L^2(\Omega)} \le C_1 \int_0^t \|g(\zeta_1)(s) - g(\zeta_2)(s)\|^2_{L^2(\Omega)} ds$$

$$+ C_2 \int_0^t \|\zeta_1(s) - \zeta_2(s)\|^2_{L^2(\Omega)} ds.$$

By Gronwall's inequality it follows that there exists a positive constant C such that for any $\zeta_1, \zeta_2 \in X$ and $t \in [0, T]$:

$$\|g(\zeta_1)(t) - g(\zeta_2)(t)\|^2_{L^2(\Omega)} \le C \int_0^t \|\zeta_1(s) - \zeta_2(s)\|^2_{L^2(\Omega)} ds$$

$$= C \int_0^t \|\zeta_1(s) - \zeta_2(s)\|^2_{L^2(\Omega)} e^{-2\lambda s} e^{2\lambda s} ds \le C\, dist(\zeta_1, \zeta_2)^2 \frac{e^{2\lambda t}}{2\lambda}.$$

It follows that

$$dist(g(\zeta_1), g(\zeta_2))^2 \le \frac{C}{2\lambda} dist(\zeta_1, \zeta_2)^2,$$

for any $\zeta_1, \zeta_2 \in X$.

If we take λ to be a constant bigger than $\frac{C}{2}$, then we get that g is a contraction on X, and consequently by Banach's fixed point result it has a unique fixed point.

So, for any $u \in \mathcal{U}_\omega$, Problem (3.1) admits a unique solution, denoted from now on y^u. Moreover, y^u satisfies

$$0 < y^u(x, t) \le \tilde{y}(x, t), \quad \text{a.e. } x \in \Omega, \text{ for any } t \in (0, T].$$

Theorem 3.1 *There exists at least one optimal control u^* for* (OH1).

Proof Let

$$\Phi(u) = \int_0^T \int_\omega [u(x, t) y^u(x, t) - c u(x, t)] dx\, dt,$$

and $R = \sup_{u \in \mathcal{U}_\omega} \Phi(u)$. Since for any $u \in \mathcal{U}_\omega$, we have that

$$-cLT \le \Phi(u) \le L \int_0^T \int_\omega \tilde{y}(x, t) dx\, dt,$$

we may infer that $R \in \mathbb{R}$. It follows that there exists $\{u_n\}_{n \in \mathbf{N}^*} \subset \mathcal{U}_\omega$ such that

$$R - \frac{1}{n} < \Phi(u_n) \le R,$$

for any $n \in \mathbb{N}^*$.

Since the sequence given by

$$f_n(x,t) = \eta(x)y^{u_n}(x,t) - y^{u_n}(x,t)\int_\Omega k(x,x')y^{u_n}(x',t)dx$$
$$- \mathbb{I}_\omega(x)u_n(x,t)y^{u_n}(x,t),$$

$(x,t) \in Q$, is bounded in $L^\infty(Q)$ and $\{u_n\}_{n\in\mathbb{N}^*}$ is bounded in $L^2(\omega \times (0,T))$, we get that on a subsequence

$$\mathbb{I}_\omega u_{n_q} \longrightarrow \mathbb{I}_\omega u^*, \quad f_{n_q} \longrightarrow f \quad \text{weakly in } L^2(Q).$$

The set $\mathcal{U}_\omega$ is convex and closed in $L^2(\omega \times (0,T))$, and so $u^* \in \mathcal{U}_\omega$.

Thanks to the compactness (in $L^2(\Omega)$) of the C_0-semigroup generated by the operator $\mathcal{A}$, defined by

$$\mathcal{A}y = d\Delta y, \qquad D(\mathcal{A}) = \{v \in H^2(\Omega) : \frac{\partial v}{\partial n}(x) = 0 \text{ on } \partial\Omega\},$$

the last convergence implies that (on a subsequence)

$$y^{u_{n_q}} \longrightarrow y \quad \text{in } C([0,T]; L^2(\Omega))$$

(see [9, 37, 164]), where y is the solution to

$$\begin{cases} \dfrac{\partial y}{\partial t}(x,t) - d\Delta y(x,t) = f(x,t), & (x,t) \in Q, \\ \dfrac{\partial y}{\partial n}(x,t) = 0, & (x,t) \in \Sigma, \\ y(x,0) = y_0(x), & x \in \Omega. \end{cases}$$

On the other hand,

$$f_{n_q} \longrightarrow \eta y - y\int_\Omega k(\cdot,x')y(x',t)dx' - \mathbb{I}_\omega u^* y$$

weakly in $L^2(Q)$. Hence, y is the unique solution to (3.1) with $u := u^*$. We immediately conclude that $\Phi(u_{n_q}) \longrightarrow \Phi(u^*)$, and so $\Phi(u^*) = R$. □

As far as the first order necessary optimality conditions are concerned, we may prove the following result.

Theorem 3.2 *If u^* is an optimal control (effort) for* (OH1) *and if p is the solution to*

$$\begin{cases} \dfrac{\partial p}{\partial t}(x,t) + d\Delta p(x,t) = -\eta(x)p(x,t) \\ \qquad +p(x,t)\displaystyle\int_\Omega k(x,x')y^{u^*}(x',t)dx' \\ \qquad +\displaystyle\int_\Omega k(x',x)y^{u^*}(x',t)p(x',t)dx' \\ \qquad +\mathbb{I}_\omega(x)u^*(x,t)(1+p(x,t)), & (x,t)\in Q, \\ \dfrac{\partial p}{\partial n}(x,t) = 0, & (x,t)\in\Sigma, \\ p(x,T) = 0, & x\in\Omega, \end{cases} \tag{3.5}$$

then

$$u^*(x,t) = \begin{cases} 0, \ \text{a.e.}\ (x,t)\in\omega\times(0,T), \ y^{u^*}(x,t)p(x,t)+y^{u^*}(x,t)-c<0, \\ L, \ \text{a.e.}\ (x,t)\in\omega\times(0,T), \ y^{u^*}(x,t)p(x,t)+y^{u^*}(x,t)-c>0. \end{cases} \tag{3.6}$$

In (3.6) for "a.e. $\langle Prop\rangle$," we mean that for almost all $x\in\Omega$ (with respect to the relevant Lebesgue measure) such that $\langle Prop\rangle$ holds true.

We may notice that (3.5) has a unique solution p, meaning that $p\in C([0,T];L^2(\Omega))\cap AC([0,T);L^2(\Omega))\cap L^2(0,T;H^1(\Omega))\cap L^2_{loc}([0,T);H^2(\Omega))$ and satisfies

$$\begin{cases} \dfrac{\partial p}{\partial t}(x,t) + d\Delta p(x,t) = -\eta(x)p(x,t) \\ \qquad +p(x,t)\displaystyle\int_\Omega k(x,x')y^{u^*}(x',t)dx' \\ \qquad +\displaystyle\int_\Omega k(x',x)y^{u^*}(x',t)p(x',t)dx' \\ \qquad +\mathbb{I}_\omega(x)u^*(x,t)(1+p(x,t)), & \text{a.e. in } Q, \\ \dfrac{\partial p}{\partial n}(x,t) = 0, & \text{a.e. in } \Sigma, \\ p(x,T) = 0, & \text{a.e. in } \Omega. \end{cases}$$

By taking into account that by the substitution $T-t:=s$, Problem (3.5) may be included in the general framework (3.3). The existence and uniqueness of the solution to (3.5) follows via Banach's fixed point theorem.

We shall use the following auxiliary result (see [23]).

Lemma 3.3 *Let v be arbitrary such that $u^*+\varepsilon v\in\mathcal{U}_\omega$ for sufficiently small $\varepsilon>0$ ($v\in L^\infty(\omega\times(0,T))$, and $0\le u^*(x,t)+\varepsilon v(x,t)\le L$, a.e. in $\omega\times(0,T)$). Then*

(i)

$$y^{u^*+\varepsilon v}\longrightarrow y^{u^*} \quad in\ L^\infty(Q),$$

and

(ii)

$$\frac{1}{\varepsilon}w_\varepsilon \longrightarrow z \quad in\ L^\infty(Q),$$

as $\varepsilon \to 0+$, *where we have denoted* $w_\varepsilon = y^{u^*+\varepsilon v} - y^{u^*}$, *and* z *is the solution to*

$$\begin{cases} \frac{\partial z}{\partial t}(x,t) - d\Delta z(x,t) = \eta(x)z(x,t) \\ \qquad -z(x,t)\int_\Omega k(x,x')y^{u^*}(x',t)dx' \\ \qquad -y^{u^*}(x,t)\int_\Omega k(x,x')z(x',t)dx' \\ \qquad -\mathbb{I}_\omega(x)[u^*(x,t)z(x,t) + v(x,t)y^{u^*}(x,t)], \ (x,t)\in Q, \\ \frac{\partial z}{\partial n}(x,t) = 0, \qquad (x,t)\in\Sigma, \\ z(x,0) = 0, \qquad x\in\Omega. \end{cases} \tag{3.7}$$

It is obvious that, via Banach's fixed point theorem, Problem (3.7) has a unique solution (notice that (3.7) may be included in the framework of (3.3)).

Proof of Theorem 3.2 The existence and uniqueness of a solution to (3.5) follows via Banach's fixed point theorem. Let now v be as in Lemma 3.3; then

$$\Phi(u^*) \geq \Phi(u^* + \varepsilon v),$$

which implies that

$$\int_0^T\int_\omega [(u^*(x,t)+\varepsilon v(x,t))y^{u^*+\varepsilon v}(x,t) - u^*(x,t)y^{u^*}(x,t) - c\varepsilon v(x,t)]dx\,dt \leq 0.$$

Consequently, we obtain that $\int_0^T\int_\omega \left[u^*\frac{w_\varepsilon}{\varepsilon} + vy^{u^*+\varepsilon v} - cv\right]dx\ dt \leq 0$, and passing to the limit ($\varepsilon \to 0+$) we obtain that

$$\int_0^T\int_\omega [u^*(x,t)z(x,t) + v(x,t)y^{u^*}(x,t) - cv(x,t)]dx\,dt \leq 0. \tag{3.8}$$

By multiplying the first equation in (3.5) by z and integrating over Q, we get that $\mathcal{T}_1 = \mathcal{T}_2$, where

$$\mathcal{T}_1 = \int_0^T\int_\Omega z(x,t)\left[\frac{\partial p}{\partial t}(x,t) + d\Delta p(x,t)\right]dx\,dt,$$

and

$$\mathcal{T}_2 = \int_0^T \int_\Omega z(x,t)\Big[-\eta(x)p(x,t) + p(x,t)\int_\Omega k(x,x')y^{u^*}(x',t)dx'$$

$$+\int_\Omega k(x',x)y^{u^*}(x',t)p(x',t)dx' + \mathbb{I}_\omega(x)u^*(x,t)(1+p(x,t))\Big]dx\,dt.$$

By integrating by parts (with respect to t) and taking into account (3.7) and (3.5), we get that

$$\mathcal{T}_1 = \int_0^T \int_\Omega p(x,t)\Big[-\eta(x)z(x,t) + z(x,t)\int_\Omega k(x,x')y^{u^*}(x',t)dx'$$

$$+y^{u^*}(x,t)\int_\Omega k(x,x')z(x',t)dx' + \mathbb{I}_\omega(x)(u^*(x,t)z(x,t) + v(x,t)y^{u^*}(x,t))\Big]dx\,dt.$$

Since $\mathcal{T}_1 = \mathcal{T}_2$, we may conclude that $\int_0^T\int_\omega u^* z dx\,dt = \int_0^T\int_\omega v y^{u^*} p dx\,dt$, and by (3.8) we infer that

$$\int_0^T\int_\omega v(x,t)[y^{u^*}(x,t)p(x,t) + y^{u^*}(x,t) - c]dx\,dt \le 0,$$

for any v such that $u^* + \varepsilon v \in \mathcal{U}_\omega$ for sufficiently small $\varepsilon > 0$.

Let us prove now that the first part of (3.6) holds, i.e., $u^*(x,t) = 0$, a.e. $(x,t) \in \mathcal{A}_1$, where

$$\mathcal{A}_1 = \{(x,t) \in \omega \times (0,T) :\ y^{u^*}(x,t)p(x,t) + y^{u^*}(x,t) - c < 0\}.$$

We argue by contradiction; assume that there exists a subset $\mathcal{A}_0$ of $\mathcal{A}_1$, with positive Lebesgue measure meas$(\mathcal{A}_0)$, such that

$$u^*(x,t) \in (0,L], \quad \text{a.e. } (x,t) \in \mathcal{A}_0.$$

Consider now an arbitrary (but fixed) $v \in L^\infty(\omega \times (0,T))$ such that

$$u^* + \varepsilon v \in \mathcal{U}_\omega,$$

for any $\varepsilon > 0$ sufficiently small, and

$$v(x,t) < 0, \text{ a.e. } (x,t) \in \mathcal{A}_0, \quad v(x,t) = 0, \text{ a.e. } (x,t) \in \omega \times (0,T) \setminus \mathcal{A}_0.$$

We obtain that

$$0 \geq \int_0^T \int_\omega v(x,t)[y^{u^*}(x,t)p(x,t) + y^{u^*}(x,t) - c]dx\,dt$$

$$= \int_{\mathcal{A}_0} v(x,t)[y^{u^*}(x,t)p(x,t) + y^{u^*}(x,t) - c]dx\,dt > 0,$$

which is absurd. So, $u^*(x,t) = 0$, a.e. $(x,t) \in \mathcal{A}_1$.

Similarly we can prove the second part of (3.6). □

Remark 3.4 If we multiply the first equation in (3.5) by y^{u^*} and integrate over $\Omega \times (0,T)$, we obtain after some calculation (and using (3.1) and (3.5)) that

$$\Phi(u^*) = -\int_\Omega y_0(x)p(x,0)dx$$
$$-\int_0^T \int_\Omega \int_\Omega k(x',x)y^{u^*}(x',t)p(x',t)y^{u^*}(x,t)dx'\,dx\,dt.$$

A special attention will be paid in Sections 3.4 and 3.5 to the case $c = 0$, which corresponds to the case when the cost paid for harvesting is a negligible part of the gain for the harvested population. Since $y^{u^*}(x,t) > 0$ a.e. in Q, we get by (3.6) that

$$u^*(x,t) = \begin{cases} 0, & \text{a.e. } (x,t) \in \omega \times (0,T),\ 1 + p(x,t) < 0, \\ L, & \text{a.e. } (x,t) \in \omega \times (0,T),\ 1 + p(x,t) > 0, \end{cases}$$

where p is the solution to

$$\begin{cases} \dfrac{\partial p}{\partial t}(x,t) + d\Delta p(x,t) = -\eta(x)p(x,t) \\ \qquad +p(x,t)\displaystyle\int_\Omega k(x,x')y^{u^*}(x',t)dx' \\ \qquad +\displaystyle\int_\Omega k(x',x)y^{u^*}(x',t)p(x',t)dx' \\ \qquad +\mathbb{I}_\omega(x)L(1+p(x,t))^+, & (x,t) \in Q, \\ \dfrac{\partial p}{\partial n}(x,t) = 0, & (x,t) \in \Sigma, \\ p(x,T) = 0, & x \in \Omega. \end{cases}$$

If $p(x,t) \neq -1$, a.e. in $\omega \times (0,T)$, then $u^*(x,t) = L \cdot H(1 + p(x,t))$, where H is the Heaviside function. This happens if, for example, $k \equiv 0$ and $\eta(x) \neq 0$ a.e. in ω, or if $k(x,x') = k_0\delta_x(x')$ with k_0 a positive constant ($\delta_x(x')$ is the Dirac mass centered at x and evaluated in x') and if $\eta(x) < 2k_0y^L(x,t)$, a.e. in $\omega \times (0,T)$, or if $\eta(x) > 2k_0y^0(x,t)$, a.e. in $\omega \times (0,T)$ (more details will be given in Section 3.3;

this is a limit case and is not however a particular case since such k does not satisfy the hypothesis (H2)).

Actually, we may conclude that the maximal profit (for a given ω) is

$$\Phi(u^*) = -\int_\Omega y_0(x)p(x,0)dx - \int_0^T \int_\Omega \int_\Omega k(x',x)y(x',t)p(x',t)y(x,t)dx'\,dx\,dt,$$

where (y, p) is the solution to

$$\begin{cases} \dfrac{\partial y}{\partial t}(x,t) - d\Delta y(x,t) = \eta(x)y(x,t) \\ \qquad -y(x,t)\displaystyle\int_\Omega k(x,x')y(x',t)dx' - \mathbb{I}_\omega(x)LH(1+p(x,t))y(x,t), \; (x,t) \in Q, \\ \dfrac{\partial p}{\partial t}(x,t) + d\Delta p(x,t) = -\eta(x)p(x,t) \\ \qquad +p(x,t)\displaystyle\int_\Omega k(x,x')y(x',t)dx' \\ \qquad +\displaystyle\int_\Omega k(x',x)y(x',t)p(x',t)dx' + \mathbb{I}_\omega(x)L(1+p(x,t))^+, & (x,t) \in Q, \\ \dfrac{\partial y}{\partial n}(x,t) = \dfrac{\partial p}{\partial n}(x,t) = 0, & (x,t) \in \Sigma, \\ y(x,0) = y_0(x), \quad p(x,T) = 0, & x \in \Omega. \end{cases}$$

Remark 3.5 A similar conclusion may be obtained in the general case if, in addition, we get that $y^{u^*}(x,t)p(x,t) + y^{u^*}(x,t) - c \neq 0$, a.e. in $\omega \times (0,T)$.

3.3 A Spatially Structured Population with Local Logistic Term

Consider now the case in which competition for resources is local (pointwise), as modelled by the following evolution equation:

$$\begin{cases} \dfrac{\partial y}{\partial t}(x,t) - d\Delta y(x,t) = \eta(x)y(x,t) - \zeta(x)y^2(x,t) \\ \qquad -\mathbb{I}_\omega(x)uy(x,t), & (x,t) \in Q, \\ \dfrac{\partial y}{\partial n}(x,t) = 0, & (x,t) \in \Sigma, \\ y(x,0) = y_0(x), & x \in \Omega. \end{cases} \tag{3.9}$$

The logistic term here, $\zeta(x)y(x,t)^2$, is local, depending only on the position, x.

This section is devoted to the following optimal harvesting problem related to a spatially structured population with diffusion and local logistic term:

$$\text{(OH2)} \qquad \underset{u \in \mathcal{U}_\omega}{Maximize} \int_0^T \int_\omega [u(x,t)y^u(x,t) - cu(x,t)]dx\,dt,$$

where $L \in (0, +\infty)$ and $c \in [0, +\infty)$ are constants and y^u is the solution to (3.9).

Here are the hypotheses that are going to be used in this section:

(H1)" $\eta, y_0 \in L^\infty(\Omega)$, $y_0(x) \geq 0$, a.e. in Ω, and $y_0 \neq 0_{L^\infty(\Omega)}$.
(H2)" $\zeta \in L^\infty(\Omega)$, $\zeta(x) \geq 0$, a.e. in Ω.

By adopting the same arguments as in the previous sections, it is possible to prove the following results:

Theorem 3.6 *There exists at least one optimal control u^* for* (OH2).

Theorem 3.7 *If u^* is an optimal control for* (OH2) *and if p is the solution to*

$$\begin{cases} \dfrac{\partial p}{\partial t}(x,t) + d\Delta p = -\eta(x)p + 2\zeta(x)y^{u^*}p + \mathbb{I}_\omega(x)u^*(1+p), & (x,t) \in Q, \\ \dfrac{\partial p}{\partial n}(x,t) = 0, & (x,t) \in \Sigma, \\ p(x,T) = 0, & x \in \Omega, \end{cases} \tag{3.10}$$

then

$$u^*(x,t) = \begin{cases} 0, & a.e.\ (x,t) \in \omega \times (0,T),\ y^{u^*}(x,t)p(x,t) + y^{u^*}(x,t) - c < 0, \\ L, & a.e.\ (x,t) \in \omega \times (0,T),\ y^{u^*}(x,t)p(x,t) + y^{u^*}(x,t) - c > 0. \end{cases} \tag{3.11}$$

Remark 3.8 Problem (3.9) may be regarded as a limit version of (3.1), for

$$k(x,x') = \zeta(x)\delta_x(x').$$

Theorems 3.6 and 3.7 are not obtained as particular cases from Theorems 3.1 and 3.2. However their proofs follow in a similar manner.

An important particular case is when $c = 0$. If we multiply the first equation in (3.10) by y^{u^*} and integrate over Q, after an easy calculation (and using (3.9)), we obtain

$$\Phi(u^*) = -\int_\Omega y_0(x)p(x,0)dx - \int_0^T \int_\Omega \zeta(x)y^{u^*}(x,t)^2 p(x,t)dx\,dt,$$

where $\Phi(u) = \int_0^T \int_\Omega u(x,t)y^u(x,t)dx\,dt$, and

$$u^*(x,t) = \begin{cases} 0, & \text{a.e. } (x,t) \in \omega \times (0,T),\ 1 + p(x,t) < 0, \\ L, & \text{a.e. } (x,t) \in \omega \times (0,T),\ 1 + p(x,t) > 0 \end{cases} \tag{3.12}$$

(by (3.11) and because $y^{u^*}(x,t) > 0$ a.e. in Q). This however implies that p is the unique solution to

$$\begin{cases} \dfrac{\partial p}{\partial t}(x,t) + d\Delta p = -\eta(x)p + 2\zeta(x)y^{u^*}p + \mathbb{I}_\omega(x)L(1+p)^+, & (x,t) \in Q, \\ \dfrac{\partial p}{\partial n}(x,t) = 0, & (x,t) \in \Sigma, \\ p(x,T) = 0, & x \in \Omega. \end{cases}$$

Remark 3.9 If under these last hypotheses there exists a subset $\mathcal{A} \subset \omega \times (0,T)$ of positive measure, such that $p(x,t) = -1$, a.e. in $\mathcal{A}$, and if $\zeta(x) > 0$, a.e. in ω, then by (3.10) we get that $0 = \eta(x) - 2\zeta(x)y^{u^*}(x,t)$, a.e. in $\mathcal{A}$, which implies that $y^{u^*}(x,t) = \dfrac{\eta(x)}{2\zeta(x)}$ a.e. in $\mathcal{A}$, and consequently (by (3.9))

$$-d\Delta\left(\frac{\eta(x)}{2\zeta(x)}\right) = \frac{\eta(x)}{2\zeta(x)}\left(\frac{\eta(x)}{2} - u^*(x,t)\right),$$

a.e. in $\mathcal{A}$. If $\eta(x) \neq 0$ a.e. in ω, then

$$u^*(x,t) = \frac{2\zeta(x)}{\eta(x)}d\Delta\left(\frac{\eta(x)}{\zeta(x)}\right) + \frac{\eta(x)}{2},$$

a.e. in $\mathcal{A}$. Hence, if $\eta(x) \neq 0$ a.e. in ω and if

$$\frac{2\zeta(x)}{\eta(x)}d\Delta\left(\frac{\eta(x)}{\zeta(x)}\right) + \frac{\eta(x)}{2} > L,$$

a.e. in ω or if

$$\frac{2\zeta(x)}{\eta(x)}d\Delta\left(\frac{\eta(x)}{\zeta(x)}\right) + \frac{\eta(x)}{2} < 0,$$

a.e. in ω, then we get that $p(x,t) \neq -1$, a.e. in $\omega \times (0,T)$, and so u^* is a bang–bang control, and $u^*(x,t) = L \cdot H(1 + p(x,t))$, a.e. in $\omega \times (0,T)$.

If $\dfrac{\eta}{\zeta}$ is a constant function and if $\eta < 0$ or $\eta > 2L$, a.e. in ω, then we have the same conclusion. Moreover, it follows (as in the previous section) that

$$\Phi(u^*) = -\int_\Omega y_0(x)p(x,0)dx - \int_0^T\int_\Omega \zeta(x)y(x,t)^2 p(x,t)dx\,dt,$$

where (y,p) is the solution to

$$\begin{cases} \dfrac{\partial y}{\partial t} - d\Delta y = \eta(x)y - \zeta(x)y^2 - \mathbb{I}_\omega(x)LH(1+p)y, & (x,t) \in Q, \\ \dfrac{\partial p}{\partial t} + d\Delta p = -\eta(x)p + 2\zeta(x)yp + \mathbb{I}_\omega(x)L(1+p)^+, & (x,t) \in Q, \\ \dfrac{\partial y}{\partial n}(x,t) = \dfrac{\partial p}{\partial n}(x,t) = 0, & (x,t) \in \Sigma, \\ y(x,0) = y_0(x), \quad p(x,T) = 0, & x \in \Omega. \end{cases}$$

Remark 3.10 In the particular case when $c = 0$ and $\zeta \equiv 0$ (a particular case also for the problem in Section 3.2: for $k \equiv 0$), we may conclude that

$$\Phi(u^*) = -\int_\Omega y_0(x)p(x,0)dx,$$

where p is the unique solution to

$$\begin{cases} \dfrac{\partial p}{\partial t}(x,t) + d\Delta p(x,t) = -\eta(x)p(x,t) + \mathbb{I}_\omega(x)L(1+p(x,t))^+, & (x,t) \in Q, \\ \dfrac{\partial p}{\partial n}(x,t) = 0, & (x,t) \in \Sigma, \\ p(x,T) = 0, & x \in \Omega. \end{cases}$$

If in addition $\eta(x) \neq 0$, a.e. in ω, then there is a unique optimal control u^*, which is given by (3.12).

3.4 Localizing a Suboptimal Subregion ω where the Control Acts

We investigate here the optimal harvesting for a particular model of population dynamics and corresponding to the case when $k \equiv 0$ (there is no logistic competition). We wish to identify an optimal choice of the subregion ω where the control is localized. Consider a $2D$ domain $\Omega \subset \mathbb{R}^2$, and let φ be an implicit function associated with ω, the subregion of Ω in which we wish to apply our control ($\omega = \{x \in \Omega;\ \varphi(x) > 0\}$). We shall use the level set approach to treat this last problem (see also [80, 116]).

Our optimal control problem here includes both the magnitude of the harvesting effort $u \in \mathcal{U}_\omega = \{v \in L^\infty(\omega \times (0,T)) :\ 0 \leq v(x,t) \leq L,\ \text{a.e.}\}$ and the choice of the subregion ω:

$$\underset{\varphi}{Maximize}\ \underset{u \in \mathcal{U}_\omega}{Maximize}\ \Psi_\varphi(u),$$

where

$$\Psi_\varphi(u) = \int_0^T \int_\omega u(x,t)y^u(x,t)dx\,dt - c_1 \int_\Omega \delta(\varphi(x))|\nabla\varphi(x)|dx - c_2 \int_\Omega H(\varphi(x))dx,$$

and y^u is the solution to

$$\begin{cases} \dfrac{\partial y}{\partial t}(x,t) - d\Delta y(x,t) = \eta(x)y(x,t) - \mathbb{I}_\omega(x)u(x,t)y(x,t), & (x,t) \in Q, \\ \dfrac{\partial y}{\partial n}(x,t) = 0, & (x,t) \in \Sigma, \\ y(x,0) = y_0(x), & x \in \Omega. \end{cases}$$

Assume that the hypotheses in Section 3.2 are satisfied. Using the remarks in Sections 3.2 and 3.3, we may rewrite the above problem as

$$\underset{\varphi}{Minimize} \left\{ \int_\Omega y_0(x)p(x,0)dx + c_1 \int_\Omega \delta(\varphi(x))|\nabla\varphi(x)|dx + c_2 \int_\Omega H(\varphi(x))dx \right\},$$

where p is the solution to

$$\begin{cases} \dfrac{\partial p}{\partial t} + d\Delta p = -\eta(x)p + LH(\varphi(x))(1+p)H(1+p), & (x,t) \in Q, \\ \dfrac{\partial p}{\partial n}(x,t) = 0, & (x,t) \in \Sigma, \\ p(x,T) = 0, & x \in \Omega. \end{cases}$$

For computational purposes this problem can be "approximated" by the following one, where the Heaviside function H is substituted by its mollified version $H_\varepsilon(t) = \dfrac{1}{2}\left(1 + \dfrac{2}{\pi}\arctan\left(\dfrac{t}{\varepsilon}\right)\right)$ and its derivative by the mollified function $\delta_\varepsilon(t) = \dfrac{\varepsilon}{\pi(\varepsilon^2 + t^2)}$.

So, for a suitably small but fixed $\varepsilon > 0$, we shall deal in what follows with the following "approximating" (regional harvesting) problem:

(OH regional) $\quad \underset{\varphi}{Minimize}\ J(\varphi),$

where $\varphi : \overline{\Omega} \longrightarrow \mathbb{R}$ is a smooth function and

$$J(\varphi) = \int_\Omega y_0(x)p^\varphi(x,0)dx + c_1 \int_\Omega \delta_\varepsilon(\varphi(x))|\nabla\varphi(x)|dx + c_2 \int_\Omega H_\varepsilon(\varphi(x))dx,$$

and p^{φ} is the solution to

$$\begin{cases} \dfrac{\partial p}{\partial t} + d\Delta p = -\eta(x)p + LH_{\varepsilon}(\varphi(x))(1+p)H_{\varepsilon}(1+p), & (x,t) \in Q, \\ \dfrac{\partial p}{\partial n}(x,t) = 0, & (x,t) \in \Sigma, \\ p(x,T) = 0, & x \in \Omega. \end{cases} \tag{3.13}$$

The following result gives the directional derivative of J:

Theorem 3.11 *For any smooth functions $\varphi, \psi : \overline{\Omega} \longrightarrow \mathbb{R}$ (with $\nabla\varphi(x) \neq 0$, a.e. $x \in \Omega$), we have that*

$$dJ(\varphi)(\psi) = \int_{\Omega} \delta_{\varepsilon}(\varphi(x))[-c_1 \operatorname{div}\left(\frac{\nabla\varphi(x)}{|\nabla\varphi(x)|}\right) + c_2$$

$$- L\int_0^T (1 + p^{\varphi}(x,t))H_{\varepsilon}(1 + p^{\varphi}(x,t))r^{\varphi}(x,t)dt]\psi(x)dx$$

$$+c_1 \int_{\partial\Omega} \frac{\delta_{\varepsilon}(\varphi(x))}{|\nabla\varphi(x)|}\frac{\partial\varphi}{\partial n}(x)\psi(x)d\sigma,$$

where r^{φ} is the solution to

$$\begin{cases} \dfrac{\partial r}{\partial t} - d\Delta r = \eta(x)r - LH_{\varepsilon}(\varphi(x))H_{\varepsilon}(1+p^{\varphi})r \\ \qquad -LH_{\varepsilon}(\varphi(x))(1+p^{\varphi})\delta_{\varepsilon}(1+p^{\varphi})r, & (x,t) \in Q, \\ \dfrac{\partial r}{\partial n}(x,t) = 0, & (x,t) \in \Sigma, \\ r(x,0) = y_0(x), & x \in \Omega. \end{cases} \tag{3.14}$$

Proof It is possible to prove in a standard manner that

$$\frac{1}{\theta}[p^{\varphi+\theta\psi} - p^{\varphi}] \longrightarrow q^{\psi} \quad \text{in } C([0,T]; L^{\infty}(\Omega)),$$

as $\theta \to 0$, where q^{ψ} is the solution to the problem

$$\begin{cases} \dfrac{\partial q}{\partial t} + d\Delta q = -\eta(x)q + L\delta_{\varepsilon}(\varphi(x))(1+p^{\varphi})H_{\varepsilon}(1+p^{\varphi})\psi(x) \\ \qquad +LH_{\varepsilon}(\varphi(x))H_{\varepsilon}(1+p^{\varphi})q \\ \qquad +LH_{\varepsilon}(\varphi(x))(1+p^{\varphi})\delta_{\varepsilon}(1+p^{\varphi})q, & (x,t) \in Q, \\ \dfrac{\partial q}{\partial n}(x,t) = 0, & (x,t) \in \Sigma, \\ q(x,T) = 0, & x \in \Omega. \end{cases} \tag{3.15}$$

For any φ we have that

$$\lim_{\theta\to 0}\frac{1}{\theta}[J(\varphi+\theta\psi)-J(\varphi)] = \int_\Omega y_0(x)q^\psi(x,0)dx + c_1\int_\Omega \delta_\varepsilon'(\varphi(x))\psi(x)|\nabla\varphi(x)|dx$$

$$+ c_1\int_\Omega \delta_\varepsilon(\varphi(x))\frac{\nabla\varphi(x)\cdot\nabla\psi(x)}{|\nabla\varphi(x)|}dx + c_2\int_\Omega \delta_\varepsilon(\varphi(x))\psi(x)dx$$

$$= \int_\Omega y_0(x)q^\psi(x,0)dx + c_1\int_\Omega \mathrm{div}\left(\delta_\varepsilon(\varphi(x))\frac{\nabla\varphi(x)}{|\nabla\varphi(x)|}\psi(x)\right)dx$$

$$- c_1\int_\Omega \delta_\varepsilon(\varphi(x))\psi(x)\mathrm{div}\left(\frac{\nabla\varphi(x)}{|\nabla\varphi(x)|}\right)dx + c_2\int_\Omega \delta_\varepsilon(\varphi(x))\psi(x)dx.$$

After applying the Gauss–Ostrogradski's formula (see [39, p.13]), we obtain

$$dJ(\varphi)(\psi) = \int_\Omega y_0(x)q^\psi(x,0)dx - c_1\int_\Omega \delta_\varepsilon(\varphi(x))\mathrm{div}\left(\frac{\nabla\varphi(x)}{|\nabla\varphi(x)|}\right)\psi(x)dx$$
$$+ c_2\int_\Omega \delta_\varepsilon(\varphi(x))\psi(x)dx + c_1\int_{\partial\Omega}\frac{\delta_\varepsilon(\varphi(x))}{|\nabla\varphi(x)|}\frac{\partial\varphi}{\partial n}(x)\psi(x)d\sigma. \tag{3.16}$$

By multiplying the first equation in (3.15) by r and integrating over Q, we get, after an easy calculation, and using (3.14) as well, that

$$\int_\Omega y_0(x)q^\psi(x,0)dx$$
$$= -\int_0^T\int_\Omega Lr^\varphi(x,t)\delta_\varepsilon(\varphi(x))(1+p^\varphi(x,t))H_\varepsilon(1+p^\varphi(x,t))\psi(x)dx\,dt. \tag{3.17}$$

Now, from (3.16) and (3.17), we get the conclusion of Theorem 3.11. □

Remark 3.12 The gradient descent with respect to φ is

$$\begin{cases} \dfrac{\partial\varphi}{\partial\theta}(x,\theta) = \delta_\varepsilon(\varphi(x,\theta))[c_1\,\mathrm{div}\left(\dfrac{\nabla\varphi(x,\theta)}{|\nabla\varphi(x,\theta)|}\right) - c_2 \\ \qquad +L\displaystyle\int_0^T (1+p^\varphi(x,t))H_\varepsilon(1+p^\varphi(x,t))r^\varphi(x,t)dt], \; x\in\Omega,\ \theta>0, \\ \dfrac{\delta_\varepsilon(\varphi(x,\theta))}{|\nabla\varphi(x,\theta)|}\dfrac{\partial\varphi}{\partial n}(x,\theta) = 0, \qquad x\in\partial\Omega,\ \theta>0 \end{cases} \tag{3.18}$$

(θ is an artificial time).

3.5 Computational Issues

In the first paragraph, we describe the conceptual algorithm for the optimal control procedure related to the problem investigated in Section 3.4. In the second paragraph, we describe the numerical methods adopted for the solution of Equations (3.13) and (3.14). In the last paragraph, we report and discuss the computational results.

3.5.1 Conceptual Algorithm

From Theorem 3.11 we derive the following conceptual iterative algorithm, a semi-implicit gradient descent one, to improve at each step the region where the harvesting effort acts in order to obtain a smaller value for J.

STEP 0: set $k := 0$, $J^{(0)} := 10^6$ and $\theta_0 > 0$ a small constant
initialize $\varphi^{(0)} = \varphi^{(0)}(x, 0)$.

STEP 1: compute $p^{(k+1)}$ the solution of (3.13) corresponding to $\varphi^{(k)}(\cdot, 0)$

$$\text{compute } J^{(k+1)} = \int_\Omega y_0(x) p^{(k+1)}(x, 0) dx$$
$$+ c_1 \int_\Omega \delta_\varepsilon(\varphi^{(k)}(x, 0)) |\nabla \varphi^{(k)}(x, 0)| dx + c_2 \int_\Omega H_\varepsilon(\varphi^{(k)}(x, 0)) dx.$$

Step 2: if $\left| J^{(k+1)} - J^{(k)} \right| < \varepsilon_1$ or $J^{(k+1)} \geq J^{(k)}$ then **STOP**
else go to **Step 3**.

Step 3: compute $r^{(k+1)}$ the solution of problem (3.14) corresponding to
$\varphi^{(k)}(\cdot, 0)$ and $p^{(k+1)}$.

Step 4: compute $\varphi^{(k+1)}$ using (3.18) and the initial condition $\varphi^{(k+1)}(x, 0) = \varphi^{(k)}(x, \theta_0)$ and a semi-implicit timestep scheme.

Step 5: if $\|\varphi^{(k+1)} - \varphi^{(k)}\|_{L^2} < \varepsilon_2$ then **STOP**
else $k := k + 1$
go to **Step 1**.

$\varepsilon_1 > 0$ in Step 2 and $\varepsilon_2 > 0$ in Step 5 are prescribed convergence parameters. For more information about gradient (descent) methods, see [30, §2.3].

3.5.2 Numerical Methods

Equations (3.13), (3.14), and (3.18) are discretized by finite elements in space and finite differences in time.

3.5.2.1 Space Discretization

We first apply a standard Galerkin procedure to the weak formulations of systems (3.13) and (3.14). The unit square domain Ω is discretized by a uniform grid of 64×64 bilinear finite elements ($\mathbf{Q}^1$), yielding a total amount of $N = 4225$ discretization nodes. The stiffness matrix is computed exactly, whereas the mass matrix is obtained by applying the mass lumping technique.

In the following, we describe in detail this procedure applied to Equation (3.14), being the treatment of Equations (3.13) and (3.18) analogous.

Let us first denote by V the Sobolev space $H^1(\Omega)$, by $(\cdot, \cdot)$ the L^2-inner product, and by $a(\cdot, \cdot) : V \times V \longrightarrow \mathbb{R}$ the bilinear form

$$a(u, v) := \int_{\Omega} d\nabla u \cdot \nabla v dx, \quad \text{for any } u, v \in V.$$

For the sake of simplifying the notations, let us also set $p = p^{\varphi}$. Then, using the Green formula, the variational formulation of problem (3.14) reads as follows:

given $\varphi \in V$ and p, find r such that

$$\left\{ \frac{\partial}{\partial t}(r(t), \phi) + a(r(t), \phi) = (G(r(t), p(t), \varphi), \phi), \quad \text{for any } \phi \in V, \right. \tag{3.19}$$

subject to the initial condition $r(x, 0) = y_0(x)$, and where

$$\begin{aligned} G(r(x,t), p(x,t), \varphi(x)) = \eta(x) r(x,t) - L H_{\varepsilon}(\varphi(x)) H_{\varepsilon}(1 + p(x,t)) r(x,t) \\ - L H_{\varepsilon}(\varphi(x))(1 + p(x,t)) \delta_{\varepsilon}(1 + p(x,t)) r(x,t). \end{aligned}$$

Let us now introduce the $\mathbf{Q}^1$ finite element space V^h, associated with a Cartesian grid discretizing the unit square domain Ω. For a detailed description of this space, see Appendix B.

The space semi-discretization of problem (3.19) then reads as follows:
find $r_h : (0, T] \rightarrow V^h$ such that

$$\left\{ \frac{\partial}{\partial t}(r_h(t), \phi) + a(r_h(t), \phi) = (G(r_h(t), p_h(t), \varphi_h), \phi), \quad \text{for any } \phi \in V^h, \right.$$

and a corresponding initial condition, where $p_h : (0, T] \to V^h$ and $\varphi_h \in V^h$ are the finite element approximations of p and φ. The previous variational equation is equivalent to

$$\left\{ \frac{\partial}{\partial t}(r_h(t), \phi_i) + a(r_h(t), \phi_i) = (G(r_h(t), p_h(t), \varphi_h), \phi_i), \quad i = 1, \ldots, N, \right. \tag{3.20}$$

where $\phi_i, i = 1, \ldots, N$, are the *hat basis functions* of the $\mathbf{Q}^1$ finite element space V^h (see Appendix B for the definition of this basis).

Expanding the unknowns $r_h(x, t)$, $p_h(x, t)$, and $\varphi_h(x)$ with respect to the basis $\{\phi_i\}_{i=1}^N$, we have

$$r_h(x,t) = \sum_{j=1}^N R_j(t)\phi_j(x), \quad p_h(x,t) = \sum_{j=1}^N P_j(t)\phi_j(x), \quad \varphi_h(x) = \sum_{j=1}^N \Phi_j\phi_j(x), \tag{3.21}$$

where $R_j : [0, T] \to \mathbb{R}$ are unknown scalar functions to be determined, and we assume that $P_j : [0, T] \to \mathbb{R}$ and $\Phi_j \in \mathbb{R}$ are given. Indeed they are obtained proceeding analogously as solutions of problems (3.13) and (3.18).

In the next steps, we follow closely the approximation of the nonlinear problem presented in Appendix B. Hence, denoting by A the stiffness matrix associated with the bilinear form $a(\cdot, \cdot)$ and by M the lumped mass matrix (see Appendix B for the construction of such matrices) and substituting the expressions (3.21) into system (3.20), the semi-discrete problem then reads as follows:
find $\mathbf{R} : (0, T] \to \mathbb{R}^N$ such that

$$\left\{ \mathrm{M}\frac{d\mathbf{R}}{dt}(t) + \mathrm{A}\mathbf{R}(t) = \mathrm{M}\mathbf{G}(\mathbf{R}(t), \mathbf{P}(t), \mathbf{\Phi}), \right. \tag{3.22}$$

where

$$\begin{aligned}
\mathbf{R}(t) &= [\ldots, R_j(t), \ldots]^T, \\
\mathbf{P}(t) &= [\ldots, P_j(t), \ldots]^T, \\
\mathbf{\Phi} &= [\ldots, \Phi_j, \ldots]^T, \\
\mathbf{G}(\mathbf{R}(t), \mathbf{P}(t), \mathbf{\Phi}) &= [\ldots, G(R_j(t), P_j(t), \Phi_j), \ldots]^T.
\end{aligned}$$

3.5.2.2 Time Discretization

After space discretization by finite elements, we obtain the semi-discrete problem (3.22) that consists of a system of ordinary differential equations (ODEs). We solve these ODEs by employing a first order semi-implicit finite difference scheme, where the linear diffusion and reaction terms are approximated by Backward Euler, whereas the nonlinear reaction terms are approximated by Forward Euler.

In formulae, at the generic time step n,
given $\mathbf{R}^n$, we have to find $\mathbf{R}^{n+1}$ such that

$$\left\{ \mathrm{M}\frac{\mathbf{R}^{n+1} - \mathbf{R}^n}{\Delta t} + \mathrm{A}\mathbf{R}^{n+1} = \mathrm{M}\mathbf{G}(\mathbf{R}^n, \mathbf{P}^{n+1}, \boldsymbol{\Phi}). \right. \tag{3.23}$$

As a result, at each time step it is required the solution of the linear system of algebraic equations

$$\left(\frac{1}{\Delta t}\mathrm{M} + \mathrm{A}\right)\mathbf{R}^{n+1} = \frac{1}{\Delta t}\mathrm{M}\mathbf{R}^n + \mathrm{M}\mathbf{G}(\mathbf{R}^n, \mathbf{P}^{n+1}, \boldsymbol{\Phi}),$$

which in our specific case has the dimension of 4225 degrees of freedom. The linear system is solved by the Gaussian elimination with the built-in function of MATLAB. The time step size is $\Delta t = 0.05$.

For further details on the numerical approximation of parabolic equations, we refer, e.g., to [168].

3.5.3 *Numerical Simulations*

We report in this section the results obtained with the MATLAB code

```
run_opt_harv.m,
```

available on the Birkhäuser website provided in the Preface and also at the link
`http://www.mat.unimi.it/users/scacchi/ACS_codes/run_opt_harv.m.`

We consider a normal initial population density $y_0(x_1, x_2) = \frac{1}{2\pi}e^{-\frac{x_1^2+x_2^2}{2}}$, $(x_1, x_2) \in \Omega$. Let the diffusion coefficient be $d = 1$, the final time $T = 1$, $L = 1$, and the regularization parameter $\varepsilon = 1$. For the convergence tests we consider $\varepsilon_1 = \varepsilon_2 = 0.001$. In the following, in every figure, the light gray/blue area represents the subregion ω that provides a small value for J.

3.5.3.1 Test 1

Let the natural growth rate of the population be constant, for example, $\eta(x_1, x_2) = 3$, $(x_1, x_2) \in \Omega$. We consider $\varphi^{(0)}(x_1, x_2, 0) = (-x_1+0.25)(-x_2+0.5)+\frac{1}{4}\sin(x_1 - 0.4)\sin(0.25 - x_2)$, $(x_1, x_2) \in \Omega$. We penalize the length of $\partial\omega$ by $c_1 = 0.6$ and $c_1 = 0.7$ and the area of ω by $c_2 = 1$. The corresponding results are shown in Figure 3.1.

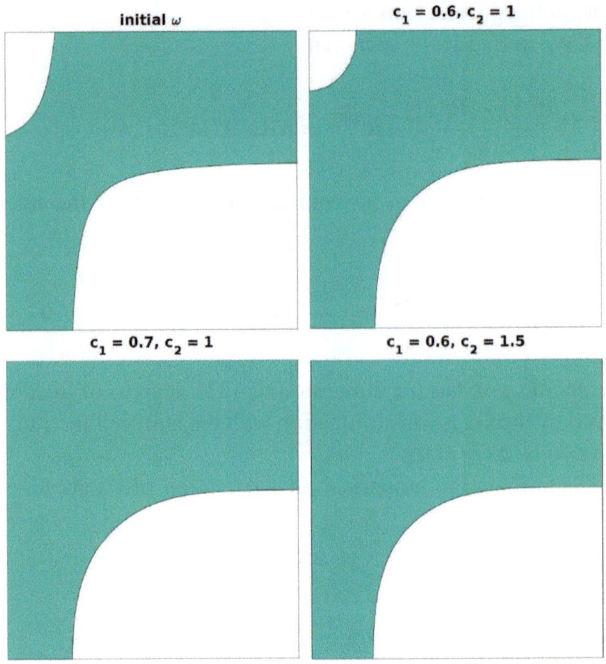

Fig. 3.1 Test 1. We solve the optimal harvesting problem described in Section 3.4, with the algorithms and discretization parameters reported in Section 3.5. Initial configuration of ω and suboptimal configurations depending on the choices of parameters c_1 and c_2

3.5.3.2 Test 2

We consider the growth rate of the population $\eta(x_1, x_2) = e^{x_1+x_2}$ and $(x_1, x_2) \in \Omega$. We penalize the length of $\partial\omega$ and the area of ω by varying c_1 in the set $\{0.1, 0.2, 0.3, 0.4, 0.5\}$, respectively, by $c_2 = 0.5$. The initialization of φ is made by $\varphi^{(0)}(x_1, x_2, 0) = \sin(5\pi x_1)\sin(5\pi x_2)$, $(x_1, x_2) \in \Omega$, a function that produces an initial checkerboard shape. The results are shown in Figure 3.2.

Let us point out that if we try to increase the value of c_2, the level set function returning by the computer program is everywhere negative. The interpretation is that the penalty applied to the area is too big. In all these examples, the convergence was obtained by the first test in Step 2.

3.6 Final Comments

The methods we have developed in the previous sections may be used even for the general optimal control problems of the form

$$\textbf{(OC)} \qquad \underset{u \in \mathcal{U}}{Maximize} \int_0^T G(u(t), y^u(t))dt,$$

Fig. 3.2 Test 2. We solve the optimal harvesting problem described in Section 3.4, with the algorithms and discretization parameters reported in Section 3.5. Initial configuration of ω and suboptimal configurations depending on the choices of parameters c_1 and c_2

where $\mathcal{U}$ is a convex and closed subset of $L^2(0, T; X)$ with X a real Hilbert space (a functional space) and y^u is the unique (Carathéodory) solution to

$$\begin{cases} y'(t) = f(u(t), y(t)), \ t \in (0, T), \\ y(0) = y_0. \end{cases} \tag{3.24}$$

We assume that (3.24) is the abstract form of a reaction–diffusion system where the control acts in a subset ω of the whole domain Ω (and so $\mathcal{U}$ depends on ω) and that for any $u \in \mathcal{U}$, (3.24) has a unique Carathéodory solution. Here $f : X \times Y \to Y$ and $G : X \times Y \to \mathbb{R}$ are continuously differentiable functions (Y is another real Hilbert space); we will denote by f_u, f_y, G_u, and G_y the Gâteaux derivatives with respect to u and y.

Under appropriate assumptions (OC) admits at least one optimal control u^*. Assume that there exists a unique (Carathéodory) solution p to

$$\begin{cases} p'(t) = -f_y(u^*(t), y^{u^*}(t))^* p(t) + G_y(u^*(t), y^{u^*}(t)), \ t \in (0, T), \\ p(T) = 0. \end{cases} \tag{3.25}$$

Let B^* denote the adjoint of a linear and bounded operator B.

A formal calculation allows us to conclude that

$$-G_u(u^*(t), y^{u^*}(t)) + f_u(u^*(t), y^{u^*}(t))^* p(t) \in N_{\mathcal{U}}(u^*(t))$$

(see [9, p.59–62]). Here $N_{\mathcal{U}}(w) \subset L^2(0, T; X)$ is the normal cone at $\mathcal{U}$ in w.

If we may conclude in addition that $u^*(t) = \mathcal{F}(y^{u^*}(t), p(t))$, then (y^{u^*}, p) is the solution to

$$\begin{cases} y'(t) = f(\mathcal{F}(y(t), p(t)), y(t)), & t \in (0, T), \\ p'(t) = -f_y(\mathcal{F}(y(t), p(t)), y(t))^* p(t) + G_y(\mathcal{F}(y(t), p(t)), y(t)), & t \in (0, T), \\ y(0) = y_0, \qquad p(T) = 0. \end{cases}$$

If in addition we have that $f(u, y) = \mathcal{A}(u)y$ (f is linear with respect to y), then $f_y(u, y) = \mathcal{A}(u)$. Multiplying (3.25) by y^{u^*}, we may conclude after an easy formal calculation (involving (3.24) as well) that the maximal value of the cost function is

$$\Phi(u^*) = -y_0 \cdot p(0) - \int_0^T y(t) \cdot G_y(\mathcal{F}(y(t), p(t)), y(t))dt,$$

where (y, p) is the solution to

$$\begin{cases} y'(t) = \mathcal{A}(\mathcal{F}(y(t), p(t))y(t), & t \in (0, T), \\ p'(t) = -\mathcal{A}(\mathcal{F}(y(t), p(t)))^* p(t) + G_y(\mathcal{F}(y(t), p(t)), y(t)), & t \in (0, T), \\ y(0) = y_0, \qquad p(T) = 0. \end{cases}$$

Appropriate maximization problems with respect to ω (when $\mathcal{U}$ depends on ω) may be formulated and treated in the same manner as in Section 3.4.

Remark 3.13 An extremely important problem is that when the harvesting effort is localized in ω and is equal to a constant L. The problem to be investigated does not depend anymore on a dual state p and may be formulated as

$$\underset{\varphi}{Maximize} \{L \int_0^T \int_\Omega H(\varphi) y^\varphi dx\, dt - c_1 \int_\Omega \delta(\varphi)|\nabla \varphi| dx - c_2 \int_\Omega H(\varphi) dx\},$$

where y^φ is the solution to

$$\begin{cases} \frac{\partial y}{\partial t} - d\Delta y(x, t) = \eta(x) y(x, t) - LH(\varphi(x)) y(x, t), & (x, t) \in Q, \\ \frac{\partial y}{\partial n}(x, t) = 0, & (x, t) \in \Sigma, \\ y(x, 0) = y_0(x), & x \in \Omega. \end{cases}$$

The method presented in Section 3.4 may be used to treat such a problem as well.

Chapter 4
Optimal Harvesting: Age Dependence

4.1 Introduction and Setting of the Problem

This chapter is devoted to some optimal harvesting problems related to continuous age-dependent population dynamics. Age is an important parameter to take into account when describing a population dynamics. There is an extensive literature devoted to age-dependent population dynamics. For basic results and different approaches and applications, we refer to the monographs [8, 9, 82, 93, 132–134, 147, 153, 193].

There is a high interest in the control of age-dependent population dynamics. This concerns optimal control problems, stabilizability, and controllability problems as well. A special attention is paid to the optimal harvesting of age-dependent populations. Some of the very first investigations devoted to this problem were done by M. Brokate in [60, 61]. There is a huge interest and literature concerning the optimal control of population dynamics. We refer here to the monographs devoted to this issue [131, 145, 176] and to [8, 9, 41, 104, 105] for the control of age-dependent population dynamics.

In the second section of this chapter, we shall deduce the basic properties of the solutions to a linear model that describes the dynamics of a single population species with age structure. This section is based on Section 4.1 in [9]. The third section concerns the optimal harvesting problem related to a linear population dynamics. The section slightly extends the results in Section 4.2 in [9]. The fourth section is devoted to the properties of the solution to a nonlinear age-dependent population dynamics in a periodic environment and a related regional optimal harvesting problem. In the fifth section, some computational issues are discussed. Numerical results for an optimal control problem investigated in the previous section are given.

S. Aniţa et al., *Mathematical Modeling and Control in Life and Environmental Sciences*, Modeling and Simulation in Science, Engineering and Technology,
https://doi.org/10.1007/978-3-031-49971-5_4

4.2 A Linear Age-Dependent Population Dynamics

Throughout this chapter, T, A are positive constants, $Q_T = (0, A) \times (0, T)$, and one denotes

$$Dy(a,t) = \lim_{h \to 0} \frac{y(a+h, t+h) - y(a,t)}{h}.$$

Note that if y is differentiable at (a, t), then

$$Dy(a,t) = \frac{\partial y}{\partial a}(a,t) + \frac{\partial y}{\partial t}(a,t).$$

Let $y(a, t)$ be the number of individuals of age $a \in [0, A]$ at time $t \in [0, T]$ of a certain population species, and A the maximal age for this population species. The so-called vital rates are μ and β; μ is the mortality rate and depends on both age and time, and β is the birth rate and depends on a and t as well. Actually the birth process is described by the "birth law" (or "renewal law")

$$y(0,t) = \int_0^A \beta(a,t) y(a,t) da, \quad t \in [0, T],$$

where $y(0, t)$ is the number of newborn individuals at moment t, and $\beta(a, t)y(a, t)$ is the number of newborn individuals at time t with parents of age a. The initial age distribution of individuals is $y_0(a)$, $a \in [0, A]$.

The balance law shows that the number of individuals of age a at moment t that die in the time interval $[t, t + dt]$ is

$$y(a,t) - y(a+dt, t+dt) = \mu(a,t) y(a,t) dt.$$

Dividing by dt, we get that

$$Dy(a,t) + \mu(a,t) y(a,t) = 0.$$

If a certain infusion of population ($f(a, t) \geq 0$ is the inflow of individuals of age a at time t) occurs, then the population dynamics is described by the equation

$$Dy(a,t) + \mu(a,t) y(a,t) = f(a,t).$$

We obtain the following continuous linear model for a single age-structured population dynamics (see [8, 132, 193])

$$
\begin{cases}
Dy(a,t) + \mu(a,t)y(a,t) = f(a,t), \ (a,t) \in Q_T, \\
y(0,t) = \int_0^A \beta(a,t)y(a,t)da, \qquad t \in (0,T), \\
y(a,0) = y_0(a), \qquad a \in (0,A).
\end{cases} \tag{4.1}
$$

Assume that the following general (but natural) hypotheses are satisfied:

(H1) $\beta \in L^\infty(Q_T)$, $\mu \in L^1_{loc}([0,A) \times [0,T])$,

$$\beta(a,t) \geq 0, \quad \mu(a,t) \geq 0, \quad \text{a.e. } (a,t) \in Q_T.$$

(H2) $y_0 \in L^1(0,A)$, $f \in L^1(Q_T)$,

$$y_0(a) \geq 0, \ \text{a.e. } a \in (0,A), \quad f(a,t) \geq 0, \ \text{a.e. } (a,t) \in Q_T.$$

Definition 4.1 We say that y is a solution to (4.1) if:

(i) $y \in L^\infty(0,T;L^1(0,A))$, y is absolutely continuous along almost every characteristic line of equation $a - t = k$, $(a,t) \in \overline{Q}_T$, $k \in \mathbb{R}$.
(ii) y satisfies

$$
\begin{cases}
Dy(a,t) + \mu(a,t)y(a,t) = f(a,t), \ \text{a.e. } (a,t) \in Q_T, \\
y(0,t) = \int_0^A \beta(a,t)y(a,t)da, \qquad \text{a.e. } t \in (0,T), \\
y(a,0) = y_0(a), \qquad \text{a.e. } a \in (0,A).
\end{cases}
$$

Notice that the meaning of $y(0,t)$ and $y(a,0)$ is obvious if we take into consideration the absolute continuity of y along almost every characteristic line, i.e.,

$$y(0,t) = \lim_{h \to 0+} y(h, t+h), \ \text{a.e. } t \in (0,T),$$

$$y(a,0) = \lim_{h \to 0+} y(a+h, h), \ \text{a.e. } a \in (0,A).$$

Theorem 4.2 *Problem (4.1) has a unique solution. The solution is nonnegative.*

Proof If y is a solution to (4.1), then integrating along the characteristic lines, we get that

$$y(a,t)=\begin{cases} \exp\left\{-\int_0^a \mu(s,t-a+s)ds\right\} b(t-a) \\ \qquad + \int_0^a \exp\left\{-\int_s^a \mu(\tau,t-a+\tau)d\tau\right\} f(s,t-a+s)ds, \\ \qquad\qquad \text{a.e. } (a,t)\in Q_T,\ a<t, \\ \exp\left\{-\int_0^t \mu(a-t+s,s)ds\right\} y_0(a-t) \\ \qquad + \int_0^t \exp\left\{-\int_s^t \mu(a-t+\tau,\tau)d\tau\right\} f(a-t+s,s)ds, \\ \qquad\qquad \text{a.e. } (a,t)\in Q_T,\ a>t. \end{cases} \tag{4.2}$$

Here

$$b(t)=\int_0^A \beta(a,t)y(a,t)da, \quad \text{a.e. } t\in(0,T). \tag{4.3}$$

Since $\beta \in L^\infty(Q_T)$ and $y \in L^\infty(0,T;L^1(0,A))$, we get that $b \in L^\infty(0,T)$.

If we assume that $b \in L^\infty(0,T)$ is given, then y defined by (4.2) is the unique solution to

$$\begin{cases} Dy(a,t)+\mu(a,t)y(a,t)=f(a,t), & (a,t)\in Q_T, \\ y(0,t)=b(t), & t\in(0,T), \\ y(a,0)=y_0(a), & a\in(0,A), \end{cases} \tag{4.4}$$

meaning that y satisfies (i) in Definition 4.1 and

$$\begin{cases} Dy(a,t)+\mu(a,t)y(a,t)=f(a,t), & \text{a.e. } (a,t)\in Q_T, \\ \lim\limits_{h\to 0+} y(h,t+h)=b(t), & \text{a.e. } t\in(0,T), \\ \lim\limits_{h\to 0+} y(a+h,h)=y_0(a), & \text{a.e. } a\in(0,A). \end{cases}$$

A solution y to (4.1) satisfies (4.4), where b is defined by (4.3). Hence, b satisfies

$$b(t)=\int_0^t \beta(a,t)\exp\left\{-\int_0^a \mu(s,t-a+s)ds\right\} b(t-a)da$$
$$+\int_0^t \beta(a,t)\int_0^a \exp\left\{-\int_s^a \mu(\tau,t-a+\tau)d\tau\right\} f(s,t-a+s)ds\, da$$

$$+\int_t^A \beta(a,t)\exp\left\{-\int_0^t \mu(a-t+s,s)ds\right\} y_0(a-t)da$$

$$+\int_t^A \beta(a,t)\int_0^t \exp\left\{-\int_s^t \mu(a-t+\tau,\tau)d\tau\right\} f(a-t+s,s)ds\,da,$$

for almost any $0 < t < \min\{T, A\}$, and

$$b(t) = \int_0^A \beta(a,t)\exp\left\{-\int_0^a \mu(s,t-a+s)ds\right\} b(t-a)da$$

$$+\int_0^A \beta(a,t)\int_0^a \exp\left\{-\int_s^a \mu(\tau,t-a+\tau)d\tau\right\} f(s,t-a+s)ds\,da,$$

for almost any $A < t < T$ (if $A < T$).

It follows that b satisfies the following Volterra equation:

$$b(t) = F(t) + \int_0^t K(t,t-s)b(s)ds, \qquad \text{a.e. } t \in (0,T), \tag{4.5}$$

also known as the "renewal equation," where

$$K(t,a) = \begin{cases} \beta(a,t)\exp\left\{-\int_0^a \mu(s,t-a+s)ds\right\}, & \text{a.e. } (a,t) \in Q_T,\ a < t, \\ 0, & \text{elsewhere}, \end{cases} \tag{4.6}$$

and

$$F(t) = \begin{cases} \int_0^t \beta(a,t)\int_0^a \exp\left\{-\int_s^a \mu(\tau,t-a+\tau)d\tau\right\} f(s,t-a+s)ds\,da \\ \quad +\int_t^A \beta(a,t)\int_0^t \exp\left\{-\int_s^t \mu(a-t+\tau,\tau)d\tau\right\} f(a-t+s,s)ds\,da \\ \quad +\int_t^A \beta(a,t)\exp\left\{-\int_0^t \mu(a-t+s,s)ds\right\} y_0(a-t)da, \\ \qquad \text{a.e. } 0 < t < \min\{T,A\}, \\ \int_0^A \beta(a,t)\int_0^a \exp\left\{-\int_s^a \mu(\tau,t-a+\tau)d\tau\right\} f(s,t-a+s)ds\,da, \\ \qquad \text{a.e. } A < t < T. \end{cases} \tag{4.7}$$

The assumptions on β, μ, y_0, f imply that

$$K \in L^\infty(Q_T), \quad K(t,a) \geq 0, \quad \text{a.e. } (a,t) \in Q_T,$$

$$F \in L^\infty(0,T), \quad F(t) \geq 0, \quad \text{a.e. } t \in (0,T).$$

We shall prove now via Banach's fixed point theorem that (4.5) has a unique solution $b \in L^\infty(0,T)$.

Consider the following norm of $L^\infty(0,T)$:

$$\|g\| = Ess\ sup_{t\in(0,T)}(e^{-\lambda t}|g(t)|), \quad g \in L^\infty(0,T),$$

where λ is a positive constant to be precised later. This norm is equivalent to the usual norm on $L^\infty(0,T)$.

We define the operator $\mathcal{T} : L^\infty(0,T) \longrightarrow L^\infty(0,T)$ by

$$(\mathcal{T}g)(t) = F(t) + \int_0^t K(t, t-s)g(s)ds, \quad \text{a.e. } t \in (0,T).$$

For any $g_1, g_2 \in L^\infty(0,T)$, we have that for almost any $t \in (0,T)$:

$$|(\mathcal{T}g_1)(t) - (\mathcal{T}g_2)(t)| = \left| \int_0^t K(t,t-s)(g_1(s) - g_2(s))e^{-\lambda s}e^{\lambda s}ds \right|$$

$$\leq \|K\|_{L^\infty(Q_T)}\|g_1 - g_2\| \int_0^t e^{\lambda s}ds = \|K\|_{L^\infty(Q_T)}\|g_1 - g_2\|\frac{1}{\lambda}(e^{\lambda t} - 1),$$

and consequently,

$$\|\mathcal{T}g_1 - \mathcal{T}g_2\| \leq \frac{\|K\|_{L^\infty(Q_T)}}{\lambda}\|g_1 - g_2\|.$$

So, for any $\lambda > \|K\|_{L^\infty(Q_T)}$, we have that $\mathcal{T}$ is a contraction on $(L^\infty(Q_T), \|\cdot\|)$, and via Banach's fixed point theorem, we get that there exists a unique solution $b \in L^\infty(0,T)$ to (4.5) and that

$$\lim_{n\to+\infty} b_n = b \quad \text{in } L^\infty(0,T),$$

where

$$\begin{cases} b_0(t) = F(t), & t \in (0,T), \\ b_n(t) = F(t) + \int_0^t K(t,t-s)b_{n-1}(s)ds, & t \in (0,T),\ n \in \mathbb{N}^*. \end{cases} \tag{4.8}$$

Since F and K are nonnegative, we get by induction that $b_n(t) \geq 0$, and consequently, $b(t) \geq 0$, a.e. $t \in (0,T)$.

We conclude that (4.1) has a unique solution y. This solution is given by (4.2), where b is the unique solution to (4.5). On the other hand, by (4.2) and using the nonnegativity of b, y_0 and f, we conclude that $y(a,t) \geq 0$, a.e. $(a,t) \in Q_T$. Moreover, by (4.2), we get that $\mu y \in L^1(Q_T)$. □

Remark 4.3 If $f(a,t) > 0$, a.e. $(a,t) \in Q_T$, then by (4.2) we get that $y(a,t) > 0$, a.e. $(a,t) \in Q_T$.

If the mortality rate satisfies additionally that:

(H3) $\int_0^A \mu(a, t-A+a)da = +\infty$, a.e. $t \in (0,T)$, where μ is extended by 0 on $(0,A) \times (-\infty, 0)$

then the solution y to (4.1) satisfies

$$\lim_{h \to 0+} y(A-h, t-h) = 0, \quad \text{a.e. } t \in (0,T)$$

(i.e., $y(A,t) = 0$, a.e. $t \in (0,T)$). Hence, $(H3)$ is equivalent to the fact that A is the maximal age for this population species.

Indeed, for almost any $t \in (0, \min\{T, A\})$ and $h > 0$ sufficiently small, this yields

$$y(A-h, t-h) = \exp\left\{-\int_0^{t-h} \mu(A-t+s, s)ds\right\} y_0(A-t)$$

$$+\int_0^{t-h} \exp\left\{-\int_s^{t-h} \mu(A-t+\tau, \tau)d\tau\right\} f(A-t+s, s)ds$$

$$\longrightarrow \exp\left\{-\int_0^{t} \mu(A-t+s, s)ds\right\} y_0(A-t)$$

$$+\int_0^{t} \exp\left\{-\int_s^{t} \mu(A-t+\tau, \tau)d\tau\right\} f(A-t+s, s)ds$$

(as $h \to 0+$)

$$= \exp\left\{-\int_{A-t}^{A} \mu(\tau, t-A+\tau)d\tau\right\} y_0(A-t)$$

$$+\int_0^{t} \exp\left\{-\int_{A-t+s}^{A} \mu(\tau, t-A+\tau)d\tau\right\} f(A-t+s, s)ds = 0$$

(by $(H3)$).

On the other hand for almost any $t \in (\min\{T, A\}, T)$ (if $A < T$) and for any $h > 0$ sufficiently small

$$y(A-h, t-h) = \exp\left\{-\int_0^{A-h} \mu(s, t-A+s, s)ds\right\} b(t-A)$$

$$+\int_0^{A-h} \exp\left\{-\int_s^{A-h} \mu(\tau, t-A+\tau)d\tau\right\} f(s, t-A+s)ds$$

$$\longrightarrow \exp\left\{-\int_0^{A} \mu(s, t-A+s)ds\right\} b(t-A)$$

$$+\int_0^{A} \exp\left\{-\int_s^{A} \mu(\tau, t-A+\tau)d\tau\right\} f(s, t-A+s)ds = 0,$$

as $h \to 0+$.

Remark 4.4 If besides $(H1)$, $(H2)$ we have that:

(j) $y_0 \in L^2(0, A)$, $f \in L^2(Q_T)$, then the solution y to (4.1) belongs to $L^\infty(0, T; L^2(0, A))$.
(jj) $y_0 \in L^\infty(0, A)$, $f \in L^\infty(Q_T)$, then the solution y to (4.1) belongs to $L^\infty(Q_T)$.

Remark 4.5 If we assume that besides the hypotheses $(H1)$, $(H2)$, $(H3)$ we have that $f \equiv 0$, $\beta \in C([0, T]; L^\infty(0, A))$, and $\mu \in C([0, T]; L^\infty(0, \tilde{A}))$, for all $\tilde{A} \in (0, A)$, then the solution to (4.5) satisfies $b \in C([0, T])$ (see Theorem 2.1-p.33 in [132]).

Remark 4.6 Let $\{f^n\}_{n\in\mathbb{N}^*}$ such that $(H2)$ is satisfied for any $f := f^n$, $n \in \mathbb{N}^*$. If $f^n \longrightarrow f$ in $L^1(Q_T)$, then

$$y^n \longrightarrow y \quad \text{in } L^\infty(0, T; L^1(0, A)),$$

where y^n is the solution to (4.1) corresponding to $f := f^n$.

Indeed, we get that $F^n \longrightarrow F$ in $L^\infty(0, T)$, where F^n is defined by (4.7) when $f := f^n$, and consequently, $b_{f^n} \longrightarrow b$ in $L^\infty(0, T)$, where b_{f^n} is given by (4.3) where $y := y^n$ (and is the solution to (4.5) with $F := F^n$). Using now (4.2), we get the conclusion.

Let us establish next a comparison result for the solutions to (4.1).

Theorem 4.7 *If* $(\beta_j, \mu_j, y_{0j}, f_j)$ *satisfy* $(H1)$, $(H2)$ *(where* $\beta := \beta_j$, $\mu := \mu_j$, $y_0 := y_{0j}$, $f := f_j$, $j \in \{1, 2\}$*) and*

$$\beta_1(a, t) \le \beta_2(a, t), \quad \mu_1(a, t) \ge \mu_2(a, t), \quad a.e.\ (a, t) \in Q_T,$$

$$y_{01}(a) \le y_{02}(a),\ a.e.\ a \in (0, A), \quad f_1(a, t) \le f_2(a, t),\ a.e.\ (a, t) \in Q_T,$$

then

$$y_1(a,t) \le y_2(a,t), \quad a.e.\ (a,t) \in Q_T,$$

where y_j is the solution to (4.1) *corresponding to* $(\beta_j, \mu_j, y_{0j}, f_j)$.

Proof Denote by b^j the solution to (4.5) corresponding to $(\beta_j, \mu_j, y_{0j}, f_j)$ ($j \in \{1, 2\}$), i.e., b^j is the solution to

$$b^j(t) = F_j(t) + \int_0^t K_j(t, t-s) b^j(s) ds, \quad t \in (0, T), \tag{4.9}$$

where F_j and K_j are defined by (4.6) and (4.7), respectively, with $\beta := \beta_j$, $\mu := \mu_j$, $y_0 := y_{0j}$, $f := f_j$.

It is obvious that the hypotheses of Theorem 4.7 imply that

$$K_1(t,a) \le K_2(t,a), \quad \text{a.e. } (a,t) \in Q_T,$$

$$F_1(t) \le F_2(t), \quad \text{a.e. } t \in (0,T),$$

and so by (4.9), we get by mathematical induction that

$$b_n^1(t) \le b_n^2(t), \quad \text{a.e. } t \in (0,T), \text{ for any } n \in \mathbb{N},$$

where b_n^j is the solution to (4.8) corresponding to $F := F_j$ and $K := K_j$.

Passing to the limit, one obtains that $b^1(t) \le b^2(t)$, a.e. $t \in (0, T)$. By (4.2), we get the conclusion of the theorem. □

Consider now (4.1) in the limit case $T = +\infty$. Denote $Q = Q_\infty$.

Definition 4.8 We say that y is a solution to (4.1) with $T = +\infty$ if:

(l) $y \in L^\infty_{loc}([0, +\infty); L^1(0, A))$, y is absolutely continuous along almost every characteristic line of equation $a - t = k$, $(a,t) \in \overline{Q}$, $k \in \mathbb{R}$.

(ll) y satisfies

$$\begin{cases} Dy(a,t) + \mu(a,t)y(a,t) = f(a,t), \text{ a.e. } (a,t) \in Q, \\ y(0,t) = \int_0^A \beta(a,t) y(a,t) da, \qquad \text{a.e. } t \in (0,T), \\ y(a,0) = y_0(a), \qquad \text{a.e. } a \in (0, A). \end{cases}$$

If $\beta \in L^\infty_{loc}([0, A] \times [0, +\infty))$, $\mu \in L^1_{loc}([0, A) \times [0, +\infty))$,

$$\beta(a,t) \ge 0, \quad \mu(a,t) \ge 0, \quad \text{a.e. } (a,t) \in Q,$$

and $y_0 \in L^1(0, A)$, $f \in L^1_{loc}([0, A] \times [0, +\infty))$,

$$y_0(a) \geq 0, \text{ a.e. } a \in (0, A), \quad f(a, t) \geq 0, \text{ a.e. } (a, t) \in Q,$$

then (4.1) with $T = +\infty$ admits a unique solution y. The solution is nonnegative.

Moreover, the comparison result holds in this limit case too.

The proofs are based on the fact that actually y is a solution to (4.1) with $T = +\infty$, if and only if it is a solution to (4.1) for any $T \in (0, +\infty)$.

Before ending this section, we note that by (4.2) the solution y to (4.1) is also a weak solution to (4.1) in the following sense:

$$\int_0^T \int_0^A [-D\varphi(a, t) + \mu(a, t)\varphi(a, t) - \beta(a, t)\varphi(0, t)]y(a, t)da\, dt$$

$$= \int_0^A \varphi(a, 0)y_0(a)da + \int_0^T \int_0^A f(a, t)\varphi(a, t)da\, dt$$

for any φ satisfying:

- φ is absolutely continuous along almost every characteristic line of equation $a - t = k$, $(a, t) \in \overline{Q}_T$, $k \in \mathbb{R}$.
- $\varphi \in L^\infty(Q_T)$, $D\varphi \in L^1(Q_T)$, $D\varphi - \mu\varphi + \beta\varphi(0, \cdot) \in L^\infty(Q_T)$,

$$\varphi(A, t) = 0, \text{ a.e. } t \in (0, T), \quad \varphi(a, T) = 0, \text{ a.e. } a \in (0, A).$$

Denote by Φ the set of all functions φ with the above-mentioned properties.

The next result states the existence of a unique weak solution to (4.1) (see Theorem 2.1.4 in [8]).

Theorem 4.9 *Problem (4.1) admits a unique weak solution.*

Proof Let y_1, y_2 be two weak solutions to (4.1). Their difference $v = y_1 - y_2$ is a weak solution to (4.1) corresponding to $f := 0$.

We shall use now the following auxiliary result established in [8] (Lemma 2.1.5). □

Lemma 4.10 *Let $G \in L^\infty(Q_T)$. The following problem*

$$\begin{cases} D\varphi(a, t) - \mu(a, t)\varphi(a, t) = G(a, t) - \beta(a, t)\varphi(0, t), & (a, t) \in Q_T, \\ \varphi(A, t) = 0, & t \in (0, T), \\ \varphi(a, T) = 0, & a \in (0, A) \end{cases} \tag{4.10}$$

has a unique solution φ.

If moreover, $G(a, t) \geq 0$, a.e. in Q_T, then $\varphi(a, t) \leq 0$, a.e. $(a, t) \in Q_T$.

A function $\varphi \in \Phi$ is a solution to (4.10) if

$$\begin{cases} D\varphi(a,t) - \mu(a,t)\varphi(a,t) = G(a,t) - \beta(a,t)\varphi(0,t), & \text{a.e. } (a,t) \in Q_T, \\ \varphi(A,t) = 0, & \text{a.e. } t \in (0,T), \\ \varphi(a,T) = 0, & \text{a.e. } a \in (0,A). \end{cases}$$

The proof of this lemma is similar to that of Theorem 4.2. We integrate along the characteristic lines and reduce the investigation to a certain Volterra equation (for $B(t) = \varphi(0,t)$).

Proof of Theorem 4.9 (continued) Consider the problem

$$\begin{cases} D\varphi(a,t) - \mu(a,t)\varphi(a,t) = sign\ v(a,t) - \beta(a,t)\varphi(0,t), & (a,t) \in Q_T, \\ \varphi(A,t) = 0, & t \in (0,T), \\ \varphi(a,T) = 0, & a \in (0,A), \end{cases}$$

where

$$sign\ r = \begin{cases} 1, & \text{if } r > 0, \\ 0, & \text{if } r = 0, \\ -1, & \text{if } r < 0. \end{cases}$$

The solution φ to this problem belongs to Φ. If we use the test function φ in the definition of the weak solution v, we get that

$$\int_0^T \int_0^A |v(a,t)| da\, dt = 0,$$

and so $v(a,t) = 0$ and $y_1(a,t) = y_2(a,t)$, a.e. $(a,t) \in Q_T$. □

4.3 The Optimal Harvesting Problem

This section concerns an optimal harvesting problem governed by a linear age-structured population dynamics where the control acts only for ages $a \in \omega = (a_1, a_2) \subset (0, A)$ $(0 \le a_1 < a_2 \le A)$. Our purpose is to prove the existence of an optimal harvesting effort (control), obtain first order necessary conditions of optimality, and use them to build a numerical algorithm.

The problem to be investigated here has a great importance, both from an economical point of view (because indicates the optimal strategy) and from biological point of view.

Assume that besides $(H1)$ and $(H3)$ the following hypothesis holds:

(H3.2) $y_0 \in L^2(0, A)$, $f \in L^2(Q_T)$,

$$y_0(a) \geq 0, \text{ a.e. } a \in (0, A), \quad f(a,t) \geq 0, \text{ a.e. } (a,t) \in Q_T.$$

Let $\omega = (a_1, a_2)$ be a prescribed open subinterval of $(0, A)$ and denote by $\mathbb{I}_\omega$ its characteristic function. For any $w \in L^\infty(\omega)$ we define

$$\mathbb{I}_\omega(a)w(a) = \begin{cases} w(a), & \text{if } a \in \omega, \\ 0, & \text{if } a \in (0, A) \setminus \omega. \end{cases}$$

Consider now the first problem to be investigated in this section

$$\textbf{(OH)} \qquad \underset{u \in \mathcal{U}_\omega}{\text{Maximize}} \int_0^T \int_0^A \mathbb{I}_\omega(a)g(a)u(a,t)y^u(a,t)da\,dt,$$

where $\mathcal{U}_\omega = \{w \in L^\infty(\omega \times (0,T)) : 0 \leq w(a,t) \leq L, \text{ a.e. in } \omega \times (0,T)\}$ and y^u is the solution to

$$\begin{cases} Dy(a,t) + \mu(a,t)y(a,t) = f(a,t) - \mathbb{I}_\omega(a)u(a,t)y(a,t), & (a,t) \in Q_T, \\ y(0,t) = \displaystyle\int_0^A \beta(a,t)y(a,t)da, & t \in (0,T), \\ y(a,0) = y_0(a), & a \in (0,A). \end{cases} \tag{4.11}$$

Here $L > 0$, and $\mathcal{U}_\omega$ is the set of constrained controls. $u \in \mathcal{U}_\omega$ is the control, also called the harvesting effort. u plays the role of an additional mortality rate.

$g(a)$ is the price for an individual of age a, and $g \in C([0, A])$, $g(a) \geq 0$, for any $a \in [0, A]$. The integral $\int_0^T \int_0^A \mathbb{I}_\omega(a)g(a)u(a,t)y^u(a,t)da\,dt$ represents the total value of the harvest.

Our goal is to find the control (the harvesting effort) that gives the maximal value of the harvest, i.e., the optimal control (or the optimal harvesting effort).

So, we deal here with an optimal harvesting problem where the control acts only for ages $a \in \omega$. A cost related to this action in ω will be taken into account in the second part of this section. On the contrary, we assume that the cost of the harvesting effort u is negligible.

Since $\mu := \mu + \mathbb{I}_\omega u$ satisfies $(H1)$ (for any $u \in \mathcal{U}_\omega$), we conclude that System (4.11) has a unique solution y^u. This solution is nonnegative.

4.3.1 Existence of an Optimal Control

Let us recall first the following corollary of Mazur's theorem (see [42]):

Lemma 4.11 *Let $\{x_n\}_{n\in\mathbb{N}^*}$ be a sequence in a real Banach space X, which is weakly convergent to $x \in X$. Then there exists a sequence $\{\tilde{x}_n\}_{n\in\mathbb{N}^*} \subset X$, $\tilde{x}_n \in conv\,\{x_k;\ k \geq n+1\}$, $n \in \mathbb{N}^*$, such that $\{\tilde{x}_n\}_{n\in\mathbb{N}}$ converges (strongly) to x.*

This will allow us to establish the next existence result.

Theorem 4.12 *Problem (OH) admits at least one optimal control.*

Proof Define

$$\Psi(u) = \int_0^T \int_0^A \mathbb{I}_\omega(a) g(a) u(a,t) y^u(a,t)da\,dt, \quad u \in \mathcal{U}_\omega,$$

and let

$$d = \sup_{u\in\mathcal{U}_\omega} \Psi(u).$$

By Theorem 4.7, we get that

$$0 \leq \Psi(u) \leq L \int_0^T \int_0^A g(a) y^0(a,t)da\,dt < +\infty$$

(y^0 is the solution of (4.11) corresponding to $u := 0$), for any $u \in \mathcal{U}_\omega$, and consequently, $d \in [0, +\infty)$.

Let $\{u_n\}_{n\in\mathbb{N}^*} \subset \mathcal{U}_\omega$ be a sequence of controllers satisfying

$$d - \frac{1}{n} < \Psi(u_n) \leq d.$$

The same Theorem 4.7 allows us to conclude that

$$0 \leq y^{u_n}(a,t) \leq y^0(a,t), \quad \text{a.e. in } Q_T,$$

and so, on a subsequence, also denoted by $\{y^{u_n}\}$, we have

$$y^{u_n} \longrightarrow y^* \quad \text{weakly in } L^2(Q_T).$$

Using Lemma 4.11, we obtain a sequence $\{\tilde{y}_n\}$ satisfying

$$\tilde{y}_n = \sum_{i=n+1}^{k_n} \lambda_i^n y^{u_i}, \quad \lambda_i^n \geq 0, \quad \sum_{i=n+1}^{k_n} \lambda_i^n = 1$$

($k_n \geq n+1$), and

$$\tilde{y}_n \longrightarrow y^* \quad \text{in } L^2(Q_T).$$

Let the controls $\tilde{u}_n$ be defined as follows:

$$\tilde{u}_n(a,t) = \begin{cases} \dfrac{\sum_{i=n+1}^{k_n} \lambda_i^n y^{u_i}(a,t) u_i(a,t)}{\sum_{i=n+1}^{k_n} \lambda_i^n y^{u_i}(a,t)}, & \text{if } \sum_{i=n+1}^{k_n} \lambda_i^n y^{u_i}(a,t) \neq 0, \\ 0, & \text{if } \sum_{i=n+1}^{k_n} \lambda_i^n y^{u_i}(a,t) = 0, \end{cases}$$

for $(a,t) \in \omega \times (0,T)$. These controls satisfy $\tilde{u}_n \in \mathcal{U}_\omega$ and

$$\tilde{y}_n(a,t) = y^{\tilde{u}_n}(a,t), \quad \text{a.e. in } Q_T.$$

We can take a subsequence (also denoted by $\{\tilde{u}_n\}$) such that

$$\begin{cases} \tilde{u}_n \longrightarrow u^*, & \text{weakly in } L^2(\omega \times (0,T)), \\ \Psi(\tilde{u}_n) = \sum_{i=n+1}^{k_n} \lambda_i^n \Psi(u_i) \longrightarrow d, & \text{as } n \to +\infty, \end{cases}$$

and we may infer that

$$d = \lim_{n \to +\infty} \Psi(\tilde{u}_n) = \int_0^T \int_0^A \mathbb{I}_\omega(a) g(a) u^*(a,t) y^*(a,t) da\, dt. \tag{4.12}$$

On the other hand, the convergences $\tilde{u}_n \longrightarrow u^*$ weakly in $L^2(\omega \times (0,T))$ and $y^{\tilde{u}_n} \longrightarrow y^*$ in $L^2(Q_T)$ imply that y^* is a weak solution to (4.11) corresponding to $u := u^*$. Since the weak solution to (4.11) is unique, we get that

$$y^*(a,t) = y^{u^*}(a,t), \quad \text{a.e. } (a,t) \in Q_T.$$

By (4.12), we get that $d = \Psi(u^*)$ and consequently u^* is an optimal control for problem (OH). □

4.3.2 The Maximum Principle

Theorem 4.13 *Assume in addition that $f(a,t) > 0$, a.e. in Q_T. If u^* is an optimal control for Problem (OH) and p is the solution of*

$$\begin{cases} Dp - \mu p = \mathbb{I}_\omega u^*(g+p) - \beta(a,t)p(0,t), & (a,t) \in Q_T, \\ p(A,t) = 0, & t \in (0,T), \\ p(a,T) = 0, & a \in (0,A), \end{cases} \tag{4.13}$$

then we have

$$u^*(a,t) = \begin{cases} 0, \ \text{a.e.}\ (a,t) \in \omega \times (0,T), \ g(a) + p(a,t) < 0, \\ L, \ \text{a.e.}\ (a,t) \in \omega \times (0,T), \ g(a) + p(a,t) > 0. \end{cases} \tag{4.14}$$

Proof The existence and uniqueness of the solution p to (4.13) follow via Lemma 4.10.

For any $v \in L^\infty(Q_T)$, such that $u^* + \varepsilon v \in \mathcal{U}_\omega$ for any $\varepsilon > 0$ sufficiently small, we get

$$\int_0^T \int_0^A \mathbb{I}_\omega g u^* y^{u^*} da\, dt \geq \int_0^T \int_0^A \mathbb{I}_\omega g(u^* + \varepsilon v) y^{u^*+\varepsilon v} da\, dt,$$

and this implies that

$$\int_0^T \int_0^A \mathbb{I}_\omega g u^* \frac{y^{u^*+\varepsilon v} - y^{u^*}}{\varepsilon} da\, dt + \int_0^T \int_0^A \mathbb{I}_\omega g v y^{u^*+\varepsilon v} da\, dt \leq 0. \tag{4.15}$$

The convergence $y^{u^*+\varepsilon v} \longrightarrow y^{u^*}$ in $L^\infty(0,T;L^1(0,A))$, as $\varepsilon \to 0+$ follows by Remark 4.6. □

Lemma 4.14 *The following convergence holds*

$$\frac{1}{\varepsilon}[y^{u^*+\varepsilon v} - y^{u^*}] \longrightarrow z \quad \text{in}\ L^\infty(0,T;L^1(0,A)), \ \text{as}\ \varepsilon \to 0+,$$

where z is the solution of

$$\begin{cases} Dz(a,t) + \mu(a,t)z(a,t) = -\mathbb{I}_\omega v y^{u^*} - \mathbb{I}_\omega u^* z, & (a,t) \in Q_T, \\ z(0,t) = \displaystyle\int_0^A \beta(a,t)z(a,t)da, & t \in (0,T), \\ z(a,0) = 0, & a \in (0,A). \end{cases} \tag{4.16}$$

Proof (Lemma 4.14) Existence and uniqueness of z, solution of (4.16), follow by Theorem 4.2. Denote by

$$w_\varepsilon(a,t) = \frac{1}{\varepsilon}\big[y^{u^*+\varepsilon v}(a,t) - y^{u^*}(a,t)\big] - z(a,t), \quad (a,t) \in Q_T.$$

It is obvious that w_ε is the solution of

$$\begin{cases} Dw(a,t) + \mu(a,t)w(a,t) = -\mathbb{I}_\omega u^* w - \mathbb{I}_\omega v\big[y^{u^*+\varepsilon v} - y^{u^*}\big], & (a,t) \in Q_T, \\ w(0,t) = \displaystyle\int_0^A \beta(a,t)w(a,t)da, & t \in (0,T), \\ w(a,0) = 0, & a \in (0,A). \end{cases}$$

Since $y^{u^*+\varepsilon v} - y^{u^*} \longrightarrow 0$ in $L^\infty(0,T;L^1(0,A))$, as $\varepsilon \to 0+$, we infer via Remark 4.6 that $w_\varepsilon \to 0$ in $L^\infty(0,T;L^1(0,A))$, as $\varepsilon \to 0+$, and this concludes the proof of the lemma. □

Proof of Theorem 4.13 (continued) By passing to the limit in (4.15), we may conclude

$$\int_0^T \int_0^A \mathbb{I}_\omega g(u^* z + v y^{u^*})da\,dt \le 0. \tag{4.17}$$

By multiplying $(4.13)_1$ by z and integrating over $[0,T] \times [0,A]$, we get

$$\int_0^T \int_0^A (Dp - \mu p)z\,da\,dt = \int_0^T \int_0^A [\mathbb{I}_\omega u^*(g+p)z - \beta(a,t)p(0,t)z(a,t)]da\,dt.$$

By using (4.13) and $(4.16)_2$, after some calculation, we obtain

$$-\int_0^T \int_0^A p(Dz + \mu z)da\,dt = \int_0^T \int_0^A \mathbb{I}_\omega u^*(g+p)z\,da\,dt. \tag{4.18}$$

If we use (4.18), we get that

$$\int_0^T \int_0^A \mathbb{I}_\omega p v y^{u^*}\, da\, dt = \int_0^T \int_0^A \mathbb{I}_\omega g u^* z\, da\, dt,$$

and by (4.17) it follows that

$$\int_0^T \int_0^A \mathbb{I}_\omega v(a,t)(g(a)+p(a,t))y^{u^*}(a,t)da\, dt \le 0,$$

for any $v \in L^\infty(\omega \times (0,T))$, such that $u^* + \varepsilon v \in \mathcal{U}_\omega$, for any $\varepsilon > 0$ sufficiently small. This implies (arguing by contradiction) that

$$u^*(a,t) = \begin{cases} 0, \text{ a.e. } (a,t) \in \omega \times (0,T), \ (g(a)+p(a,t))y^{u^*}(a,t) < 0, \\ L, \text{ a.e. } (a,t) \in \omega \times (0,T), \ (g(a)+p(a,t))y^{u^*}(a,t) > 0. \end{cases}$$

Since $f(a,t) > 0$, a.e. in Q_T, we conclude that $y^{u^*}(a,t) > 0$, a.e. in Q_T, and consequently, (4.14) holds. □

Remark 4.15 As a consequence of Theorem 4.13, we obtain that p (the solution of (4.13)) is a solution of

$$\begin{cases} Dp - \mu p = L\mathbb{I}_\omega(g+p)^+ - \beta(a,t)p(0,t), \ (a,t) \in Q_T, \\ p(A,t) = 0, \qquad t \in (0,T), \\ p(a,T) = 0, \qquad a \in (0,A). \end{cases} \tag{4.19}$$

A function $p \in \Phi$ (with Φ defined in Section 4.2) is a solution to (4.19) if

$$\begin{cases} Dp - \mu p = L\mathbb{I}_\omega(g+p)^+ - \beta(a,t)p(0,t), \quad \text{a.e. } (a,t) \in Q_T, \\ p(A,t) = 0, \qquad \text{a.e. } t \in (0,T), \\ p(a,T) = 0, \qquad \text{a.e. } a \in (0,A). \end{cases}$$

Actually (4.19) has a unique solution.

If f satisfies the weaker assumption $f(a,t) \ge 0$, a.e. in Q_T and y_0 and β satisfy the stronger ones: $y_0(a) > 0$, a.e. $a \in (0,A)$, $\beta(a,t) > 0$, a.e. in Q_T, then it is obvious that $y^{u^*}(a,t) > 0$, a.e. in Q_T, and so, the conclusion of Theorem 4.13 remains true, as well as the conclusion in the first part of this remark.

Remark 4.16 The adjoint state p (which is also the solution of (4.19)) does not depend on f neither on y_0.

An uniqueness result for the optimal control is the next one:

Remark 4.17 Let u^* be an optimal control for (OH). Assume in addition that $f(a,t) > 0$, a.e. in Q_T, and that

$$\text{(H3.4)} \qquad \begin{cases} g \equiv 1, \\ \mu(a,t) > 0, \quad \text{a.e. in } \omega \times (0,T), \\ \text{and, for almost any } t \in (0,T),\ \dfrac{\beta}{\mu}(\cdot,t) \\ \text{is not a strictly positive constant on any subset} \\ \text{of positive measure of } \omega. \end{cases}$$

Under these additional conditions, Equation $(4.13)_1$ implies that the set

$$\mathcal{D} = \{(a,t) \in \omega \times (0,T) : \ p(a,t) = -1\}$$

has Lebesgue measure zero (p is the solution of (4.13)), and u^* is a bang-bang control (u^* takes only a finite number of values, almost everywhere).

In addition, the optimal control is unique. Denote by u^* this optimal control.

The last remark allows us to formulate the following result for the general case $f(a,t) \geq 0$, a.e. in Q_T:

Theorem 4.18 *If $(H1)$, $(H3.2)$, $(H3)$, $(H3.4)$ hold, then the control u^* is optimal also for problem (OH) corresponding to nonnegative inflow f.*

Proof Consider $f_n \in L^1(Q_T)$, such that $f_n(a,t) > 0$, a.e. in Q_T and $f_n \to f$ in $L^1(Q_T)$ (as $n \to +\infty$). Denote by $\Psi_n(u)$ the cost function corresponding to $f := f_n$ in Problem (OH), and by y_n^u the solution to (4.11) corresponding to u and $f := f_n$.

Remark 4.6 allows us to conclude that for any $u \in \mathcal{U}_\omega$,

$$y_n^u \longrightarrow y^u$$

in $L^\infty(0,T; L^1(0,A))$, as $n \to +\infty$. This implies

$$\Psi_n(u) \longrightarrow \Psi(u), \quad \text{as } n \to +\infty.$$

Since for any $u \in \mathcal{U}_\omega$ we have $\Psi_n(u^*) \geq \Psi_n(u)$, and by using the last convergence, we conclude that u^* is an optimal control for (OH). □

Remark 4.19 If $(H1)$, $(H3.2)$, $(H3)$, $(H3.4)$ hold, then we determine first the solution p of (4.19). An optimal control is given now by (4.14).

Assume now that we are interested to determine the interval $\omega = (a_1, a_2)$ for which we get the maximal yield (the value of the harvest—the cost associated to the action in the age interval (a_1, a_2)). If we multiply (4.11) by p and integrate over

$[0, T] \times [0, A]$, we get after some calculation and taking into account (4.13) that

$$\int_0^T \int_0^A \mathbb{I}_\omega g u^* y^{u^*} da\, dt = -\int_0^A y_0(a) p(a, 0) da - \int_0^T \int_0^A f(a, t) p(a, t) da\, dt.$$

If we denote by p^{a_1,a_2} the solution to (4.19), then we can formulate our problem as

$$(\mathbf{OH_{regional}}) \quad \underset{a_1,a_2 \in [0,A], a_1 < a_2}{\text{Minimize}} \left\{ \int_0^A y_0(a) p^{a_1,a_2}(a, 0) da \right.$$

$$\left. + \int_0^T \int_0^A f(a, t) p^{a_1,a_2}(a, t) da\, dt + \mathcal{G}(a_1, a_2) \right\}.$$

Here $\mathcal{G}(a_1, a_2)$ is the cost to be paid to act in the age interval (a_1, a_2); $\mathcal{G}$ is a continuous function.

The control here is the couple (a_1, a_2). Since for any control (a_1, a_2) we can compute the solution p^{a_1,a_2} to (4.19), then comparing the values of the cost functional in $(OH_{regional})$ we get the optimal one.

4.4 Optimal Harvesting for a Nonlinear Population Dynamics with Periodic Rates

Let $T_0 \in (0, +\infty)$. Consider the following controlled nonlinear model of age-dependent population dynamics with time-periodic vital rates (of period T_0)

$$\begin{cases} Dy(a, t) + \mu(a, t) y(a, t) + \mathcal{M}(Y(t)) y(a, t) = -u(t) y(a, t), & (a, t) \in Q, \\ Y(t) = \int_0^A y(a, t) da, & t > 0, \\ y(0, t) = \int_0^A \beta(a, t) y(a, t) da, & t > 0, \\ y(a, 0) = y_0(a), & a \in (0, A), \end{cases} \tag{4.20}$$

where $Q = (0, A) \times (0, +\infty)$ (here $T = +\infty$). Here $Y(t)$ is the total population at time t and $\mathcal{M}(Y(t)) y(a, t)$ is a logistic term. Actually, $\mathcal{M}(Y(t))$ is an additional mortality rate (besides the natural one μ) and is due to the competition for resources.

Assume that:

(H4.1) $\beta \in C([0, A] \times [0, +\infty))$, $\beta(a, t) \geq 0$, for all $(a, t) \in Q$,

$$\beta(a,t) = \beta(a,t+T_0), \quad \text{for all } (a,t) \in Q,$$

and there exist $\delta, \tau > 0$ and $a_0 \in (0, A)$ such that $a_0 + T_0 \le A$ and $\beta(a,\tau) \ge \delta$ for any $a \in (a_0, a_0 + T_0)$.
This assumption means that the fertile period is longer than or equal to the period T_0.

(H4.2) $\mu = \mu_0 + \overline{\mu}$, where

$$\mu_0 \in C([0, A)),\ \mu_0(a) \ge 0, \text{ for all } a \in [0, A),\ \int_0^A \mu_0(a)da = +\infty,$$
$$\overline{\mu} \in C([0, A] \times [0, +\infty)),\ \overline{\mu}(a,t) \ge 0, \quad \text{for all } (a,t) \in Q,$$

$$\overline{\mu}(a,t) = \overline{\mu}(a,t+T_0), \quad \text{for all } (a,t) \in Q.$$

(H4.3) $y_0 \in L^\infty(0, A)$, $y_0(a) > 0$, a.e. $a \in (0, A)$.

(H4.4) $\mathcal{M} : [0, +\infty) \longrightarrow [0, +\infty)$ is a continuously differentiable function, $\mathcal{M}'(r) > 0$, for all $r \ge 0$, and

$$\mathcal{M}(0) = 0, \quad \text{and there exists } \lim_{r \to +\infty} \mathcal{M}(r) = +\infty.$$

u is a harvesting effort (a control), $u \in L^\infty([0, +\infty))$, $u(t) \ge 0$, a.e. $t > 0$ and $u(t) = u(t + T_0)$, a.e. $t > 0$.

Definition 4.20 By a solution to (4.20), we mean a function y satisfying:

(i_n) $y \in L^\infty_{loc}([0, +\infty); L^1(0, A))$, y is absolutely continuous along almost every characteristic line of equation $a - t = k$, $(a,t) \in \overline{Q}$, $k \in \mathbb{R}$.

(ii_n) y satisfies

$$\begin{cases} Dy(a,t) + \mu(a,t)y(a,t) + \mathcal{M}(Y(t))y(a,t) = -u(t)y(a,t), & \text{a.e. } (a,t) \in Q, \\ Y(t) = \int_0^A y(a,t)da, & \text{a.e. } t > 0, \\ y(0,t) = \int_0^A \beta(a,t)y(a,t)da, & \text{a.e. } t > 0, \\ y(a,0) = y_0(a), & \text{a.e. } a \in (0, A). \end{cases}$$

First of all, it follows immediately that (4.20) has at most one solution. Next, let us notice that there exists a separable solution y to (4.20), i.e., which can be written as

$$y(a,t) = \tilde{y}(a,t)h(t), \quad (a,t) \in \overline{Q},$$

where $\tilde{y}$ is the solution to (4.20) with $\mathcal{M} := 0$ and h is the Carathéodory solution to the following Cauchy problem:

$$\begin{cases} h'(t) + \mathcal{M}(\tilde{Y}(t)h(t))h(t) = 0, & t > 0, \\ h(0) = 1, \end{cases}$$

where $\tilde{Y}(t) = \int_0^A \tilde{y}(a,t)da$, $t > 0$.

Any solution h satisfies $0 < h(t) \leq 1$, for any $t \in [0, +\infty)$, and h is a solution to the following integral equation:

$$h(t) = 1 - \int_0^t \mathcal{M}(\tilde{Y}(s)h(s))h(s)ds, \quad t \geq 0.$$

By Banach's fixed point theorem, we get that this integral equation has a unique solution $h \in C([0, \tilde{T}])$ for any $\tilde{T} > 0$, and consequently, the Cauchy problem has a unique solution.

We get the existence and uniqueness of a solution to (4.20).

We notice that actually $\tilde{y}(a,t) = \overline{y}(a,t)e^{-\int_0^t u(\tau)d\tau}$, $(a,t) \in [0, A] \times [0, +\infty)$, where $\overline{y}$ is the solution to (4.20) with $\mathcal{M} := 0$ and $u := 0$.

Since $\overline{y} \in C([A, +\infty); L^1(0, A))$ (by Remark 4.5), it follows that $\tilde{Y} \in C([A, +\infty))$.

Since $\mu = \mu_0 + \overline{\mu}$ with $\mu_0 \in C([0, A))$, $\mu_0(a) \geq 0$, for any $a \in [0, A)$, $\int_0^A \mu_0(a)da = +\infty$, and $\overline{\mu} \in C([0, A] \times [0, +\infty))$, $\overline{\mu}(a,t) \geq 0$, for any $(a,t) \in Q$, we additionally get that $\overline{y} \in C([A, +\infty); L^\infty(0, A))$, and consequently, $\tilde{y} \in C([A, +\infty); L^\infty(0, A))$.

Consider now the following nonlinear periodic model:

$$\begin{cases} Dy(a,t) + \mu(a,t)y(a,t) + \mathcal{M}(Y(t))y(a,t) = -u(t)y(a,t), & (a,t) \in Q, \\ Y(t) = \int_0^A y(a,t)da, & t > 0, \\ y(0,t) = \int_0^A \beta(a,t)y(a,t)da, & t > 0, \\ y(a,t) = y(a, t + T_0), & (a,t) \in Q. \end{cases} \tag{4.21}$$

Definition 4.21 By a solution to (4.21), we mean a function y satisfying:

(j_n) $y \in L^\infty_{loc}([0, +\infty); L^1(0, A))$, y is absolutely continuous along almost every characteristic line of equation $a - t = k$, $(a,t) \in \overline{Q}$, $k \in \mathbb{R}$.

(jj_n) y satisfies

$$\begin{cases} Dy(a,t)+\mu(a,t)y(a,t)+\mathcal{M}(Y(t))y(a,t)=-u(t)y(a,t), \text{ a.e. } (a,t)\in Q, \\ Y(t)=\int_0^A y(a,t)da, \qquad\qquad \text{a.e. } t>0, \\ y(0,t)=\int_0^A \beta(a,t)y(a,t)da, \qquad\qquad \text{a.e. } t>0, \\ y(a,t)=y(a,t+T_0), \qquad\qquad \text{a.e. } (a,t)\in Q. \end{cases}$$

An ergodicity result established in [187] (see also Theorem 2.3-p.36 in [9, 132]-p.174) implies that there exists a unique $(\alpha^*, y^*) \in \mathbb{R} \times C([0,+\infty); L^\infty(0,A))$ such that y^* is a nonnegative solution to

$$\begin{cases} Dy^*(a,t)+\mu(a,t)y^*(a,t)+\alpha^* y^*(a,t)=0, \ (a,t)\in Q, \\ y^*(0,t)=\int_0^A \beta(a,t)y^*(a,t)da, \qquad t>0, \\ y^*(a,t)=y^*(a,t+T_0), \qquad (a,t)\in Q \end{cases} \tag{4.22}$$

(the solution to (4.22) is defined as for (4.21) with $\mathcal{M} := 0$, $u := 0$) and satisfies Ess $\sup\{y^*(a,t) : (a,t)\in Q\} = 1$.

We recall the following result (see Theorem 4.18-p.174 in [9]).

Lemma 4.22 *The following convergence holds*

$$\lim_{t\to+\infty} \|y(t) - y_{T_0}(t)\|_{L^\infty(0,A)} = 0,$$

where y is the solution to (4.20) and y_{T_0} is a nonnegative solution to (4.21).

If $T_0\alpha^* \le \int_0^T u(t)dt$, then $y_{T_0} \equiv 0$ is the unique nonnegative solution to (4.21).

The case $T_0\alpha^* > \int_0^{T_0} u(t)dt$ corresponds to the situation when the population persists, and the case $T_0\alpha^* \le \int_0^{T_0} u(t)dt$ corresponds to the case when the population goes towards extinction. This second situation happens if $\alpha^* \le 0$ (so, the population would go to extinction even without harvesting) or if $\alpha^* > 0$, but the harvesting effort is too high to get population persistence.

It is obvious that

$$\int_t^{t+T_0}\int_0^A u(s)y(a,s)da\,ds \longrightarrow \int_0^{T_0}\int_0^A u(s)y_{T_0}(a,s)da\,ds,$$

as $t \to +\infty$ (the total harvest on $[t, t + T_0]$ tends to the total harvest on any time interval of length T_0 for y_{T_0}). It follows that

$$\frac{1}{N}\int_0^{NT_0}\int_0^A u(s)y(a,s)da\,ds \longrightarrow \int_0^{T_0}\int_0^A u(s)y_{T_0}(a,s)da\,ds,$$

as $N \to +\infty$.

We are interested to maximize the limit of the total harvest on a time interval $[t, t + T_0]$ (when the harvesting effort u belongs to a certain set of controls):

$$\underset{u}{\text{Maximize}} \lim_{t\to+\infty}\int_t^{t+T_0}\int_0^A u(s)y(a,s)da\,ds,$$

which is equivalent to

$$\underset{u}{\text{Maximize}} \int_0^{T_0}\int_0^A u(s)y_{T_0}(a,s)da\,ds.$$

It is clear now that we can "approximate" this by the following optimal harvesting problem:

$$\underset{u}{\text{Maximize}} \frac{1}{N}\int_0^{NT_0}\int_0^A u(s)y(a,s)da\,ds,$$

where N is a sufficiently large positive integer.

In many concrete situations, it is of interest to use controllers (in our case harvesting efforts) with a simple structure. It would be of great interest to take $u(t) = Lw(t)$, where L is a positive constant and $w(t) \in \{0, 1\}$ (for any t, $w(t)$ is either 0 or 1). When $w(t) = 1$, the harvesting is permitted, while when $w(t) = 0$, we have prohibition. Moreover, we wish that an interval of prohibition to be followed by an interval when harvesting is allowed and the sum of the lengths of the two intervals to be T_0 (usually 1 year).

We have of course to find the positions and the lengths of the prohibition intervals of time (and implicitly of the intervals when harvesting is allowed) and so we deal with a regional control problem.

Any such control (harvesting effort) can be written as

$$Lw(t) = LH\left(\gamma + \sin\left(\frac{2\pi t}{T_0} + \theta\right)\right),$$

where H is the Heaviside function, $\gamma \in [-1, 1]$, $\theta \in [0, 2\pi]$.

γ "dictates" the length of the prohibition intervals, and θ "dictates" the position of the prohibition intervals.

We can formulate now the following optimal harvesting problem:

$(\mathbf{OH^N_{regional}})$

$$\underset{\gamma\in[-1,1],\theta\in[0,2\pi]}{\text{Maximize}} \ \frac{1}{N}\int_0^{NT_0}\int_0^A LH\left(\gamma+\sin\left(\frac{2\pi s}{T_0}+\theta\right)\right) y^{\gamma,\theta}(a,s)da\,ds,$$

where $y^{\gamma,\theta}$ is the solution to

$$\begin{cases} Dy(a,t)+\mu(a,t)y(a,t)+\mathcal{M}(Y(t))y(a,t) \\ \qquad = -LH\left(\gamma+\sin\left(\dfrac{2\pi t}{T_0}+\theta\right)\right) y(a,t), \ (a,t)\in Q, \\ Y(t)=\displaystyle\int_0^A y(a,t)da, & t>0, \\ y(0,t)=\displaystyle\int_0^A \beta(a,t)y(a,t)da, & t>0, \\ y(a,0)=y_0(a), & a\in(0,A). \end{cases} \tag{4.23}$$

We leave as an exercise to the reader to prove that for a given N, problem $(OH^N_{regional})$ admits at least one optimal control $(\gamma^*,\theta^*)\in[-1,1]\times[0,2\pi]$.

Since the control is a couple of numbers, we can calculate for any such couple (γ,θ) the value of the cost functional and compare these values to decide which one provides the maximal one.

4.5 Computational Issues

In the first paragraph, we describe the numerical method to approximate the solution to (4.19) corresponding to $\omega = (a_1,a_2)$. This allows to investigate numerically problem $(OH_{regional})$.

In the second paragraph, we show the numerical method for the approximation of y^{a_1,a_2}, the solution to (4.23). Numerical results for the regional optimal control problem $(OH^N_{regional})$ are given as well.

4.5.1 Numerical Method to Approximate the Dual State p

We first introduce the grids

$$a_i=(i-1)h, \qquad i=1,2,\ldots,M,$$

and

$$t_j = (j-1)h, \qquad j = 1, 2, \ldots, N,$$

where $h > 0$ is the grid step and

$$M = 1 + A/h, \qquad N = 1 + T/h.$$

For the approximation, we use the notation

$$\psi(a_i, t_j) \approx \psi_i^{(j)}.$$

We use a semi-implicit Euler scheme to approximate the solution to (4.19):

$$\frac{p_i^{(j)} - p_{i-1}^{(j-1)}}{h} - \mu_{i-1}^{(j-1)} p_{i-1}^{(j-1)} = L\mathbb{I}_{\omega,i} \Pi(g_i^{(j)} + p_i^{(j)}) - \beta_i^{(j)} p_1^{(j)}.$$

Here $\Pi(r) = r^+ = \max\{r, 0\}$.

The following conditions are satisfied as well

$$p_M^{(j)} = 0, \qquad p_i^{(N)} = 0.$$

Note that in the above considered scheme we have taken $\mu_{i-1}^{(j-1)} p_{i-1}^{(j-1)}$ instead of $\mu_i^{(j)} p_i^{(j)}$ (corresponding to an explicit Euler scheme) to avoid the situation $\mu_M^{(j)} = +\infty$.

For any control (a_1, a_2), one computes the corresponding approximation of the solution to (4.19) and the corresponding value of the cost functional for $(OH_{regional})$. Next one decides which control (a_1, a_2) provides the optimal value of the cost functional.

4.5.2 *Numerical Method for the Optimal Harvesting Problem with Periodic Rates*

We discuss here a numerical method for the regional optimal harvesting problem $(OH_{regional}^N)$. We consider the grids

$$a_i = (i-1)h, \qquad i = 1, 2, \ldots, M,$$

and

$$t_j = (j-1)h, \qquad j = 1, 2, \ldots, N_t,$$

where $h > 0$ is the grid step and

$$M = 1 + A/h, \qquad N_t = 1 + NT_0/h.$$

For an arbitrary but fixed $(\gamma, \theta) \in [-1, 1] \times [0, 2\pi]$, we use a semi-implicit Euler scheme to approximate the solution $y^{\gamma,\theta}$ to (4.23):

$$\frac{y_i^{(j)} - y_{i-1}^{(j-1)}}{h} + \mu_i^{(j)} y_i^{(j)} + \mathcal{M}(Y^{(j-1)}) y_i^{(j)} = -LH\left(\gamma + \sin\left(\frac{2\pi t_j}{T_0} + \theta\right)\right) y_i^{(j)},$$

where

$$Y^{(j-1)} = \int_0^A y^{(j-1)}(a)da$$

is obtained using the trapezoidal rule.

The nonlocal boundary condition becomes

$$y_1^{(j)} = \int_0^A \beta^{(j)}(a) y^{(j)}(a)da,$$

where the integral is approximated using the trapezoidal rule.

We consider a rate β that satisfies the natural condition $\beta(0, t) = 0$ (meaning that the birth rate for the newborn individuals is 0), and so $y_1^{(j)} = \int_0^A \beta^{(j)}(a) y^{(j)}(a)da$ works properly.

The initial condition is

$$y_i^{(1)} = y_0(a_i).$$

For any control $(\gamma, \theta) \in [-1, 1] \times [0, 2\pi]$, one computes the approximation of $y^{\gamma,\theta}$ and the corresponding approximation of the value of the cost functional. Next one decides the control that provides the optimal value of the cost functional and the optimal intervals to allow harvesting and for prohibition.

The numerical tests below were obtained for $T_0 = 0.25$, $N = 20$, $A = 1$, $\mathcal{M}(r) = r$, $L = 5$, $C = 60$,

$$\beta(a, t) = Ca^2(A - a)\left(1 + \sin\frac{\pi a}{A}\right)(0.6 + \sin(8\pi t))^+,$$

$$\mu(a, t) = \frac{e^{-a}}{A - a} + e^{-4a}(2 + \cos(8\pi t)),$$

$$y_0(a) = 2e^{-0.5a^2}.$$

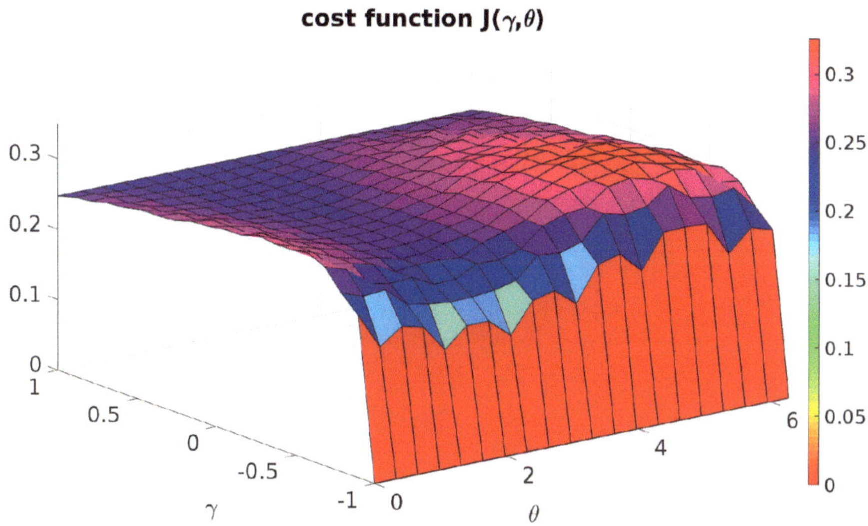

Fig. 4.1 We solve the optimal harvesting problem described in Section 4.4, with the algorithms and discretization parameters reported in Section 4.5. Plot of the cost function

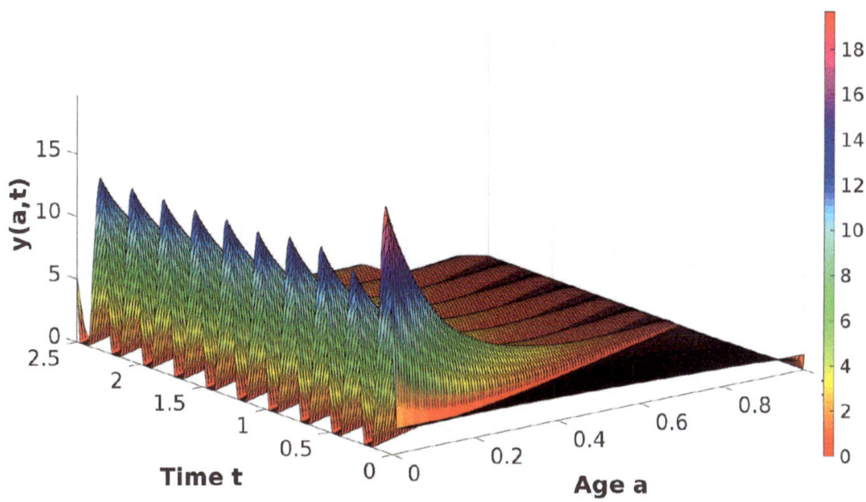

Fig. 4.2 We solve the optimal harvesting problem described in Section 4.4, with the algorithms and discretization parameters reported in Section 4.5. Plot of y^{γ^*,θ^*}

Figure 4.1 shows the graph of the cost function as a function of (γ, θ), and we see that the optimal control is

$$\gamma^* = -0.58, \quad \theta^* = 4.96.$$

In Figure 4.2, the plot of y^{γ^*,θ^*} has been reported.

Part III
Controlling Epidemics in Agriculture

Chapter 5
Controlling the Spread of an Epidemic in Agriculture: The Case of *Xylella fastidiosa*

5.1 Introduction

The etiological agent of olive trees disease known as olive quick decline syndrome (OQDS) is the plant pathogenic bacterium *Xylella fastidiosa*, which is a vector borne bacterium.

The main vector of *Xylella fastidiosa* in Southern Italy has been identified in the so-called meadow spittlebug, i.e., the *Philaenus spumarius*, a xylem sap-feeding specialist. Their juvenile form (nymphs) develops on weeds or ornamental plants, confined in a foam produced by themselves for protection from predators and temperature, while their adult forms move to olive tree canopies.

X. fastidiosa transmission is the result of four events (see, e.g., [2, 170] and references therein):

(a) Acquisition from a source plant
(b) Attachment and retention to vector's foregut
(c) Inoculation into a new host
(d) Development of the infection in a plant after inoculation

A successful transmission also includes bacterial multiplication.

Once a plant is infected, bacteria multiply within the xylem vessels inducing the production of a gel in the plant xylem, which occludes the xylem vessels, thus inhibiting the flux of water through the lymph vessels eventually blocking the nutrition of the plant. Typical symptoms are leaf scorch, dieback of twigs, branches, and even of the whole plant [78].

Sanitation of infected olive trees is unfeasible; the scope of our research is the mathematical modelling of the population dynamics of the ecosystem in the presence of the infection. The availability of a sound mathematical model may lead to predictive analysis of the relevant populations, so to suggest possible eradication strategies, and possible optimal control strategies.

S. Aniţa et al., *Mathematical Modeling and Control in Life and Environmental Sciences*, Modeling and Simulation in Science, Engineering and Technology,
https://doi.org/10.1007/978-3-031-49971-5_5

In a series of papers by the same authors of this monograph, a simple epidemiological model, describing the Olive Quick Decline Syndrome (OQDS) outbreak, which has occurred in Southern Italy since 2013, was presented.

The OQDS main players considered in the model are its insect vectors, *Philaenus spumarius*, and the host plants (olive trees and weeds) of the insects and of the bacterium. A spatially homogeneous model was based on a system of ordinary differential equations, the analysis of which have provided interesting results about possible equilibria of the epidemic system and their stability [48, 62]. Although the model presented there was mathematically rather simplified, its analysis has highlighted threshold parameters that could be the target of control strategies within an integrated pest management framework. A key issue emerged there; for the eventual eradication of the disease, it is not necessary for the removal of the productive resource represented by the olive trees. Indeed, numerical simulations support the outcomes of the mathematical analysis according to which a removal of a suitable amount of weed biomass (a reservoir of *Xylella fastidiosa*) from olive orchards and surrounding areas resulted the most efficient strategy to control the spread of the OQDS. In addition, as expected, the adoption of more resistant olive tree cultivars has been shown to be a good strategy, though less cost-effective, in controlling the pathogen.

Later [25], for a more realistic description, and a clearer interpretation of the proposed control measures, a spatial structure of the epidemic system has been included. In order to keep mathematical technicalities at a minimum level, in this case, only two players have been included in the spatially structured model, trees and insects, while the weed biomass has been taken as a given quantity.

As possible control actions, insects traps, weed cut, choice of the cultivar, and reduction of contact rates have been taken into account. In the numerical simulations, particular attention has been paid to weeds cut and choice of the cultivar, the other control measures being more difficult to implement. It is worth mentioning that, in recent investigations presented in [177], the authors, by means of a cellular automaton simulator, have confirmed the relevance of the olive cultivar as a possible effective control strategy.

A relevant contribution of our approach consists of the restriction of measures of intervention (control) only to a subregion of the whole habitat of interest. It has been shown that such a practice can still lead to the eradication of an epidemic outbreak.

Numerical simulations are presented, which confirm the analytical results. In the numerical simulations, the relevant parameters have been taken from [62].

5.2 Our Model

Since the feeding behavior and metabolic processes are qualitatively similar for both nymphs and adults [135], only one stage of active vectors of the infection has been considered.

In the spatially structured model, the population of vector insects and the population of infected olive trees have been included; weeds will be taken into account as a given environmental parameter.

Let $C_I(x,t)$ denote the spatial density of the total population of insects, $s_1(x,t)$ the spatial density of susceptible insects, and $i_1(x,t)$ the spatial density of infected insects, so that

$$C_I(x,t) = s_1(x,t) + i_1(x,t).$$

Let $C_T(x,t)$ denote the spatial density of the total biomass of olive trees, $s_2(x,t)$ the spatial density of the biomass of healthy trees, and $i_2(x,t)$ the spatial density of the biomass of infected trees, so that

$$C_T(x,t) = s_2(x,t) + i_2(x,t).$$

- VECTOR—*Philaenus spumarius*

Due to the fact that bacteria reside only in the foregut of an adult insect, the latter generate only healthy offsprings (see, e.g., [2, 170], and references therein). The parameter r denotes the birth rate of new insects.

Reproduction, however, can occur only if bushes and other plants, whether healthy or infected, are present, explaining the presence of a "carrying capacity" $M(x)$ in the logistic term, describing the biomass of all such plants, that we will simply call *weeds*.

Both healthy and infected insects experience natural mortality at a rate δ; κ is a scale parameter.

Finally, we assume that insects may diffuse in the habitat (with constant diffusion coefficient to avoid purely technical complications) [142].

Concerning the "local incidence" for insects, at point $x \in \overline{\Omega}$, and time $t \geq 0$, as in previous models [11, 67], we assume that it is due to bites to infected olive trees at any point $x' \in \Omega$ of the habitat, within a spatial neighborhood of x represented by a suitable probability kernel $k(x, x')$, depending on the specific structure of the local ecosystem (see also [179]); as a trivial simplification, one may assume $k(x, \cdot)$ as a Gaussian density centered at x; hence, the "local incidence rate (i.r.)" for insects, at point $x \in \Omega$, and time $t \geq 0$, is taken as

$$(i.r.)_I(x,t) = s_1(x,t) \int_\Omega k(x,x') i_2(x',t) dx'.$$

Hence, the spatial dynamics of susceptible insects is expressed by the following equation:

$$\begin{aligned} \frac{\partial s_1}{\partial t}(x,t) &= d\Delta s_1(x,t) + rC_I(x,t)[M(x) - \kappa s_1(x,t)] - \delta s_1(x,t) \\ &\quad -\beta s_1(x,t) \int_\Omega k(x,x') i_2(x',t) dx', \end{aligned} \tag{5.1}$$

while the spatial dynamics of infective insects is expressed by the following equation:

$$\frac{\partial i_1}{\partial t}(x,t) = d\Delta i_1(x,t) - \delta i_1(x,t) - r\kappa i_1(x,t)C_I(x,t) + \beta s_1(x,t)\int_\Omega k(x,x')i_2(x',t)dx'. \tag{5.2}$$

Both Equations (6.1) and (6.2) act in a spatial domain $\Omega \subset \mathbb{R}^2$, at times $t \in (0,+\infty)$. They are subject to suitable initial conditions. We assume homogeneous Neumann boundary conditions:

- OLIVE TREES

For the olive trees, it is better to refer to their canopies, so that we may consider pruning and regrowth. Healthy trees (canopy) are produced at constant net regrowth rate q, get infected by contact with infected insects at rate ζ, or by human activities such as budding and grafting, at rate b. For trees, in view of their long survival, we may neglect natural mortality. Infected trees experience disease-related mortality μ and human-induced mortality ℓ due to pruning and logging.

Hence, the spatial dynamics of trees is expressed by the following equations:

$$\begin{aligned}\frac{\partial s_2}{\partial t}(x,t) &= (q-\ell)s_2(x,t) - s_2(x,t)\frac{C_T(x,t)}{C^0} \\ &-(\zeta i_1(x,t) + b\ell i_2(x,t))s_2(x,t) + \alpha i_2(x,t),\end{aligned} \tag{5.3}$$

$$\begin{aligned}\frac{\partial i_2}{\partial t}(x,t) &= -\mu i_2(x,t) - \ell i_2(x,t) - i_2(x,t)\frac{C_T(x,t)}{C^0} \\ &+(\zeta i_1(x,t) + b\ell i_2(x,t))s_2(x,t) - \alpha i_2(x,t).\end{aligned} \tag{5.4}$$

Both Equations (5.3) and (5.4) act in the spatial domain $\Omega \subset \mathbb{R}^2$, at times $t \in (0,+\infty)$. They are subject to suitable initial conditions. The description of the model parameters is presented in Table 5.1.

Assumptions

- $\Omega \subset \mathbb{R}^2$ is a bounded domain with a sufficiently smooth boundary.
- $M \in L^\infty(\Omega)$, $M(x) \geq 0$, a.e. in Ω.
- $k \in L^\infty(\Omega \times \Omega)$, $k(x,x') \geq 0$, a.e. in $\Omega \times \Omega$, normalized to β.
- $d,\ r,\ \kappa, \delta,\ q,\ \ell,\ C^0,\ \mu,\ \zeta,\ b,\ \alpha,\ \beta$ are positive constants.

Table 5.1 Model parameters

Symbol	Description
r	Insects birth rate
κ	Insect intraspecific competition rate
δ	Insects mortality rate
q	Healthy trees (canopy) regrowth rate
C^0	Trees carrying capacity parameter
ℓ	Elimination rate of trees by pruning or logging
b	Infection rate of trees by infected tools
μ	Infected trees mortality rate
α	Infected trees recovery rate
β	Insects infection rate by infected trees
ζ	Trees infection rate by infected insects

5.3 Control, Eradicability

By summing Equations (6.1), (6.2), we obtain an equation for the total insect population $C_I(x,t)$

$$\frac{\partial C_I}{\partial t}(x,t) = d\Delta C_I(x,t) + rC_I(x,t)[M(x) - \kappa C_I(x,t)] - \delta C_I(x,t).$$

For M constant, pure Neumann boundary conditions allow constant in space and time solutions, satisfying the following equilibrium equation:

$$rC_I^*(M - \kappa C_I^*) - \delta C_I^* = 0,$$

which we may rewrite as

$$[r(M - \kappa C_I^*) - \delta]C_I^* = 0.$$

This equation may admit two solutions

$$C_I^* = 0,$$

and the solution of

$$r(M - \kappa C_I^{**}) - \delta = 0.$$

This shows that for $M \leq \dfrac{\delta}{r}$ no other nonnegative solution is feasible, but the trivial one. Otherwise, another equilibrium is feasible

$$C_I^{**} = \frac{1}{\kappa}\left(M - \frac{\delta}{r}\right), \tag{5.5}$$

provided

$$M > \frac{\delta}{r}.$$

From the above simple reasoning, we may conjecture that a way to eradicate the disease is to eliminate the insect population by a significant reduction of the carrying capacity M, which may be obtained by eliminating weeds in the relevant olive orchards. Equation (5.5) shows the quantitative role of the scale parameter κ; a smaller value of κ allows a larger value of C_I^{**}, so that we may say that $\frac{M}{\kappa}$ plays the role of an "effective carrying capacity" of insects. Numerical simulations, reported in Section 5.4, illustrate these facts.

A more accurate analysis follows. Assume that certain constant controls are considered in a nonempty open subset $\omega \subset \Omega$.

Possible regional controls (acting in ω) are the following ones:

C1: Traps: $\delta \rightarrow \delta + \gamma_1$; increase insect death rate, by insecticides and/or treated nets.
C2: Weed cut: $M \rightarrow M(1 - \gamma_{21})$; decrease carrying capacity of insects, by eliminating weeds in the olive orchards.
C3: Choice of the cultivar: $\zeta \rightarrow \zeta(1 - \gamma_{22})$; decrease infection rate of trees, acting on the cultivar.
C4: Reduce contacts: $\beta \rightarrow \beta(1 - \gamma_{23})$; decrease contact rate of trees towards insects, by installing treated nets.
C5: Clean tools: $b \rightarrow b(1 - \gamma_3)$; decrease infection rate from tools to trees, by disinfection.

For various reasons, including cost reduction, we may consider the possibility of implementing the proposed control measures only on a suitable subregion $\omega \subset \Omega$ of the whole habitat, including areas already affected by the epidemic, augmented by a preventive confinement area.

Let $\mathbb{I}_\omega$ denote the characteristic function of ω. So that the controlled system is

$$\frac{\partial s_1}{\partial t}(x,t) = d\Delta s_1(x,t) + rC_I(x,t)[M(x)(1-\gamma_{21}\mathbb{I}_\omega(x)) - \kappa s_1(x,t)]$$
$$-(\delta+\gamma_1\mathbb{I}_\omega(x))s_1(x,t) - \beta(1-\gamma_{23}\mathbb{I}_\omega(x))s_1(x,t)\int_\Omega k(x,x')i_2(x',t)dx', \quad (5.6)$$

$$\frac{\partial i_1}{\partial t}(x,t) = d\Delta i_1(x,t) - (\delta + \gamma_1\mathbb{I}_\omega(x))i_1(x,t) - r\kappa i_1(x,t)C_I(x,t)$$
$$+\beta(1-\gamma_{23}\mathbb{I}_\omega(x))s_1(x,t)\int_\Omega k(x,x')i_2(x',t)dx', \quad (5.7)$$

$$\frac{\partial s_2}{\partial t}(x,t) = (q-\ell)s_2(x,t) - s_2(x,t)\frac{C_T(x,t)}{C^0}$$
$$-(\zeta(1-\gamma_{22}\mathbb{I}_\omega(x))i_1(x,t) + b(1-\gamma_3\mathbb{I}_\omega(x))\ell i_2(x,t))s_2(x,t) + \alpha i_2(x,t), \quad (5.8)$$

$$\frac{\partial i_2}{\partial t}(x,t)=-\mu i_2(x,t)-\ell i_2(x,t)-i_2(x,t)\frac{C_T(x,t)}{C^0}$$
$$+(\zeta(1-\gamma_{22}\mathbb{I}_\omega(x))i_1(x,t)+b(1-\gamma_3\mathbb{I}_\omega(x))\ell i_2(x,t))s_2(x,t)-\alpha i_2(x,t). \quad (5.9)$$

We assume that there is no flux of insects through the boundary of Ω (the domain is isolated):

$$\frac{\partial s_1}{\partial n}(x,t)=\frac{\partial i_1}{\partial n}(x,t)=0, \quad x\in\partial\Omega,\ t>0, \quad (5.10)$$

and that the following initial conditions are satisfied

$$s_j(x,0)=s_{j0}(x),\ i_j(x,0)=i_{j0}(x), \quad x\in\Omega,\ j\in\{1,2\}. \quad (5.11)$$

Assumptions

- $s_{10},\ i_{10}\in L^\infty(\Omega), \quad s_{10}(x),\ i_{10}(x)\geq 0$, a.e. $x\in\Omega$.
- $s_{20},\ i_{20}\in L^\infty(\Omega), \quad s_{20}(x),\ i_{20}(x)\geq 0$, a.e. $x\in\Omega$.
- $\gamma_1\geq 0,\gamma_{21},\ \gamma_{22},\ \gamma_{23}\in[0,1],\ \gamma_3\in[0,1]$ are constants.

Under these hypotheses, it follows via a fixed point argument and using the comparison for the solutions to the parabolic equations, that there exists a unique solution (s_1,i_1,s_2,i_2) to (5.6)–(5.11), i.e., for any $T\in(0,+\infty)$:

$s_1,\ i_1\ \in\ C([0,T];L^2(\Omega))\cap AC((0,T];L^2(\Omega))\cap L^2(0,T;H^1(\Omega))\cap L^2_{loc}((0,T];H^2(\Omega))$,
$s_2,\ i_2\in L^2(\Omega\times(0,T))$, for almost any $x\in\Omega$ $s_2(x,\cdot),\ i_2(x,\cdot)\in AC([0,T])$ and

$$\begin{cases}
\frac{\partial s_1}{\partial t}(x,t)=d\Delta s_1(x,t)+rC_I(x,t)[M(x)(1-\gamma_{21}\mathbb{I}_\omega(x))-\kappa s_1(x,t)] \\
\qquad -(\delta+\gamma_1\mathbb{I}_\omega(x))s_1(x,t) \\
\qquad -\beta(1-\gamma_{23}\mathbb{I}_\omega(x))s_1(x,t)\int_\Omega k(x,x')i_2(x',t)dx', & \text{a.e. } (x,t)\in Q, \\
\frac{\partial i_1}{\partial t}(x,t)=d\Delta i_1(x,t)-(\delta+\gamma_1\mathbb{I}_\omega(x))i_1(x,t)-r\kappa i_1(x,t)C_I(x,t) \\
\qquad +\beta(1-\gamma_{23}\mathbb{I}_\omega(x))s_1(x,t)\int_\Omega k(x,x')i_2(x',t)dx', & \text{a.e. } (x,t)\in Q, \\
\frac{\partial s_2}{\partial t}(x,t)=(q-\ell)s_2(x,t)-s_2(x,t)\frac{C_T(x,t)}{C^0} \\
\qquad -(\zeta(1-\gamma_{22}\mathbb{I}_\omega(x))i_1(x,t)+b(1-\gamma_3\mathbb{I}_\omega(x))\ell i_2(x,t))s_2(x,t) \\
\qquad +\alpha i_2(x,t), & \text{a.e. } (x,t)\in Q, \\
\frac{\partial i_2}{\partial t}(x,t)=-\mu i_2(x,t)-\ell i_2(x,t)-i_2(x,t)\frac{C_T(x,t)}{C^0} \\
\qquad +(\zeta(1-\gamma_{22}\mathbb{I}_\omega(x))i_1(x,t)+b(1-\gamma_3\mathbb{I}_\omega(x))\ell i_2(x,t))s_2(x,t) \\
\qquad -\alpha i_2(x,t), & \text{a.e. } (x,t)\in Q, \\
\frac{\partial s_1}{\partial n}(x,t)=\frac{\partial i_1}{\partial n}(x,t)=0, & \text{a.e. } (x,t)\in\Sigma, \\
s_j(x,0)=s_{j0}(x),\ i_j(x,0)=i_{j0}(x)\ (j\in\{1,2\}), & \text{a.e. } x\in\Omega.
\end{cases}$$

Here $Q = \Omega \times (0, +\infty)$ and $\Sigma = \partial\Omega \times (0, +\infty)$. For a similar argument but for a different system, we refer to [14].

By using the comparison results for parabolic equations and for first order ODEs, it follows that

$$s_j,\ i_j \in L^\infty_{loc}(\overline{\Omega} \times [0, +\infty)),$$
$$s_j(x,t),\ i_j(x,t) \geq 0,\ \text{a.e. } (x,t) \in Q,$$

$j \in \{1, 2\}$ (see [11, 14, 27]).

As far as trees are concerned, we may easily obtain an upper bound for the total canopy. By adding Equations (5.8) and (5.9) and taking into account the initial conditions, we get that $C_T = s_2 + i_2$ satisfies

$$\begin{cases} \dfrac{\partial C_T}{\partial t} = (q - \ell)C_T - \dfrac{1}{C^0}C_T^2 - (\mu + q)i_2, & (x,t) \in Q, \\ C_T(x,0) = C_{T0}(x) = s_{20}(x) + i_{20}(x), & x \in \Omega. \end{cases}$$

By comparison results for the solutions to first order ODEs, we get that

$$C_T(x,t) \leq \tilde{C}_T(t), \quad \text{a.e. } (x,t) \in Q,$$

where $\tilde{C}_T$ is the solution to the next Bernoulli equation (which can be explicitly solved)

$$\begin{cases} \dfrac{d\tilde{C}_T}{dt} = (q - \ell)\tilde{C}_T - \dfrac{1}{C^0}\tilde{C}_T^2, \ t > 0, \\ \tilde{C}_T(0) = \|C_{T0}\|_\infty + 1. \end{cases}$$

Here and throughout this chapter, $\|\cdot\|_p$ denotes the usual norm of $L^p(\Omega)$.

It is obvious that $\tilde{C}_T$ is space independent.

Moreover:

(j) If $q \leq \ell$, then $\tilde{C}_T(t) \longrightarrow 0$, as $t \to +\infty$.
(jj) If $q > \ell$, then $\tilde{C}_T(t) \longrightarrow C^0(q - \ell)$, as $t \to +\infty$.

We may notice that the quantity $C^0(q - \ell)^+$ is the carrying capacity of the tree population.

As far as the insect population is concerned, we may add Equations (5.6) and (5.7) and take into account the boundary and initial conditions, so to obtain that $C_I = s_1 + i_1$ satisfies

$$
\begin{cases}
\dfrac{\partial C_I}{\partial t} = d\Delta C_I + r[M(1-\gamma_{21}\mathbb{I}_\omega) - \kappa C_I]C_I \\
\qquad\qquad -(\delta + \gamma_1 \mathbb{I}_\omega)C_I, & (x,t) \in Q, \\
\dfrac{\partial C_I}{\partial n}(x,t) = 0, & (x,t) \in \Sigma, \\
C_I(x,0) = C_{I0}(x) = s_{10}(x) + i_{10}(x), & x \in \Omega.
\end{cases}
\tag{5.12}
$$

Let C_I^* be the maximal nonnegative solution to

$$
\begin{cases}
-d\Delta C_I - [rM(1-\gamma_{21}\mathbb{I}_\omega) - (\delta + \gamma_1\mathbb{I}_\omega)]C_I + r\kappa C_I^2 = 0, \ x \in \Omega, \\
\dfrac{\partial C_I}{\partial n}(x) = 0, & x \in \partial\Omega
\end{cases}
\tag{5.13}
$$

(the existence of the maximal nonnegative solution will be proved in the Proposition below), and let $\tilde{\lambda}_1$ be the principal eigenvalue to

$$
\begin{cases}
-d\Delta\varphi - [rM(1-\gamma_{21}\mathbb{I}_\omega) - (\delta + \gamma_1\mathbb{I}_\omega)]\varphi = \lambda\varphi, \ x \in \Omega, \\
\dfrac{\partial \varphi}{\partial n}(x) = 0, & x \in \partial\Omega.
\end{cases}
\tag{5.14}
$$

By Rayleigh's formula (see Appendix A), we get that

$$
\tilde{\lambda}_1 = \min_{\substack{\varphi \in H^1(\Omega) \\ \|\varphi\|_2 \neq 0}} \frac{\int_\Omega [d|\nabla\varphi|^2 - [rM(1-\gamma_{21}\mathbb{I}_\omega) - (\delta + \gamma_1\mathbb{I}_\omega)]|\varphi|^2]dx}{\int_\Omega |\varphi|^2 dx}
$$

and that the space of eigenvectors corresponding to $\tilde{\lambda}_1$ has the dimension one and there exists an eigenvector $\tilde{\varphi}_1$ such that $\tilde{\varphi}_1(x) > 0$, for any $x \in \Omega$.

It is obvious that $\tilde{\lambda}_1$ depends on Ω, ω, and on the controls γ_1 and γ_{21}, and by Rayleigh's formula, we deduce that it is an increasing function of γ_1, of γ_{21} and ω—by inclusion.

By arguing as in [27], we shall prove the following result.

Proposition 5.1 *Under the above assumptions:*

(i) If $\tilde{\lambda}_1 \geq 0$, then (5.13) admits only the trivial solution.
Moreover, for any initial condition, the solution to (5.12) satisfies

$$
C_I(\cdot,t) \longrightarrow 0 \quad in\ L^\infty(\Omega),
$$

as $t \to +\infty$.

(ii) If $\tilde{\lambda}_1 < 0$, then (5.13) has two nonnegative solutions: the trivial one and C_I^* (which is strictly positive on Ω).

Moreover, if C_{I0} is not identically 0, then the solution to (5.12) satisfies

$$C_I(\cdot, t) \longrightarrow C_I^* \quad in\ L^\infty(\Omega),$$

as $t \to +\infty$.

Proof

(i) Assume first that $\tilde{\lambda}_1 > 0$. Let $\tilde{\varphi}_1$ an eigenvector to (5.14) with positive values in Ω. Actually, $\tilde{\varphi}_1 \in C(\overline{\Omega})$ (via a regularity result for elliptic equations), and there exist two positive constants θ_0 and θ_1 such that

$$0 < \theta_0 \leq \tilde{\varphi}_1(x) \leq \theta_1, \quad \text{for all } x \in \overline{\Omega}$$

(see [154]).

By the comparison result for parabolic equations, we obtain that

$$0 \leq C_I(x,t) \leq \overline{C}(x,t), \quad \text{a.e. } x \in \Omega, \text{ for all } t \geq 0,$$

where $\overline{C}$ is the solution to

$$\begin{cases} \dfrac{\partial \overline{C}}{\partial t} = d\Delta \overline{C} + r[M(1-\gamma_{21}\mathbb{I}_\omega)]\overline{C} - (\delta + \gamma_1 \mathbb{I}_\omega)\overline{C}, & (x,t) \in Q, \\ \dfrac{\partial \overline{C}}{\partial n}(x,t) = 0, & (x,t) \in \Sigma, \\ \overline{C}(x,0) = c_0 \tilde{\varphi}_1(x), & x \in \Omega. \end{cases} \tag{5.15}$$

Here c_0 is a positive constant such that

$$C_{I0}(x) \leq c_0 \tilde{\varphi}_1(x), \quad \text{a.e. } x \in \Omega.$$

The solution to (5.15) is

$$\overline{C}(x,t) = e^{-\tilde{\lambda}_1 t} c_0 \tilde{\varphi}_1(x), \quad \text{a.e. } x \in \Omega, \text{ for all } t \geq 0,$$

and so

$$C_I(\cdot,t) \longrightarrow 0 \quad \text{in } L^\infty(\Omega),$$

as $t \to +\infty$ (exponentially).

Assume that $\tilde{\lambda}_1 = 0$. Let h be the solution to the following problem:

$$\begin{cases} h'(t) = -r\kappa\theta_0 h(t)^2, & t > 0, \\ h(0) = c_0, \end{cases}$$

and denote $\overline{C}(x,t) = h(t)\tilde{\varphi}_1(x)$, $x \in \overline{\Omega}$, $t \geq 0$. By calculating h, we obtain that $\lim_{t\to+\infty} h(t) = 0$.

It is obvious that $\overline{C}$ is the solution to the following problem:

$$\begin{cases} \dfrac{\partial \overline{C}}{\partial t} = d\Delta\overline{C} + r[M(1-\gamma_{21}\mathbb{I}_\omega)]\overline{C} - (\delta+\gamma_1\mathbb{I}_\omega)\overline{C} & \\ \qquad\qquad -r\kappa\overline{C}^2 + r\kappa h^2\tilde{\varphi}_1(\tilde{\varphi}_1-\theta_0), & (x,t)\in Q, \\ \dfrac{\partial \overline{C}}{\partial n}(x,t) = 0, & (x,t)\in\Sigma, \\ \overline{C}(x,0) = c_0\tilde{\varphi}_1(x), & x\in\Omega. \end{cases}$$

Since $r\kappa h^2\tilde{\varphi}_1(\tilde{\varphi}_1-\theta_0)$ is a nonnegative function, we get via the comparison result for parabolic equations that

$$0 \leq C_I(x,t) \leq \overline{C}(x,t) = h(t)\tilde{\varphi}_1(x), \quad \text{a.e. } x\in\Omega, \text{ for all } t \geq, 0$$

and since $\lim_{t\to+\infty} h(t) = 0$, we get the conclusion. The convergence to 0 has this time the rate of $\frac{1}{r\kappa\theta_0 t}$.

The convergence $C_I(\cdot,t) \longrightarrow 0$ in $L^\infty(\Omega)$, as $t\to+\infty$, implies for $C_I(x,0) = C_I^*(x)$ that $C_I^* \equiv 0$:

(ii) Assume that $\tilde{\lambda}_1 < 0$.

Let us prove first that (5.13) has two nonnegative solutions: the trivial one and another one denoted C_I^*. In fact we shall prove that there exists a positive constant $\tilde{\theta}$ such that $C_I^*(x) \geq \tilde{\theta}$, a.e. $x\in\Omega$ (actually for any $x\in\overline{\Omega}$, because $C_I^*\in C(\overline{\Omega})$).

Indeed, let us consider $c_1 > 0$ a sufficiently small constant and $c_2 > 0$ a sufficiently large constant, satisfying that

$$c_1\tilde{\varphi}_1(x) \leq C_{I0}(x) \leq c_2\tilde{\varphi}_1(x), \quad \text{a.e. } x\in\Omega.$$

Let h_1 be the solution to

$$\begin{cases} h_1'(t) = -\tilde{\lambda}_1 h_1(t) - r\kappa\theta_1 h_1(t)^2, & t>0, \\ h_1(0) = c_1 \end{cases} \tag{5.16}$$

(the saturated solution to (5.16) is defined at least on $[0,+\infty)$), and h_2 be the solution to

$$\begin{cases} h_2'(t) = -\tilde{\lambda}_1 h_2(t) - r\kappa\theta_0 h_2(t)^2, & t>0, \\ h_2(0) = c_2 \end{cases} \tag{5.17}$$

(the saturated solution to (5.17) is defined at least on $[0,+\infty)$).

For c_1 sufficiently small, we have that $h_1'(0) = c_1(-\tilde{\lambda}_1 - r\kappa\theta_1 c_1) > 0$. It follows that $h_1'(t) > 0$, for any $t \geq 0$; otherwise, there exists $t_1 > 0$ such that $h_1'(t_1) = 0$ and $h_1'(t) > 0$, for any $t \in [0, t_1)$.

The uniqueness of the solution to the ODE in (5.16), together with the initial condition $h(t_1) = h_1(t_1)$, implies that $h_1(t) = h_1(t_1)$, for any $t \in [0, +\infty)$ and so we get a contradiction.

So, $h_1'(t) > 0$, for any $t \in [0, +\infty)$, and so h_1' is a strictly increasing function on $[0, +\infty)$. By analogy, it follows that h_2 is strictly decreasing on $[0, +\infty)$.

Let C_1 be the solution to

$$\begin{cases} \dfrac{\partial C_1}{\partial t} = d\Delta C_1 + [r\,M(1-\gamma_{21}\mathbb{I}_\omega) - (\delta + \gamma_1 \mathbb{I}_\omega)]C_1 - r\,\kappa C_1^2, & (x,t) \in Q, \\ \dfrac{\partial C_1}{\partial n}(x,t) = 0, & (x,t) \in \Sigma, \\ C_1(x,0) = c_1 \tilde{\varphi}_1(x), & x \in \Omega, \end{cases} \tag{5.18}$$

and let $\tilde{C}_1(x,t) = h_1(t)\tilde{\varphi}_1(x)$, $x \in \Omega,\ t \geq 0$. It is obvious that $\tilde{C}_1$ is the solution

$$\begin{cases} \dfrac{\partial \tilde{C}_1}{\partial t} = d\Delta \tilde{C}_1 + [r\,M(1-\gamma_{21}\mathbb{I}_\omega) - (\delta + \gamma_1 \mathbb{I}_\omega)]\tilde{C}_1 - r\,\kappa C_1^2, & \\ \qquad\qquad + r\,\kappa h_1^2 \tilde{\varphi}_1(\tilde{\varphi}_1 - \theta_1), & (x,t) \in Q, \\ \dfrac{\partial \tilde{C}_1}{\partial n}(x,t) = 0, & (x,t) \in \Sigma, \\ \tilde{C}_1(x,0) = c_1 \tilde{\varphi}_1(x), & x \in \Omega. \end{cases}$$

Since $r\,\kappa h_1^2 \tilde{\varphi}_1(\tilde{\varphi}_1 - \theta_1) \leq 0$, a.e. $(x,t) \in Q$, we get, via the comparison result for parabolic equations, that

$$h_1(t)\tilde{\varphi}_1(x) = \tilde{C}_1(x,t) \leq C_1(x,t), \quad \text{a.e. } x \in \Omega,$$

for any $t \geq 0$. This implies that

$$h_1(0)\tilde{\varphi}_1(x) = C_1(x,0) \leq C_1(x,t), \quad \text{a.e. } x \in \Omega,$$

for any $t \geq 0$. By using again the comparison result for parabolic equations, we get that

$$C_1(x,s) \leq C_1(x,s+t), \quad \text{a.e. } x \in \Omega, \ \text{for all } s,t \geq 0.$$

Hence, $t \mapsto C_1(\cdot, t)$ is nondecreasing on $[0, +\infty)$.

Let now C_2 be the solution to

$$\begin{cases} \dfrac{\partial C_2}{\partial t} = d\Delta C_2 + [r\,M(1-\gamma_{21}\mathbb{I}_\omega) - (\delta + \gamma_1\mathbb{I}_\omega)]C_2 - r\,\kappa C_2^2, & (x,t) \in Q, \\ \dfrac{\partial C_2}{\partial n}(x,t) = 0, & (x,t) \in \Sigma, \\ C_2(x,0) = c_2\tilde{\varphi}_1(x), & x \in \Omega. \end{cases} \tag{5.19}$$

It follows as before that

$$h_2(0)\tilde{\varphi}_1(x) = C_2(x,0) \geq C_2(x,t), \quad \text{a.e. } x \in \Omega,$$

for any $t \geq 0$. By using again the comparison result for parabolic equations, we get that

$$C_2(x,s) \geq C_2(x,s+t), \quad \text{a.e. } x \in \Omega, \text{ for all } s,t \geq 0,$$

and so $t \mapsto C_2(\cdot,t)$ is nonincreasing on $[0,+\infty)$.

This also yields

$$c_1\tilde{\varphi}_1(x) \leq C_1(x,t) \leq C_I(x,t) \leq C_2(x,t) \leq c_2\tilde{\varphi}_1(x), \quad \text{a.e. } x \in \Omega,$$

for any $t \geq 0$. We conclude that

$$C_1(\cdot,t) \longrightarrow C_1^*,\ C_2(\cdot,t) \longrightarrow C_2^* \quad \text{in } L^p(\Omega),$$

as $t \to +\infty$ (for any $p \in [1,+\infty)$). So,

$$c_1\tilde{\varphi}_1(x) \leq C_1^*(x) \leq C_2^*(x) \leq c_2\tilde{\varphi}_1(x), \quad \text{a.e. } x \in \Omega.$$

By Equation (5.18), we get that C_1^* is a (weak) solution (in the sense of Lax–Milgram's lemma) of

$$\begin{cases} -d\Delta C_1^* - [r\,M(1-\gamma_{21}\mathbb{I}_\omega) - (\delta + \gamma_1\mathbb{I}_\omega)]C_1^* + r\,\kappa(C_1^*)^2 = 0, & x \in \Omega, \\ \dfrac{\partial C_1^*}{\partial n}(x) = 0, & x \in \partial\Omega \end{cases} \tag{5.20}$$

and by Equation (5.19), we get that C_2^* is a (weak) solution of

$$\begin{cases} -d\Delta C_2^* - [r\,M(1-\gamma_{21}\mathbb{I}_\omega) - (\delta + \gamma_1\mathbb{I}_\omega)]C_2^* + r\,\kappa(C_2^*)^2 = 0, & x \in \Omega, \\ \dfrac{\partial C_2^*}{\partial n}(x) = 0, & x \in \partial\Omega. \end{cases} \tag{5.21}$$

If we multiply Equation (5.20) by C_2^* and Equation (5.21) by C_1^* and consider their difference, we get that

$$\int_\Omega r\kappa\, C_1^* C_2^* (C_1^* - C_2^*) dx = 0.$$

We conclude that $C_1^*(x) = C_2^*(x)$, a.e. $x \in \Omega$.

So, there exists one more nonnegative solution C_I^* to (5.13) (apart from the trivial one). This solution satisfies $C_I^*(x) \geq c_1\tilde{\varphi}_1(x)$, a.e. $x \in \Omega$.

If C^* is a nonnegative and nontrivial solution to (5.13), then the solution C_I to (5.12) corresponding to the initial condition $C_I(x,0) = C^*$ (which obviously is C^*; it is time independent) satisfies that for a $t_0 > 0$, there exists $c_1 > 0$ sufficiently small such that $C_I(x,t_0) \geq c_1\tilde{\varphi}_1(x)$, a.e. $x \in \Omega$ (see [154]). On the other hand, there exists $c_2 > 0$ sufficiently large such that $C_I(x,t_0) \leq c_2\tilde{\varphi}_1(x)$, a.e. $x \in \Omega$ (because $C_I(\cdot, t_0) \in L^\infty(\Omega)$).

It is clear now that

$$C_1(x,t) \leq C_I(x,t+t_0) = C^*(x) \leq C_2(x,t) \leq c_2\tilde{\varphi}_1(x), \quad \text{a.e. } x \in \Omega,$$

for any $t \geq 0$. By passing to the limit, we get that $C^*(x) = C_I^*(x)$, a.e. $x \in \Omega$.

So, the first part of (ii) is proved.

We also conclude that 0 is an eigenvalue for the next problem

$$\begin{cases} -d\Delta\varphi - [rM(1-\gamma_{21}\mathbb{I}_\omega) - (\delta + \gamma_1\mathbb{I}_\omega)]\varphi + r\kappa C_I^*\varphi = \lambda\varphi, & x \in \Omega, \\ \dfrac{\partial\varphi}{\partial n}(x) = 0, & x \in \partial\Omega, \end{cases} \tag{5.22}$$

and that C_I^* is actually a corresponding positive eigenvector. As a consequence of Rayleigh's formula, we get that 0 is actually the principal eigenvalue for this problem.

Let us prove now the second part of (ii). By arguing as before, it is obvious that it is enough to treat the situation when $c_1\tilde{\varphi}_1(x) \leq C_{I0}(x) \leq c_2\tilde{\varphi}_1(x)$, a.e. $x \in \Omega$. Let us prove that $C_1(\cdot,t) - C_I^* \longrightarrow 0$ in $L^\infty(\Omega)$, as $t \to +\infty$. It is obvious that $0 \leq C_I^*(x) - C_1(x,t)$, a.e. $x \in \Omega$, for any $t \geq 0$; $C_I^* - C_1$ is a solution to

$$\begin{cases} \dfrac{\partial(C_I^* - C_1)}{\partial t} = d\Delta(C_I^* - C_1) + [rM(1-\gamma_{21}\mathbb{I}_\omega) - (\delta + \gamma_1\mathbb{I}_\omega)]\ (C_I^* - C_1), & \\ \qquad -r\kappa C_I^*(C_I^* - C_1) - r\kappa C_1(C_I^* - C_1), & (x,t) \in Q, \\ \dfrac{\partial(C_I^* - C_1)}{\partial n}(x,t) = 0, & (x,t) \in \Sigma, \\ (C_I^* - C_1)(x,0) = C_I^*(x) - C_{I0}(x), & x \in \Omega. \end{cases} \tag{5.23}$$

We consider now v the solution to

$$\begin{cases} \dfrac{\partial v}{\partial t} = d\Delta v + [rM(1-\gamma_{21}\mathbb{I}_\omega) - (\delta + \gamma_1 \mathbb{I}_\omega(x))]v, \\ \qquad -r\kappa C_I^* v - r\kappa c_1 \tilde{\varphi}_1 v, & (x,t) \in Q, \\ \dfrac{\partial v}{\partial n}(x,t) = 0, & (x,t) \in \Sigma, \\ v(x,0) = C_I^*(x) - C_{I0}(x), & x \in \Omega. \end{cases} \tag{5.24}$$

The comparison result for parabolic equations implies that

$$0 \le C_I^*(x) - C_1(x,t) \le v(x,t), \quad \text{a.e. } x \in \Omega, \text{ for all } t \ge 0.$$

We consider now the following eigenvalue problem:

$$\begin{cases} -d\Delta\varphi - [rM(1-\gamma_{21}\mathbb{I}_\omega) - (\delta + \gamma_1 \mathbb{I}_\omega(x))]\varphi \\ \qquad +r\kappa C_I^*\varphi + r\kappa c_1\tilde{\varphi}_1\varphi = \lambda\varphi, & x \in \Omega, \\ \dfrac{\partial \varphi}{\partial n}(x) = 0, & x \in \partial\Omega. \end{cases}$$

By Rayleigh's formula, we get that the principal eigenvalue for this problem is equal to

$$\min_{\substack{\varphi \in H^1(\Omega) \\ \|\varphi\|_2 \neq 0}} \frac{\int_\Omega [d|\nabla\varphi|^2 - [rM(1-\gamma_{21}\mathbb{I}_\omega) - (\delta+\gamma_1\mathbb{I}_\omega)]|\varphi|^2 + r\kappa C_I^*|\varphi|^2 + r\kappa c_1\tilde{\varphi}_1|\varphi|^2]dx}{\int_\Omega |\varphi|^2 dx}$$

$$\ge \min_{\substack{\varphi \in H^1(\Omega) \\ \|\varphi\|_2 \neq 0}} \frac{\int_\Omega r\kappa c_1 \tilde{\varphi}_1 |\varphi|^2 dx}{\int_\Omega |\varphi|^2 dx} > 0.$$

The first part of Proposition 5.1 implies that

$$v(\cdot,t) \longrightarrow 0 \quad \text{in } L^\infty(\Omega), \text{ as } t \to +\infty,$$

and consequently, we get that

$$C_I^* - C_1(\cdot,t) \longrightarrow 0 \quad \text{in } L^\infty(\Omega), \text{ as } t \to +\infty$$

and

$$C_1(\cdot,t) \longrightarrow C_I^* \quad \text{in } L^\infty(\Omega), \text{ as } t \to +\infty.$$

In the same manner, it follows that

$$C_2(\cdot,t) \longrightarrow C_I^* \quad \text{in } L^\infty(\Omega), \text{ as } t \to +\infty,$$

and consequently,

$$C_I(\cdot, t) \longrightarrow C_I^* \quad \text{in } L^\infty(\Omega), \text{ as } t \to +\infty$$

(because $C_1(x,t) \le C_I(x,t) \le C_2(x,t)$, a.e. $x \in \Omega$, for any $t \ge 0$). □

We may notice that Proposition 5.1 confirms, for the full reaction-diffusion system, the outcomes of the preliminary analysis presented at the beginning of Section 6.3. In particular, it is of interest to observe that, either for small values of the weed distribution $M(x)$, or a large value of the control parameter γ_{21}, or a large domain of intervention ω, $\tilde{\lambda}_1$ may become greater or equal than zero, so that the only nonnegative solution of (5.13) is the trivial one, and the whole insect population eventually goes extinct.

The case $\tilde{\lambda}_1 < 0$ requires further investigation, since an additional nontrivial value C_I^* is possible; we may then require suitable threshold conditions for the eventual extinction of the epidemic.

Let now λ_1 be the principal eigenvalue to

$$\begin{cases} -d\Delta\varphi + (\delta + \gamma_1 \mathbb{I}_\omega(x))\varphi + r\kappa C_I^*\varphi = \lambda\varphi, & x \in \Omega, \\ \dfrac{\partial\varphi}{\partial n} = 0, & x \in \partial\Omega. \end{cases}$$

By Rayleigh's formula, we get that $\lambda_1 \ge \delta$. If ω is an open subset of Ω with a C^1–boundary such that $\overline{\omega} \subset \Omega$ and $\Omega \setminus \overline{\omega}$ is a domain, then if $\gamma_1 \nearrow +\infty$, then $\lambda_1 \nearrow \lambda_1^*$, where λ_1^* is the principal eigenvalue to

$$\begin{cases} -d\Delta\varphi + \delta\varphi + r\kappa C_I^*\varphi = \lambda\varphi, & x \in \Omega \setminus \overline{\omega}, \\ \varphi = 0, & x \in \partial\omega, \\ \dfrac{\partial\varphi}{\partial n} = 0, & x \in \partial\Omega. \end{cases}$$

Notice that λ_1^* may be as large as we wish if γ_1 is sufficiently large and ω is sufficiently "large."

The following result concerns the eradicability of the disease in the most interesting situation when $\tilde{\lambda}_1 < 0$, $q > \ell$ and $C_{I0} \ne 0_{L^\infty(\Omega)}$ (see [25]).

Theorem 5.2 *If* $\tilde{\lambda}_1 < 0, q > \ell$, $C_{I0} \ne 0_{L^\infty(\Omega)}$, *and*

$$[\beta \|C_I^*\|_\infty \|k\|_2 + \zeta C^0(q-\ell)]^2 < \lambda_1[\mu + l + \alpha - b\ell C^0(q-\ell)], \tag{5.25}$$

then

$$i_1(\cdot, t) \longrightarrow 0, \quad i_2(\cdot, t) \longrightarrow 0 \quad \text{in } L^2(\Omega),$$

as $t \to +\infty$.

Notice that condition (5.25) holds if for example γ_1 is sufficiently large and if the subset ω is sufficiently "large."

Remark 5.3 We remind that if the diffusion coefficient is strictly positive, $d > 0$, any point of the domain Ω is "strongly connected" to any other point of Ω, so that actions taken in a subregion $\omega \subset \Omega$ will eventually propagate to the whole habitat.

On the other hand, due to the lack of diffusion for i_2, it is reasonable to expect the control to be more effective (leading to a faster decay of i_1 and i_2) if $\gamma_{22}, \gamma_{23}, \gamma_3$ act on the whole domain Ω. In this case, the controlled system becomes

$$\frac{\partial s_1}{\partial t}(x,t) = d\Delta s_1(x,t) + rC_I(x,t)[M(x)(1-\gamma_{21}\mathbb{I}_\omega(x)) - \kappa s_1(x,t)]$$
$$-(\delta + \gamma_1\mathbb{I}_\omega(x))s_1(x,t) - \beta(1-\gamma_{23})s_1(x,t)\int_\Omega k(x,x')i_2(x',t)dx', \quad (5.26)$$

$$\frac{\partial i_1}{\partial t}(x,t) = d\Delta i_1(x,t) - (\delta + \gamma_1\mathbb{I}_\omega(x))i_1(x,t) - r\kappa i_1(x,t)C_I(x,t)$$
$$+\beta(1-\gamma_{23})s_1(x,t)\int_\Omega k(x,x')i_2(x',t)dx', \quad (5.27)$$

$$\frac{\partial s_2}{\partial t}(x,t) = (q-\ell)s_2(x,t) - s_2(x,t)\frac{C_T(x,t)}{C^0}$$
$$-(\zeta(1-\gamma_{22})i_1(x,t) + b(1-\gamma_3)\ell i_2(x,t))s_2(x,t) + \alpha i_2(x,t), \quad (5.28)$$

$$\frac{\partial i_2}{\partial t}(x,t) = -\mu i_2(x,t) - \ell i_2(x,t) - i_2(x,t)\frac{C_T(x,t)}{C^0}$$
$$+(\zeta(1-\gamma_{22})i_1(x,t) + b(1-\gamma_3)\ell i_2(x,t))s_2(x,t) - \alpha i_2(x,t). \quad (5.29)$$

The boundary conditions and the initial conditions are the same.

There exists a unique solution (s_1, i_1, s_2, i_2) to (5.26)–(5.29) with the above-mentioned boundary and initial conditions. By using comparison results for parabolic equations and for first order ODEs, it follows that

$$s_j,\ i_j \in L^\infty_{loc}(\overline{\Omega} \times [0,+\infty)),$$
$$s_j(x,t),\ i_j(x,t) \geq 0, \text{ a.e. } (x,t) \in \Omega \times (0,+\infty),$$

$j \in \{1, 2\}$ (see [11, 14, 27]).

In this special case, the following result holds [25].

Theorem 5.4 *If* $\tilde{\lambda}_1 < 0, q > \ell$, $C_{I0} \neq 0_{L^\infty(\Omega)}$ *and*

$$[\beta(1-\gamma_{23})\|C_I^*\|_\infty\|k\|_2 + \zeta(1-\gamma_{22})C^0(q-\ell)]^2$$
$$< \lambda_1[\mu + l + \alpha - b(1-\gamma_3)\ell C^0(q-\ell)], \tag{5.30}$$

then

$$i_1(\cdot, t) \longrightarrow 0, \quad i_2(\cdot, t) \longrightarrow 0 \quad in\ L^2(\Omega),$$

as $t \to +\infty$.

Notice that (5.30) is a weaker assumption than (5.25).

5.4 Computational Issues

5.4.1 Numerical Methods

The numerical strategy adopted to approximate the controlled system (Equations 5.6–5.9) consists of the finite element method for space discretization and the finite difference method for time discretization. This procedure is the state of the art for the solution of parabolic partial differential equations (PDEs); see, e.g., [168].

Space Discretization We first apply a standard Galerkin procedure to the weak formulations of the controlled system (Equations 5.6–5.9). For the computational domain Ω, we have taken a rectangle of size $400 \times 80\ Km^2$, which mimics, for example, the whole region of Apulia in Southern Italy, from South (right-hand side of the domain) to North (left-hand side of the domain). The rectangular domain has been discretized by a uniform grid of 200×40 bilinear finite elements ($Q1$), yielding a total number of $N = 8241$ discretization nodes. The stiffness matrix is computed exactly, whereas the mass matrix is obtained by applying the mass lumping technique.

In the following, we describe in detail this procedure applied to system (5.6)–(5.9). Let us first denote by V the Sobolev space $H^1(\Omega)$, by $(\cdot, \cdot)$ the L^2 inner product and by $a(\cdot, \cdot) : V \times V \to \mathbb{R}$ the bilinear form

$$a(u, v) := \int_\Omega d\nabla u \cdot \nabla v dx, \quad \text{for any } u, v \in V.$$

Then, using the Green formula, the variational formulation of problem (5.6)–(5.9) reads:

Find s_1, i_1 and s_2, i_2 such that for all $\phi \in V$:

$$\begin{cases} \dfrac{\partial}{\partial t}(s_1(t), \phi) + a(s_1(t), \phi) = (R_{s_1}(s_1(t), i_1(t), s_2(t), i_2(t)), \phi), \\ \dfrac{\partial}{\partial t}(i_1(t), \phi) + a(i_1(t), \phi) = (R_{i_1}(s_1(t), i_1(t), s_2(t), i_2(t)), \phi), \\ \dfrac{\partial}{\partial t}(s_2(t), \phi) = (R_{s_2}(s_1(t), i_1(t), s_2(t), i_2(t)), \phi), \\ \dfrac{\partial}{\partial t}(i_2(t), \phi) = (R_{i_2}(s_1(t), i_1(t), s_2(t), i_2(t)), \phi), \end{cases} \tag{5.31}$$

subject to the initial conditions $s_1(x, 0) = s_1^0(x), i_1(x, 0) = i_1^0(x), s_2(x, 0) = s_2^0(x)$, $i_2(x, 0) = i_2^0(x)$ and where R_{s_1}, R_{i_1}, R_{s_2}, R_{i_2} are the reaction terms in (5.6)–(5.9).

Let us now introduce the $\mathbf{Q}^1$ finite element space V^h, associated to a Cartesian grid discretizing the unit square domain Ω. For a detailed description of this space, see Appendix B.

The space semi-discretization of problem (5.31) then reads:

Find $s_{1,h}(\cdot, t), i_{1,h}(\cdot, t), s_{2,h}(\cdot, t), i_{2,h}(\cdot, t) \; : \; (0, T] \rightarrow V^h$ such that for all $\phi \in V^h$:

$$\begin{cases} \dfrac{\partial}{\partial t}(s_{1,h}(t), \phi) + a(s_{1,h}(t), \phi) = (R_{s_1}(s_{1,h}(t), i_{1,h}(t), s_{2,h}(t), i_{2,h}(t)), \phi), \\ \dfrac{\partial}{\partial t}(i_{1,h}(t), \phi) + a(i_{1,h}(t), \phi) = (R_{i_1}(s_{1,h}(t), i_{1,h}(t), s_{2,h}(t), i_{2,h}(t)), \phi), \\ \dfrac{\partial}{\partial t}(s_{2,h}(t), \phi) = (R_{s_2}(s_{1,h}(t), i_{1,h}(t), s_{2,h}(t), i_{2,h}(t)), \phi), \\ \dfrac{\partial}{\partial t}(i_{2,h}(t), \phi) = (R_{i_2}(s_{1,h}(t), i_{1,h}(t), s_{2,h}(t), i_{2,h}(t)), \phi), \end{cases}$$

which is equivalent to

$$\begin{cases} \dfrac{\partial}{\partial t}(s_{1,h}(t), \phi_i) + a(s_{1,h}(t), \phi_i) = (R_{s_1}(s_{1,h}(t), i_{1,h}(t), s_{2,h}(t), i_{2,h}(t)), \phi_i), \\ \dfrac{\partial}{\partial t}(i_{1,h}(t), \phi_i) + a(i_{1,h}(t), \phi_i) = (R_{i_1}(s_{1,h}(t), i_{1,h}(t), s_{2,h}(t), i_{2,h}(t)), \phi_i), \\ \dfrac{\partial}{\partial t}(s_{2,h}(t), \phi_i) = (R_{s_2}(s_{1,h}(t), i_{1,h}(t), s_{2,h}(t), i_{2,h}(t)), \phi_i), \\ \dfrac{\partial}{\partial t}(i_{2,h}(t), \phi_i) = (R_{i_2}(s_{1,h}(t), i_{1,h}(t), s_{2,h}(t), i_{2,h}(t)), \phi_i), \end{cases} \tag{5.32}$$

for all $i = 1, \ldots, N$, where ϕ_i, $i = 1, \ldots, N$, are the *hat basis functions* of the $\mathbf{Q}^1$ finite element space V^h (see Appendix B for the definition of this basis).

Expanding the unknowns $s_{1,h}(\cdot, t)$, $i_{1,h}(\cdot, t)$, $s_{2,h}(\cdot, t)$, $i_{2,h}(\cdot, t)$ with respect to the basis $\{\phi_i\}_{i=1}^N$, we have

$$\begin{array}{ll} s_{1,h}(x,t) = \sum_{j=1}^{N} S_{1,j}(t)\phi_j(x), & i_{1,h}(x,t) = \sum_{j=1}^{N} I_{1,j}(t)\phi_j(x), \\ s_{2,h}(x,t) = \sum_{j=1}^{N} S_{2,j}(t)\phi_j(x), & i_{2,h}(x,t) = \sum_{j=1}^{N} I_{2,j}(t)\phi_j(x), \end{array} \tag{5.33}$$

where $S_{1,j},\ I_{1,j},\ S_{2,j},\ I_{2,j}\ :\ [0,T] \rightarrow \mathbb{R}$ are unknown scalar functions to be determined.

In the next steps, we follow closely the approximation of the nonlinear problem presented in Appendix B. Hence, denoting by A the stiffness matrix associated with the bilinear form $a(\cdot,\cdot)$, by M the lumped mass matrix (see the appendix for the construction of such matrices), and substituting the expressions (5.33) into system (5.32), the semi-discrete problem then reads:

Find $\mathbf{S}_1(t), \mathbf{I}_1(t), \mathbf{S}_2(t), \mathbf{I}_2(t) : (0,T] \rightarrow \mathbb{R}^N$ such that

$$\begin{cases} \mathrm{M}\dfrac{d\mathbf{S}_1}{dt}(t) + \mathrm{A}\mathbf{S}_1(t) = \mathrm{M}\mathbf{R}_{s1}(\mathbf{S}_1(t), \mathbf{I}_1(t), \mathbf{S}_2(t), \mathbf{I}_2(t)), \\ \mathrm{M}\dfrac{d\mathbf{I}_1}{dt}(t) + \mathrm{A}\mathbf{I}_1(t) = \mathrm{M}\mathbf{R}_{i1}(\mathbf{S}_1(t), \mathbf{I}_1(t), \mathbf{S}_2(t), \mathbf{I}_2(t)), \\ \mathrm{M}\dfrac{d\mathbf{S}_2}{dt}(t) = \mathrm{M}\mathbf{R}_{s2}(\mathbf{S}_1(t), \mathbf{I}_1(t), \mathbf{S}_2(t), \mathbf{I}_2(t)), \\ \mathrm{M}\dfrac{d\mathbf{I}_2}{dt}(t) = \mathrm{M}\mathbf{R}_{i2}(\mathbf{S}_1(t), \mathbf{I}_1(t), \mathbf{S}_2(t), \mathbf{I}_2(t)), \end{cases}$$

where

$$\begin{aligned} \mathbf{S}_1(t) &= [\ldots, S_{1,j}(t), \ldots]^T, \quad j = 1, \ldots, N, \\ \mathbf{I}_1(t) &= [\ldots, I_{1,j}(t), \ldots]^T, \quad j = 1, \ldots, N, \\ \mathbf{S}_2(t) &= [\ldots, S_{2,j}(t), \ldots]^T, \quad j = 1, \ldots, N, \\ \mathbf{I}_2(t) &= [\ldots, I_{2,j}(t), \ldots]^T, \quad j = 1, \ldots, N, \\ \mathbf{R}_{s1}(\mathbf{S}_1(t), \mathbf{I}_1(t), \mathbf{S}_2(t), \mathbf{I}_2(t)) &= [\ldots, R_{s1}(S_{1,j}(t), I_{1,j}(t), S_{2,j}(t), \mathbf{I}_2(t)), \ldots]^T, \\ &\qquad j = 1, \ldots, N, \\ \mathbf{R}_{i1}(\mathbf{S}_1(t), \mathbf{I}_1(t), \mathbf{S}_2(t), \mathbf{I}_2(t)) &= [\ldots, R_{i1}(S_{1,j}(t), I_{1,j}(t), S_{2,j}(t), \mathbf{I}_2(t)), \ldots]^T, \\ &\qquad j = 1, \ldots, N, \\ \mathbf{R}_{s2}(\mathbf{S}_1(t), \mathbf{I}_1(t), \mathbf{S}_2(t), \mathbf{I}_2(t)) &= [\ldots, R_{s2}(S_{1,j}(t), I_{1,j}(t), S_{2,j}(t), I_{2,j}(t)), \ldots]^T, \\ &\qquad j = 1, \ldots, N, \\ \mathbf{R}_{i2}(\mathbf{S}_1(t), \mathbf{I}_1(t), \mathbf{S}_2(t), \mathbf{I}_2(t)) &= [\ldots, R_{i2}(S_{1,j}(t), I_{1,j}(t), S_{2,j}(t), I_{2,j}(t)), \ldots]^T, \\ &\qquad j = 1, \ldots, N. \end{aligned}$$

Denoting by x_j the generic discretization node, the nonlocal component of the generic elements $R_{s1}(S_{1,j}(t), I_{1,j}(t), S_{2,j}(t), \mathbf{I}_2(t))$ and $R_{i1}(S_{1,j}(t), I_{1,j}(t), S_{2,j}(t), \mathbf{I}_2(t))$ of the vectors $\mathbf{R}_{s1}(\mathbf{S}_1(t), \mathbf{I}_1(t), \mathbf{S}_2(t), \mathbf{I}_2(t))$ and $\mathbf{R}_{i1}(\mathbf{S}_1(t), \mathbf{I}_1(t), \mathbf{S}_2(t), \mathbf{I}_2(t))$, respectively, is computed as follows:

$$\begin{aligned}
\int_\Omega k(x_j, x') i_{2,h}(x', t) dx' &= \int_\Omega k(x_j, x') \sum_{j=1}^{N} I_{2,j}(t) \phi_j(x') dx' \\
&\approx \int_\Omega \sum_{i=1}^{N} k(x_j, x_i) \phi_i(x') \sum_{j=1}^{N} I_{2,j}(t) \phi_j(x') dx' \\
&= \mathbf{K}(x_j)^T \mathrm{M} \mathbf{I}_2(t),
\end{aligned}$$

where

$$\mathbf{K}(x_j) = [k(x_j, x_1), \ldots, k(x_j, x_i), \ldots, k(x_j, x_N)]^T.$$

Time Discretization After the spatial discretization, we obtain a semi-discrete problem that consists of a system of ordinary differential equations (ODEs). We solve these ODEs by employing a first order semi-implicit finite difference scheme, where the linear diffusion terms are approximated by Backward Euler, whereas the nonlinear reaction terms are approximated by Forward Euler.

In formulae, at the generic time step n, given $\mathbf{S}_1^n, \mathbf{I}_1^n, \mathbf{S}_2^n, \mathbf{I}_2^n$, we have to find $\mathbf{S}_1^{n+1}, \mathbf{I}_1^{n+1}, \mathbf{S}_2^{n+1}, \mathbf{I}_2^{n+1}$ such that

$$\begin{cases}
\mathrm{M}\dfrac{\mathbf{S}_1^{n+1} - \mathbf{S}_1^n}{\Delta t} + \mathrm{A}\mathbf{S}_1^{n+1} = \mathrm{M}\mathbf{R}_{s1}(\mathbf{S}_1^n, \mathbf{I}_1^n, \mathbf{S}_2^n, \mathbf{I}_2^n), \\
\mathrm{M}\dfrac{\mathbf{I}_1^{n+1} - \mathbf{I}_1^n}{\Delta t} + \mathrm{A}\mathbf{I}_1^{n+1} = \mathrm{M}\mathbf{R}_{i1}(\mathbf{S}_1^n, \mathbf{I}_1^n, \mathbf{S}_2^n, \mathbf{I}_2^n), \\
\mathrm{M}\dfrac{\mathbf{S}_2^{n+1} - \mathbf{S}_2^n}{\Delta t} = \mathrm{M}\mathbf{R}_{s2}(\mathbf{S}_1^n, \mathbf{I}_1^n, \mathbf{S}_2^n, \mathbf{I}_2^n), \\
\mathrm{M}\dfrac{\mathbf{I}_2^{n+1} - \mathbf{I}_2^n}{\Delta t} = \mathrm{M}\mathbf{R}_{i2}(\mathbf{S}_1^n, \mathbf{I}_1^n, \mathbf{S}_2^n, \mathbf{I}_2^n).
\end{cases}$$

As a result, at each time step, it is required the solution of the linear system of algebraic equations

$$\begin{bmatrix}
\frac{1}{\Delta t}\mathrm{M} + \mathrm{A} & 0 & 0 & 0 \\
0 & \frac{1}{\Delta t}\mathrm{M} + \mathrm{A} & 0 & 0 \\
0 & 0 & \frac{1}{\Delta t}\mathrm{M} & 0 \\
0 & 0 & 0 & \frac{1}{\Delta t}\mathrm{M}
\end{bmatrix}
\begin{bmatrix}
\mathbf{S}_1^{n+1} \\ \mathbf{I}_1^{n+1} \\ \mathbf{S}_2^{n+1} \\ \mathbf{I}_2^{n+1}
\end{bmatrix}
=
\begin{bmatrix}
\frac{1}{\Delta t}\mathrm{M}\mathbf{S}_1^n + \mathrm{M}\mathbf{R}_{s1}(\mathbf{S}_1^n, \mathbf{I}_1^n, \mathbf{S}_2^n, \mathbf{I}_2^n) \\
\frac{1}{\Delta t}\mathrm{M}\mathbf{I}_1^n + \mathrm{M}\mathbf{R}_{i1}(\mathbf{S}_1^n, \mathbf{I}_1^n, \mathbf{S}_2^n, \mathbf{I}_2^n) \\
\frac{1}{\Delta t}\mathrm{M}\mathbf{S}_2^n + \mathrm{M}\mathbf{R}_{s2}(\mathbf{S}_1^n, \mathbf{I}_1^n, \mathbf{S}_2^n, \mathbf{I}_2^n) \\
\frac{1}{\Delta t}\mathrm{M}\mathbf{I}_2^n + \mathrm{M}\mathbf{R}_{i2}(\mathbf{S}_1^n, \mathbf{I}_1^n, \mathbf{S}_2^n, \mathbf{I}_2^n)
\end{bmatrix}$$

Table 5.2 Values of the model parameters

Symbol	Unit	Value range	Simulation value	References
r	Year^{-1}	[37,400]	200	[181, 197]
κ	–	[0.001, 0.05]	0.03	[62]
δ	Year^{-1}	[0.95, 0.99]	0.98	[62]
q	Year^{-1}	[0.2, 0.7]	0.5	[191]
C^0	–	100	100	[62]
ℓ	Year^{-1}	0.01	0.01	[62]
b	–	0.05	0.05	[62]
μ	Year^{-1}	[0.8, 1]	0.9	[174, 175]
α	Year^{-1}	[0.1, 0.5]	0.3	[62]
β	Year^{-1}	[0.5, 1]	1	[92]
ζ	Year^{-1}	[0.2, 0.8]	0.8	[50, 106]

which in our specific case has the dimension 4×8241 degrees of freedom. The linear system is solved by Gaussian elimination with the built-in MATLAB function. The time step size is $\Delta t = 0.0002$ years. For further details on the numerical discretization of parabolic problems, we refer the reader to [168].

5.4.2 *Parameter Calibration*

The values of the simulation parameters are listed in Table 5.2. The diffusion coefficient in the PDEs is set to $d = 1e - 4\ \mathrm{Km}^2/\mathrm{year}$. The total simulation time is $T = 10$ years. The initial distributions, used in all next numerical simulations, are the following:

$$\begin{cases} s_1(x,0) = 100 - 20\exp(-100(x_1-380)^2 - 100(x_2-40)^2), \\ i_1(x,0) = 20\exp(-100(x_1-380)^2 - 100(x_2-40)^2), \\ s_2(x,0) = 50, \\ i_2(x,0) = 0. \end{cases}$$

Therefore, we assume constant initial distributions for healthy insects and trees, and we localize the initial presence of infected insects on the right-hand side of the domain.

5.4.3 *Numerical Simulations*

We report in this section the results obtained with the MATLAB code

```
run_xylella.m,
```

available on the Birkhäuser website provided in the Preface and also at the link

```
http://www.mat.unimi.it/users/scacchi/ACS_codes/run_
    xylella.m
```

Three numerical experiments have been performed to show the efficacy of the controls on the weed biomass and the olive cultivar. We take the following parameter values: $M = 1$, $\kappa = 0.01$, $\zeta = 0.2$, and $d = 1e - 4$:

- **Experiment 1**: No control has been applied.
- **Experiment 2**: Control measures have been applied only in the right-hand section of the habitat, simulating the subregion of the Apulian region already affected by the Xylella epidemic, augmented by a "containment band," i.e., the subregion

$$\omega = \{x = (x_1, x_2) \in \Omega : x_1 > 250 \text{ Km}\}.$$

- **Experiment 3**: Control measures have been applied only in a band, away from the subregion already affected by the epidemic:

$$\omega = \{x = (x_1, x_2) \in \Omega : 300 < x_1 < 350 \text{ Km}\}.$$

Experiment 1
We first run the model without control; thus the control parameters ($\gamma_{21}, \gamma_1, \gamma_{23}, \gamma_{22}, \gamma_3$) are set to zero everywhere. The Xylella epidemic starts spreading as a travelling wave from the right portion of the domain, where the initial condition of the infected insects is positive; see Figure 5.1.

Experiment 2
Applying the regional control in the subregion $x_1 > 250$ Km, setting the control parameters

$$(\gamma_{21}, \gamma_1, \gamma_{23}, \gamma_{22}, \gamma_3) = (0.6, 0.2, 0.2, 0.6, 0.2),$$

the Xylella epidemic is stopped at the origin and no travelling front occurs; see Figure 5.2. This result clearly indicates that for more resistant olive cultivar we may still block an epidemic with a lower level of weeds cut in the olive orchards, even though, by using $\kappa = 0.001$, we mean a larger effective insect carrying capacity.

Experiment 3
A regional control has been implemented only in a band, away from the subregion already affected by the epidemic, i.e., in the subregion $300 < x_1 < 350$ Km, by setting the control parameters

$$(\gamma_{21}, \gamma_1, \gamma_{23}, \gamma_{22}, \gamma_3) = (0.6, 0.2, 0.2, 0.6, 0.2).$$

The *Xylella* epidemic starts to travel towards the left-hand side of the habitat; it is blocked when it reaches the right -hand side of the control region (see Figure 5.3);

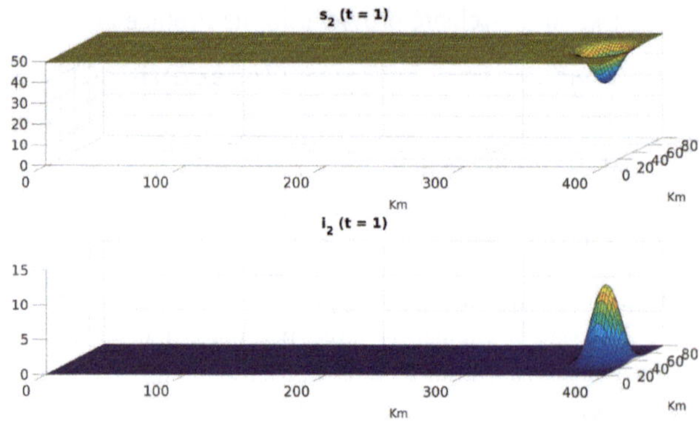

(a) $t = 1$ *year*. Top figure: healthy trees (s_2); bottom figure: infective trees (i_2).

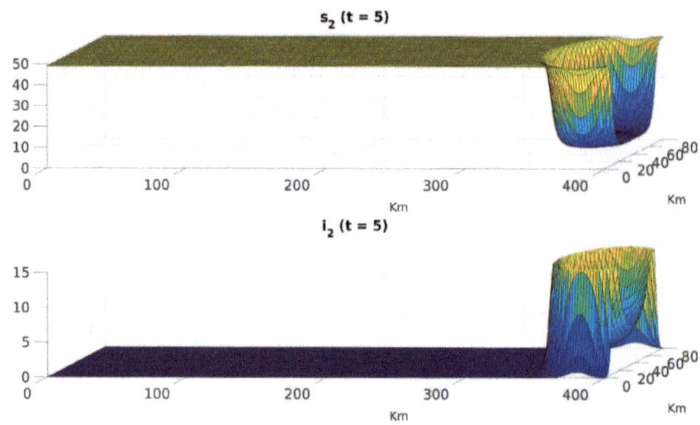

(b) $t = 5$ *years*. Same format as panel (a).

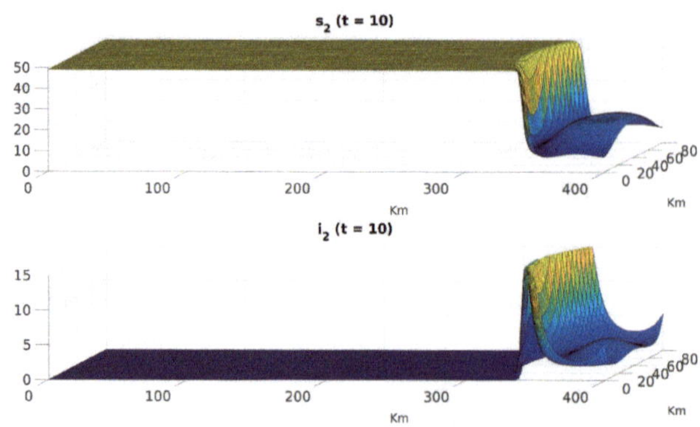

(c) $t = 10$ *years*. Same format as panel (a).

Fig. 5.1 **Experiment 1**. We solve Equations (5.6)–(5.9), with the algorithms and discretization parameters reported in Section 5.4 and the model parameters given in Table 5.2. Spatial distributions of olive trees without control: $\kappa = 0.01$, $d = 1e - 4$ Km^2/year. (**a**) $t = 1$ year. Top figure: healthy trees (s_2); bottom figure: infective trees (i_2). (**b**) $t = 5$ years. Same format as panel (**a**). (**c**) $t = 10$ years. Same format as panel (**a**)

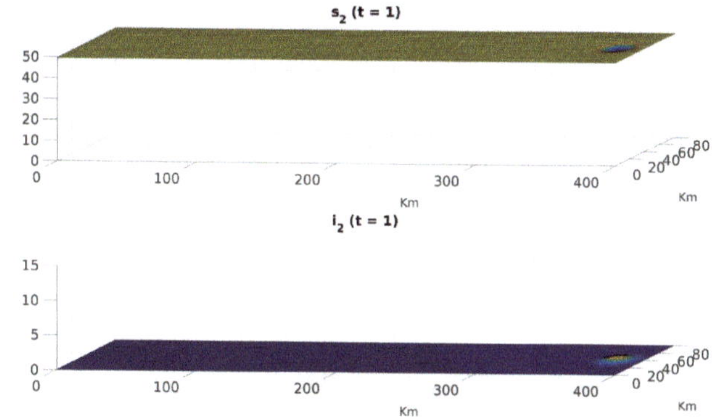

(a) $t = 1$ *year*. Top figure: healthy trees (s_2); bottom figure: infective trees (i_2).

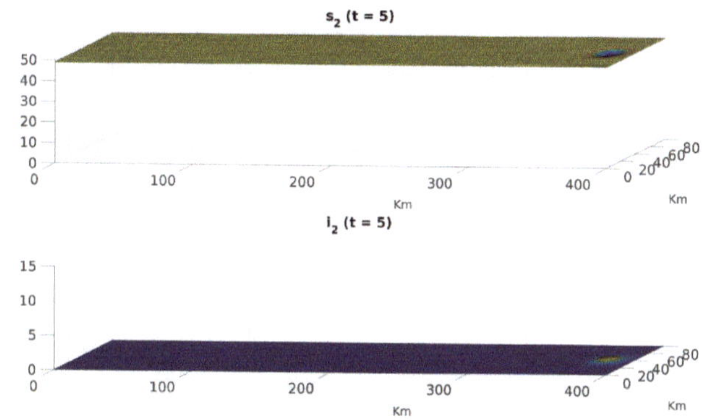

(b) $t = 5$ *years*. Same format as panel (a).

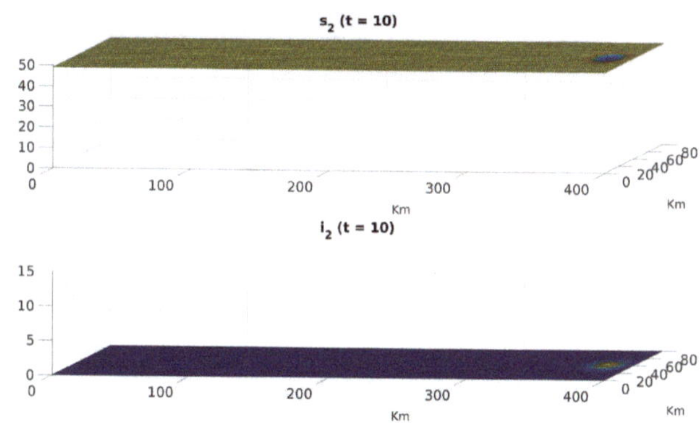

(c) $t = 10$ *years*. Same format as panel (a).

Fig. 5.2 **Experiment 2**. We solve Equations (5.6)–(5.9), with the algorithms and discretization parameters reported in Section 5.4 and the model parameters given in Table 5.2. Spatial distributions of olive trees with regional control. $\kappa = 0.01$, $d = 1e - 4$ Km^2/year, $\omega = \{x = (x_1, x_2) \in \Omega : x_1 > 250$ Km$\}$. $(\gamma_{21}, \gamma_1, \gamma_{23}, \gamma_{22}, \gamma_3) = (0.6, 0.2, 0.2, 0.6, 0.2)$. (**a**) $t = 1$ year. Top figure: healthy trees (s_2); bottom figure: infective trees (i_2). (**b**) $t = 5$ years. Same format as panel (**a**). $t = 10$ years. Same format as panel (**a**)

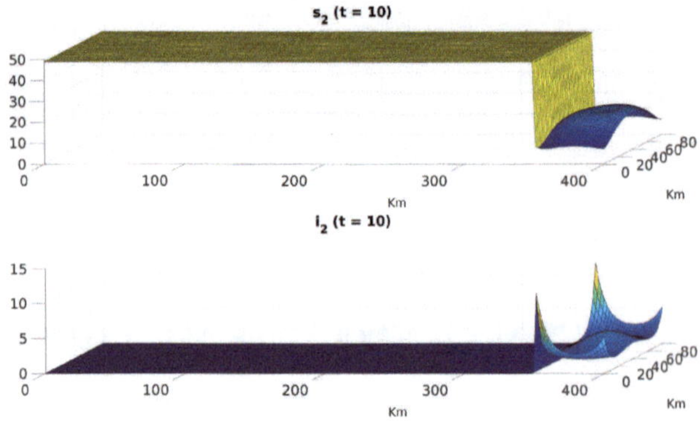

(a) $t = 10$ *years*. Top figure: healthy trees (s_2); bottom figure: infective trees (i_2).

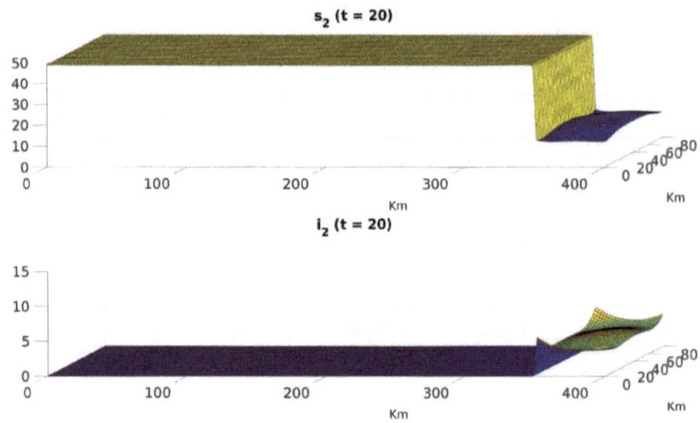

(b) $t = 5$ *years*. Same format as panel (a).

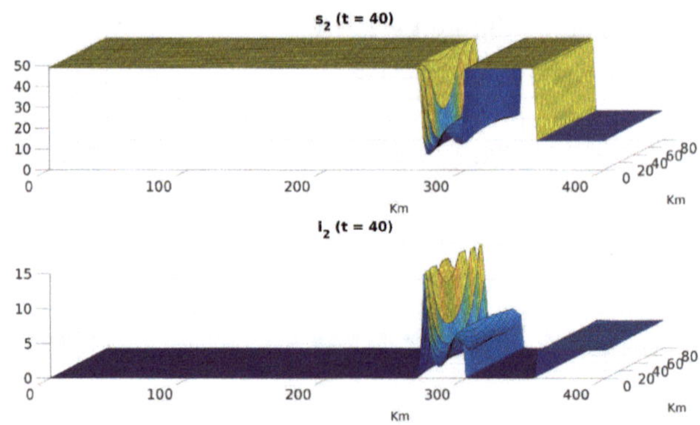

(c) $t = 10$ *years*. Same format as panel (a).

Fig. 5.3 **Experiment 3**. We solve Equations (5.6)–(5.9), with the algorithms and discretization parameters reported in Section 5.4 and the model parameters given in Table 5.2. Spatial distributions of olive trees with regional control. $\kappa = 0.01$, $d = 1e - 4$ Km^2/year, $\omega = \{x = (x_1, x_2) \in \Omega : 300 < x_1 < 350 \text{ Km}\}$. $(\gamma_{21}, \gamma_1, \gamma_{23}, \gamma_{22}, \gamma_3) = (0.6, 0.2, 0.2, 0.6, 0.2)$. (**a**) $t = 10$ years. Top figure: healthy trees (s_2); bottom figure: infective trees (i_2). (**b**) $t = 5$ years. Same format as panel (**a**). $t = 10$ years. Same format as panel (**a**)

however, with some delay, the epidemic front reappears from the left-hand side of the control region and propagates towards the left end of the whole domain. This indicates that, in order to block the epidemic, it is not sufficient to implement the controls on a band, ahead of the region already affected by the epidemic, but in the whole subregion affected by the epidemic, possibly extended by a containment band, as done in **Experiment 2**.

5.5 Concluding Remarks

The results reported here confirm that the most promising target for an effective and cost-efficient control of the *X. fastidiosa* epidemic is represented by agricultural management practices consisting of the removal of the weeds in the whole relevant habitat of the olive orchards. A further interesting strategy, as expected, is represented by the use of more resistant cultivar [62] (see also [177] and references cited therein).

Here we have extended the ordinary differential (ODE) model introduced in [62] to a partial differential (PDE) system that takes into account the spatial structure of the relevant epidemic system, subject to possible control strategies, including weeds cut, insect traps, treated nets, more resistant cultivar, etc. As emphasized in Remark 5.3, theoretical results show that eradication of an epidemic outbreak is possible by the implementation of the above-mentioned control measures only on a suitable subregion of the whole habitat, including the area already affected by the outbreak, possibly augmented by a suitable "containment band."

Unfortunately in a real practice, it might be unfeasible to act on the diffusion parameters; they are related to the ecological structure of the relevant habitat, including the behavior of the insect population (see [142] for recent biological experiments). Intervention strategies should then include possible modifications of the insects' behavior, such as by means of treated nets.

Once again, validation of the model proposed here represents a key issue: Although we have tried to render explicit the assumptions underlying our model, the latter has not yet been validated, by comparison with experimental data. Therefore, we caution that it is far from being conclusive for *X. fastidiosa* subsp. *pauca - Philaenus spumarius* olive tree epidemics. However, it is desirable that, with additional features that make it more realistic, our model might provide the foundations for designing optimal control strategies by public authorities.

Chapter 6
The Role of Predators in Controlling the Spread of an Epidemic in Agriculture: The Case of *Xylella fastidiosa*

6.1 Introduction

We continue here the investigation from Chapter 5. As discussed in the previous chapter, the main vector of *X. fastidiosa* in Southern Italy has been identified in the so-called meadow spittlebug, i.e., *Philaenus spumarius* (Hemiptera, Aphrophoridae), a xylem sap-feeding specialist.

In recent papers [94, 143, 148], *Zelus renardii* (Hemiptera, Reduviidae) has been identified as a predator of *P. spumarius* for a possible biocontrol agent of a *Xylella* epidemic.

Actually the role of a predator in controlling a vector borne epidemic crucially depends upon the functional response to predation. Possible choices adopted in the literature are the Holling type II (specialist predators) and the Holling type III (generalist predators) [190]. Similar problems have been analyzed in [26].

Indeed, as we will see later, a H. II predator of *P. spumarius* would lead to the eventual eradication of a *Xylella* epidemic. Unfortunately, *Z. renardii* has been reported to be a generalist predator, which implies the choice of an Holling type III functional response of predation. As a consequence, the introduction of *Z. renardii* as a predator of *P. spumarius* is not an efficient control strategy to eradicate a *Xylella* epidemic.

Below, both the impact of a H. II predator and the impact of a H. III predator have been analyzed.

A set of computational experiments illustrates the two cases.

6.2 The Mathematical Model

- INSECT VECTOR—*Philaenus spumarius*.

S. Aniţa et al., *Mathematical Modeling and Control in Life and Environmental Sciences*, Modeling and Simulation in Science, Engineering and Technology,
https://doi.org/10.1007/978-3-031-49971-5_6

We remind here that the spatial dynamics of the insect vector is described by the following equations:

$$\begin{aligned}\frac{\partial s_1}{\partial t}(x,t) &= d\Delta s_1(x,t) + rC_I(x,t)[M(x) - \kappa s_1(x,t)] - \delta s_1(x,t)\\ &\quad -\beta s_1(x,t)\int_\Omega k(x,x')i_2(x',t)dx',\end{aligned} \tag{6.1}$$

for the susceptible population and

$$\begin{aligned}\frac{\partial i_1}{\partial t}(x,t) &= d\Delta i_1(x,t) - \delta i_1(x,t) - r\kappa i_1(x,t)C_I(x,t)\\ &\quad +\beta s_1(x,t)\int_\Omega k(x,x')i_2(x',t)dx',\end{aligned} \tag{6.2}$$

for the infective population.

Both Equations (6.1) and (6.2) act in a spatial bounded domain $\Omega \subset \mathbb{R}^2$, at times $t \in (0, +\infty)$, subject to suitable initial and boundary conditions.

By summing Equations (6.1), (6.2), we obtain an equation for the total insect population

$$C_I(x,t) = s_1(x,t) + i_1(x,t),$$

$$\frac{\partial C_I}{\partial t}(x,t) = d\Delta C_I(x,t) + rC_I(x,t)[M(x) - \kappa C_I(x,t)] - \delta C_I(x,t). \tag{6.3}$$

- INSECT PREDATORS

The evolutionary dynamics of a population of predators, denoted by $Z(x,t)$, in the absence of preys may be described as follows:

$$\frac{\partial Z}{\partial t}(x,t) = d_2\Delta Z(x,t) + r_2 Z(x,t)\left[1 - \frac{Z(x,t)}{K(x) + cM(x)}\right], \tag{6.4}$$

for $x \in \Omega, t > 0$, subject to suitable boundary and initial conditions.

The parameter r_2 denotes the natural birth rate of new predators.

We have assumed that the carrying capacity of the predators, in addition to a general parameter $K(x)$, depends upon the additional environmental parameter $M(x), x \in \Omega$, that had been already introduced in [25, 48, 62] to describe the biomass of all other plants, such as bushes, ornamental plants, etc. that we will simply call *weeds*, either healthy or infected, which are required for the proper development of both the vector and the predator insects in the relevant ecosystem.

In order to describe mathematically the contribution of *P. spumarius* to the dynamics of the predator population, we have to take into account the functional response $g(C_I)$ to predation.

If we include the impact of predation in Equation (6.3), we have

$$\frac{\partial C_I}{\partial t}(x,t) = d\Delta C_I(x,t) + rC_I(x,t)[M(x) - \kappa C_I(x,t)] - \delta C_I(x,t) \\ - pg(C_I(x,t))Z(x,t).$$

As a consequence, Equations (6.1), (6.2) can be rewritten as follows:

$$\frac{\partial s_1}{\partial t}(x,t) = d\Delta s_1(x,t) + rC_I(x,t)[M(x) - \kappa s_1(x,t)] - \delta s_1(x,t) \\ - pg(C_I(x,t))Z(x,t)\frac{s_1(x,t)}{C_I(x,t)} - \beta s_1(x,t)\int_\Omega k(x,x')i_2(x',t)dx',$$

$$\frac{\partial i_1}{\partial t}(x,t) = d\Delta i_1(x,t) - \delta i_1(x,t) - pg(C_I(x,t))Z(x,t)\frac{i_1(x,t)}{C_I(x,t)} \\ - r\kappa i_1(x,t)C_I(x,t) + \beta s_1(x,t)\int_\Omega k(x,x')i_2(x',t)dx'.$$

In [190, p.83] (see also [26], [146, p.38]), it was evidenced that the Holling type II functional response to predation is

$$g_{II}(C_I) = \frac{\gamma_1 C_I}{1+\gamma_2 C_I},$$

with γ_1, γ_2 positive constants.

On the other hand, the Holling type III functional response to predation is

$$g_{III}(C_I) = \frac{\gamma_1 C_I^2}{1+\gamma_2 C_I^2},$$

with γ_1, γ_2 positive constants. It follows an S-shaped curve, according to which the predator dedicates its predation only to large quantities of the specific prey C_I.

The positive parameter p denotes the impact of predation on the prey. The description of the model parameters is presented in Table 6.1.

Notice that for both functional responses there is a saturation level $\frac{\gamma_1}{\gamma_2}$, meaning that there is a limitation of the number of insects predated by an individual of *Z. renardii*.

Figures 6.1 and 6.2 illustrate the two cases.

In [22], both cases have been analyzed.

Assumptions

- $\Omega \subset \mathbb{R}^2$ is a bounded domain with a sufficiently smooth boundary.
- $M \in L^\infty(\Omega)$, $M(x) \geq 0$, a.e. $x \in \Omega$.
- $K \in L^\infty(\Omega)$, $K(x) \geq k_0$, a.e. in Ω (here k_0 is a positive constant).
- $k \in L^\infty(\Omega \times \Omega)$, $k(x,x') \geq 0$, a.e. in $\Omega \times \Omega$, normalized to β.
- $d,\ d_2,\ r,\ r_2,\ \kappa,\ \delta,\ p,\ c,\ \beta$ are positive constants.

Table 6.1 Model parameters

Symbol	Description
r	*P. spumarius* birth rate
r_2	*Z. renardii* birth rate
κ	*P. spumarius* intraspecific competition rate
δ	*P. spumarius* natural mortality rate
p	A proportionality constant
$K(x)$	Baseline *Z. renardii* carrying capacity parameter
$M(x)$	Spatial distribution of the weed biomass
c	Scale parameter of the impact of the weed biomass on the carrying capacity of *Z. renardii*
β	*P. spumarius* infection rate by infected trees

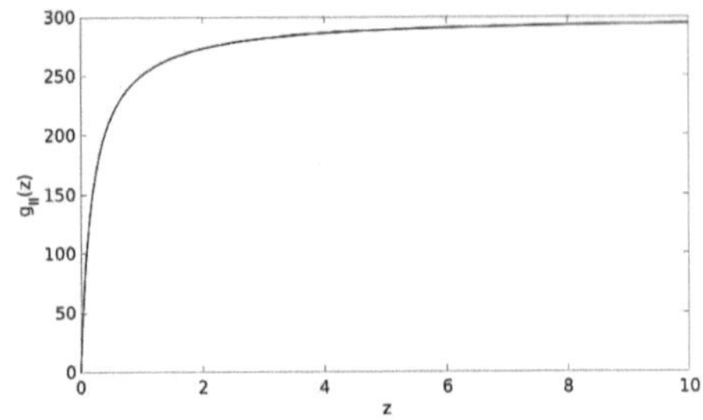

Fig. 6.1 Holling type II functional response

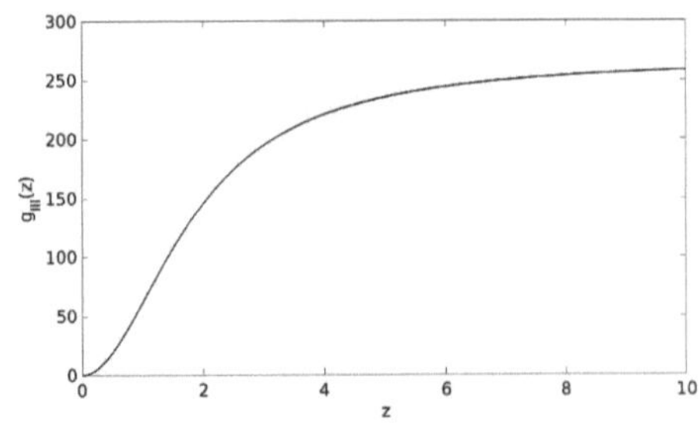

Fig. 6.2 Holling type III functional response

6.3 Control, Eradicability

Suppose that the control by predators occurs via a continuous infusion of predators; then, in the evolution equation for predators, we introduce an additional term $u(x) \geq 0$, possibly acting only in a suitable subregion $\omega \subset \Omega$ (ω is an open subset). We take $u \in L^{\infty}(\omega)$, $u(x) > 0$ a.e. $x \in \omega$. Actually, the control might also be time-dependent (seasonal), but for the time being we have taken it constant in time.

By considering only the populations of insect vectors, C_I and their predator Z, for either a H. II or a H. III functional response of predation, we have

$$\begin{cases} \dfrac{\partial C_I}{\partial t}(x,t) = d\Delta C_I(x,t) + rC_I(x,t)(M(x) - \kappa C_I(x,t)) - \delta C_I(x,t) \\ \qquad\qquad -pg(C_I(x,t))Z(x,t), \\ \dfrac{\partial Z}{\partial t}(x,t) = d_2\Delta Z(x,t) + r_2 Z(x,t)\left[1 - \dfrac{Z(x,t)}{K(x)+cM(x)}\right] \\ \qquad\qquad +g(C_I(x,t))Z(x,t) + \mathbb{I}_\omega(x)u(x), \end{cases} \tag{6.5}$$

for $(x,t) \in Q = \Omega \times (0,+\infty)$.

The above system is subject to pure Neumann boundary conditions

$$\frac{\partial C_I}{\partial n}(x,t) = \frac{\partial Z}{\partial n}(x,t) = 0, \quad (x,t) \in \Sigma = \partial\Omega \times (0,+\infty)$$

and suitable initial conditions

$$C_I(x,0) = C_{I0}(x),\ Z(x,0) = Z_0(x), \quad x \in \Omega.$$

Assume that:

- $C_{I0}, Z_0 \in L^\infty(\Omega), C_{I0}(x), Z_0(x) \geq 0$, a.e. $x \in \Omega$.

By usual methods for reaction–diffusion systems, we may state the following theorem.

Theorem 6.1 *For any $u \in L^\infty(\omega)$, $u(x) \geq 0$, a.e. $x \in \omega$, (6.5) with the above-mentioned boundary and initial conditions has a unique solution (C_I, Z).*

Moreover, $C_I(c,t), Z(x,t) \geq 0$, a.e. $x \in \Omega$, for any $t \geq 0$.

Consider now the following eigenvalue problem:

$$\begin{cases} -d\Delta\varphi - rM\varphi + \delta\varphi = \lambda\varphi,\ x \in \Omega, \\ \dfrac{\partial\varphi}{\partial n}(x) = 0, \qquad\qquad x \in \partial\Omega. \end{cases}$$

By Rayleigh's formula [55, p.516], we have that

$$\tilde{\lambda}_1 = \inf_{\varphi \in H^1(\Omega), \|\varphi\|_{L^2(\Omega)} \neq 0} \frac{\int_\Omega [d|\nabla\varphi|^2 - rM|\varphi|^2 + \delta|\varphi|^2]dx}{\int_\Omega |\varphi|^2 dx} \tag{6.6}$$

is the principal eigenvalue for the above-mentioned problem to which we may associate a positive eigenvector $\tilde{\varphi}_1$ (i.e., $\tilde{\varphi}_1(x) \geq \theta_0 > 0$, for $x \in \overline{\Omega}$).

We have shown in Chapter 5 that if c_2 is a positive real number, the function

$$C_2(x,t) = c_2\tilde{\varphi}_1(x)\exp(-\tilde{\lambda}_1 t)$$

is the unique solution to the following linear parabolic problem:

$$\begin{cases} \dfrac{\partial C}{\partial t} = d\Delta C + rMC - \delta C, \ (x,t) \in Q, \\ \dfrac{\partial C}{\partial n}(x,t) = 0, \qquad (x,t) \in \Sigma, \\ C(x,0) = c_2\tilde{\varphi}_1(x), \qquad x \in \Omega \end{cases}.$$

It is also clear that if $\tilde{\lambda}_1 > 0$, then $C_2(t) \longrightarrow 0$ in $L^\infty(\Omega)$, as $t \to +\infty$.

Now, if we take c_2 a sufficiently large constant, by comparison results for parabolic equations, we may state that

$$0 \le C_I(x,t) \le C_2(x,t), \quad \text{a.e. } x \in \Omega, \text{ for all } t \ge 0.$$

As a consequence, we get that $C_I(t) \longrightarrow 0$ in $L^\infty(\Omega)$, as $t \to +\infty$.

On the other hand, if $\tilde{\lambda}_1 = 0$, then there exist two positive constants $\tilde{\alpha}, \overline{\alpha}$ such that

$$0 \le C_I(x,t) \le \frac{\tilde{\alpha}}{t+\overline{\alpha}}\tilde{\varphi}_1(x), \quad \text{a.e. } x \in \Omega, \text{ for all } t \ge 0.$$

As a consequence, we get that $C_I(t) \longrightarrow 0$ in $L^\infty(\Omega)$, as $t \to +\infty$.

We may then state the following theorem.

Theorem 6.2 *If $\tilde{\lambda}_1 \ge 0$, then for any $u \in L^\infty(\omega)$, $u(x) \ge 0$, a.e. $x \in \omega$, we get that*

$$C_I(t) \longrightarrow 0 \quad \textit{in } L^\infty(\Omega), \ \textit{as } t \to +\infty.$$

This result expresses, in mathematical terms, the fact that for $\tilde{\lambda}_1 \ge 0$ the total population of vector insects goes extinct, independently of the presence of predators in the ecosystem.

It is clear that in the additional presence of predators, $C_I(t)$ will decay to 0 even faster.

We may further notice that, according to (6.6), we may make $\tilde{\lambda}_1 \ge 0$ by reducing M, the weed biomass, which confirms the results obtained in [25, 48, 62].

For the case $\tilde{\lambda}_1 < 0$, the role of predators becomes significant. In particular, the qualitative behavior of System (6.5) is different in the two cases of H. II and H. III predators.

6.3.1 *Holling Type II Functional Response to Predation*

Assume that $\tilde{\lambda}_1 < 0$.

For a specialist H. II functional response to predation, System (6.5) becomes

$$\begin{cases} \dfrac{\partial C_I}{\partial t}(x,t) = d\Delta C_I(x,t) + rC_I(x,t)(M(x) - \kappa C_I(x,t)) - \delta C_I(x,t) \\ \qquad\qquad -p\dfrac{\gamma_1 C_I(x,t)}{1+\gamma_2 C_I(x,t)} Z(x,t), \\ \dfrac{\partial Z}{\partial t}(x,t) = d_2\Delta Z(x,t) + r_2 Z(x,t)\left[1 - \dfrac{Z(x,t)}{K(x)+cM(x)}\right] \\ \qquad\qquad +\dfrac{\gamma_1 C_I(x,t)}{1+\gamma_2 C_I(x,t)} Z(x,t) + \mathbb{I}_\omega(x)u(x), \end{cases}$$

for $(x,t) \in Q$, subject to pure Neumann boundary conditions and the initial conditions in the previous section.

Consider the following nonlinear parabolic problem:

$$\begin{cases} \dfrac{\partial Z_1}{\partial t}(x,t) = d_2\Delta Z_1(x,t) + r_2 Z_1(x,t)\left[1 - \dfrac{Z(x,t)}{K(x)+cM(x)}\right] \\ \qquad\qquad +\mathbb{I}_\omega(x)u(x), & (x,t) \in Q, \\ \dfrac{\partial Z_1}{\partial n}(x,t) = 0, & (x,t) \in \Sigma, \\ Z_1(x,0) = 0, & x \in \Omega. \end{cases} \tag{6.7}$$

If $u(x) > 0$, a.e. $x \in \omega$, we get that $Z_1(x,t) \geq 0$, a.e. $x \in \Omega$, for any $t \geq 0$, and $Z_1(\cdot,t)$ is not identically 0. So, for any $t_0 > 0$, there exists $m_{t_0} > 0$ such that

$$Z_1(x,t_0) \geq m_{t_0}, \quad \text{a.e. } x \in \Omega$$

(see [154]).

We now show that the parabolic problem (6.7) admits a unique positive steady state Z^*, i.e., a unique positive solution of the following semilinear elliptic problem:

$$\begin{cases} d_2\Delta Z^* + r_2 Z^*\left[1 - \dfrac{Z^*}{K+cM}\right] + \mathbb{I}_\omega u = 0, & x \in \Omega, \\ \dfrac{\partial Z^*}{\partial n}(x) = 0, & x \in \partial\Omega. \end{cases} \tag{6.8}$$

Proof Consider a positive t_0 (fixed). Let $\tilde{Z}_{10}, \overline{Z}_{10} \in (0,+\infty)$, $\tilde{Z}_{10}$ sufficiently small and $\overline{Z}_{10}$ sufficiently large. The solutions $\tilde{Z}_1$ and $\overline{Z}_1$ to

$$\begin{cases} \dfrac{\partial Z}{\partial t} = d_2 \Delta Z + r_2 Z \left[1 - \dfrac{Z}{K + cM}\right] + \mathbb{I}_\omega u, & (x,t) \in \Omega \times (t_0, +\infty), \\ \dfrac{\partial Z}{\partial n}(x,t) = 0, & (x,t) \in \partial\Omega \times (t_0, +\infty), \\ Z(x, t_0) = \tilde{Z}_{10}, \; Z(x, t_0) = \overline{Z}_{10}, & x \in \Omega, \end{cases}$$

respectively, satisfy that

$$t \mapsto \tilde{Z}_1(\cdot, t) \text{ is increasing }, \quad t \mapsto \overline{Z}_1(\cdot, t) \text{ is decreasing on } [t_0, +\infty),$$

$$0 < \tilde{Z}_1(x,t) \leq Z_1(x,t) \leq \overline{Z}_1(x,t), \quad \text{a.e. } x \in \Omega, \text{ for all } t \geq t_0.$$

These properties follow as in Chapter 5. We may infer that

$$\tilde{Z}_1(t) \longrightarrow Z_1^*, \quad \overline{Z}_1(t) \longrightarrow Z_2^* \quad \text{in } L^\infty(\Omega),$$

as $t \to +\infty$, and that $0 < Z_1^*(x) \leq Z_2^*(x)$, a.e. $x \in \Omega$. Moreover, we may infer that Z_i^* are positive solutions to (6.8).

We immediately get that

$$r_2 \int_\Omega Z_1^* Z_2^* \frac{Z_2^* - Z_1^*}{K + cM} dx + \int_\omega u(Z_2^* - Z_1^*) dx = 0,$$

and so we conclude that $Z_1^*(x) = Z_2^*(x)$, a.e. $x \in \Omega$.

So, we may infer that there exists a unique positive solution to (6.8). A regularity result for the solutions to the elliptic equations implies that $Z^* \in H^2(\Omega) \subset C(\overline{\Omega})$. □

It follows that the principal eigenvalue λ_1^* to

$$\begin{cases} -d_2 \Delta \varphi - r_2 \varphi \left[1 - \dfrac{Z^*}{K + cM}\right] = \lambda \varphi, \; x \in \Omega, \\ \dfrac{\partial \varphi}{\partial n}(x) = 0, \qquad x \in \partial\Omega, \end{cases}$$

is positive. On the other hand, by comparison, we get that

$$0 \leq Z^*(x) - \tilde{Z}_1(x,t) \leq w(x,t), \quad \text{a.e. } x \in \Omega, \text{ for all } t \geq t_0,$$

where w is the solution to

$$\begin{cases} \dfrac{\partial w}{\partial t} = d_2 \Delta w + r_2 w \left[1 - \dfrac{Z^*}{K + cM}\right], & (x,t) \in \Omega \times (t_0, +\infty), \\ \dfrac{\partial w}{\partial n}(x,t) = 0, & (x,t) \in \partial\Omega \times (t_0, +\infty), \\ w(x, t_0) = Z^*(x) - \tilde{Z}_{10} \geq 0, & x \in \Omega. \end{cases}$$

Since $\lambda_1^* > 0$, we get that

$$w(t) \longrightarrow 0 \quad \text{in } L^\infty(\Omega), \text{ as } t \to +\infty$$

(at the rate of $\exp(-\lambda_1^* t)$), and so

$$\tilde{Z}_1(t) \longrightarrow Z^* \quad \text{in } L^\infty(\Omega), \text{ as } t \to +\infty.$$

Analogously, we get that

$$\overline{Z}_1(t) \longrightarrow Z^* \quad \text{in } L^\infty(\Omega), \text{ as } t \to +\infty.$$

We conclude that

$$Z_1(t) \longrightarrow Z^* \quad \text{in } L^\infty(\Omega), \text{ as } t \to +\infty.$$

It follows that, for any $\varepsilon > 0$, there exists $t(\varepsilon) \geq 0$ such that

$$Z(x,t) \geq Z_1(x,t) > Z^*(x) - \varepsilon, \quad \text{a.e. } x \in \Omega, \text{ for all } t \geq t(\varepsilon).$$

On the other hand, it can be shown that for all $\varepsilon > 0$, there exists $t_2(\varepsilon) \geq 0$, such that $C_I(x,t) \leq C^*(x) + \varepsilon$, a.e. $x \in \Omega$, for all $t \geq t_2(\varepsilon)$, where C^* is the unique positive solution to

$$\begin{cases} d\Delta C + rC(M - \kappa C) - \delta C = 0, \; x \in \Omega, \\ \dfrac{\partial C}{\partial n}(x) = 0, & x \in \partial\Omega. \end{cases}$$

Hence, for any $\varepsilon > 0$ sufficiently small, we have that

$$p\frac{\gamma_1 C_I}{1 + \gamma_2 C_I} Z \geq p\gamma_1 C_I \frac{Z^* - \varepsilon}{1 + \gamma_2 (C^* + \varepsilon)}, \quad \text{a.e. } x \in \Omega,$$

for all $t \geq \overline{t}(\varepsilon) = \max\{t(\varepsilon), t_2(\varepsilon)\}$.

We consider C_ε the solution to

$$\begin{cases} \dfrac{\partial C}{\partial t} = d\Delta C + rC(M - \kappa C) \\ \qquad -\delta C_I - \dfrac{p\gamma_1(Z^* - \varepsilon)}{1 + \gamma_2(C^* + \varepsilon)} C, \; (x,t) \in \Omega \times (\bar{t}(\varepsilon), +\infty), \\ \dfrac{\partial C}{\partial n}(x,t) = 0, \qquad (x,t) \in \partial\Omega \times (\bar{t}(\varepsilon), +\infty), \\ C(x, \bar{t}(\varepsilon)) = C_I(x, \bar{t}(\varepsilon)), \qquad x \in \Omega. \end{cases}$$

The comparison result for parabolic equations implies that

$$C_I(x,t) \le C_\varepsilon(x,t), \quad \text{a.e. } x \in \Omega, \text{ for all } t \ge \bar{t}(\varepsilon).$$

If the principal eigenvalue $\lambda_1(\varepsilon)$ to

$$\begin{cases} -d\Delta\varphi - rM\varphi + \delta\varphi + \dfrac{p\gamma_1(Z^* - \varepsilon)}{1 + \gamma_2(C^* + \varepsilon)} \varphi = \lambda\varphi, \; x \in \Omega, \\ \dfrac{\partial \varphi}{\partial n}(x) = 0, \qquad x \in \partial\Omega \end{cases}$$

(given by Rayleigh's formula),

$$\lambda_1(\varepsilon) = \inf_{\substack{\varphi \in H^1(\Omega) \\ \|\varphi\|_{L^2(\Omega)} \neq 0}} \frac{\displaystyle\int_\Omega \left[d|\nabla\varphi|^2 - rM|\varphi|^2 + \delta|\varphi|^2 + \frac{p\gamma_1(Z^* - \varepsilon)}{1 + \gamma_2(C^* + \varepsilon)} |\varphi|^2 \right] dx}{\displaystyle\int_\Omega |\varphi|^2 dx}, \tag{6.9}$$

is positive, then

$$C_\varepsilon(t) \longrightarrow 0 \quad \text{in } L^\infty(\Omega),$$

as $t \to +\infty$ (the convergence to 0 in $L^\infty(\Omega)$ follows as in Chapter 5), and so

$$C_I(t) \longrightarrow 0 \quad \text{in } L^\infty(\Omega),$$

as $t \to +\infty$.

On the other hand, by Rayleigh's formula, we get that

$$\lim_{\varepsilon \to 0+} \lambda_1(\varepsilon) = \lambda_1(0).$$

To summarize, we get the following result:

Theorem 6.3 *If $\tilde{\lambda}_1 < 0$ and if $\lambda_1(0) > 0$, then*

$$C_I(t) \longrightarrow 0 \quad \text{in } L^\infty(\Omega), \text{ as } t \to +\infty.$$

Remark 6.4 According to Theorem 6.2, if $\tilde{\lambda}_1 \geq 0$ (which may occur for a small value of M, the weed biomass), the vector population vanishes for large times, independently of the presence of predators.

While, according to Theorem 6.3, for $\tilde{\lambda}_1 < 0$ eradication of the vector population is possible upon the additional information that $\lambda_1(0) > 0$.

The form of $\lambda_1(0)$, given by Rayleigh's formula (6.9), shows that there is a monotonicity with respect to Z^*; on the other hand, Z^* is monotonically increasing with respect to $\mathbb{I}_\omega u$ (and so in particular to ω, by inclusion).

Moreover, if $Ess\ inf\ u \longrightarrow +\infty$, we may conjecture that $Ess\ inf\ Z^* \longrightarrow +\infty$ and consequently $\lambda_1(0) \longrightarrow +\infty$ (for $\omega = \Omega$, this can be proven to hold).

This means that if the inflow of specialist predators in the relevant habitat is sufficiently large (or we extend the region of intervention ω), eradication of the total vector population is possible.

We may additionally state the following.

Corollary 6.5 *Under the assumptions of Theorem 6.3, for any $\varepsilon > 0$ sufficiently small, we get that $\lambda_1(\varepsilon) > 0$ and so $C_I(t) \longrightarrow 0$ in $L^\infty(\Omega)$, as $t \to +\infty$, and the decay rate is that of* $\exp(-\lambda_1(\varepsilon)t)$.

6.3.2 *Holling Type III Functional Response to Predation*

Assume that $\tilde{\lambda}_1 < 0$.

In this case, System (6.5) has to be changed into the following one, in which an Holling type III response to predation occurs. In this case, (C_I, Z) is the solution to

$$\begin{cases} \dfrac{\partial C_I}{\partial t}(x,t) = d\Delta C_I(x,t) + rC_I(x,t)(M(x) - \kappa C_I(x,t)) - \delta C_I(x,t) \\ \qquad\qquad -p\dfrac{\gamma_1 C_I(x,t)^2}{1+\gamma_2 C_I(x,t)^2} Z(x,t), \\ \dfrac{\partial Z}{\partial t}(x,t) = d_2\Delta Z(x,t) + r_2 Z(x,t)\left[1 - \dfrac{Z(x,t)}{K(x)+cM(x)}\right] \\ \qquad\qquad +\dfrac{\gamma_1 C_I(x,t)^2}{1+\gamma_2 C_I(x,t)^2} Z(x,t) + \mathbb{I}_\omega(x)u(x), \end{cases}$$

for $(x,t) \in Q$, subject to pure Neumann boundary conditions and suitable initial conditions.

We are considering the case in which $\tilde{\lambda}_1 < 0$, as in Theorem 6.3. In that case, we have been able to show that, under the additional assumption of a large inflow of specialist predators in the relevant habitat, the vector population will eventually go extinct. For H. III predators, the result is negative, as we will show in the following theorem.

Theorem 6.6 *If $\tilde{\lambda}_1 < 0$, then there is no $u \in L^\infty(\omega)$, $u(x) \geq 0$, a.e. $x \in \omega$, such that*

$$C_I(t) \longrightarrow 0 \quad in\ L^\infty(\Omega),\ as\ t \to +\infty.$$

Proof We argue by contradiction. Suppose that there is a control u such that $C_I(t) \longrightarrow 0$ in $L^\infty(\Omega)$, as $t \to +\infty$; then for all $\varepsilon > 0$, there exists $t_3(\varepsilon) \geq 0$ such that

$$\varepsilon > C_I(x,t), \quad \text{a.e. } x \in \Omega, \text{ for all } t \geq t_3(\varepsilon).$$

Let us notice that the comparison result for parabolic equations implies that

$$0 \leq Z(x,t) \leq \overline{Z}(x,t), \quad \text{a.e. } x \in \Omega, \text{ for all } t \geq 0,$$

where $\overline{Z}$ is the unique solution to

$$\begin{cases} \dfrac{\partial Z}{\partial t} = d_2 \Delta Z + r_2 Z(1 - \dfrac{Z}{K + cM}) + \dfrac{\gamma_1}{\gamma_2} Z + \mathbb{I}_\omega u, & (x,t) \in Q, \\ \dfrac{\partial Z}{\partial n}(x,t) = 0, & (x,t) \in \Sigma, \\ Z(x,0) = Z_0(x), & x \in \Omega. \end{cases}$$

We have that $\overline{Z} \in L^\infty(Q)$, and so there exists M^0, a positive constant such that

$$\left| \frac{\gamma_1 C_I}{1 + \gamma_2 C_I^2} Z \right| \leq M^0, \quad \text{a.e. } (x,t) \in Q,$$

because the function $s \mapsto \dfrac{\gamma_1 s}{1 + \gamma_2 s^2}$ (defined on $[0, +\infty)$) is bounded. We may conclude (via the comparison principle) that

$$C_I(x,t) \geq C_{I1}(x,t), \quad \text{a.e. } x \in \Omega, \text{ for all } t \geq t_3(\varepsilon),$$

where C_{I1} is the unique solution to

$$\begin{cases} \dfrac{\partial C}{\partial t} = d\Delta C + rC(M - \kappa C) - \delta C - p\varepsilon M^0 C, & x \in \Omega,\ t > t_3(\varepsilon), \\ \dfrac{\partial C}{\partial n}(x,t) = 0, & x \in \Omega,\ t > t_3(\varepsilon), \\ C(x, t_3(\varepsilon)) = C_I(x, t_3(\varepsilon)), & x \in \Omega. \end{cases}$$

Since $C_I(t) \longrightarrow 0$ in $L^\infty(\Omega)$, as $t \to 0$, we get that $\tilde{\lambda}_1 + p\varepsilon M^0$, the principal eigenvalue to

$$\begin{cases} -d\Delta\varphi - rM\varphi + \delta\varphi + p\varepsilon M^0\varphi = \lambda\varphi, & x \in \Omega, \\ \dfrac{\partial\varphi}{\partial n}(x) = 0, & x \in \partial\Omega, \end{cases}$$

is positive. Since $\tilde{\lambda}_1 + p\varepsilon M^0 > 0$, for any $\varepsilon > 0$, we conclude that $\tilde{\lambda}_1 \geq 0$, which leads to a contradiction.

□

So, we may conclude that there is no control u such that $C_I(t) \longrightarrow 0$ in $L^\infty(\Omega)$, as $t \to +\infty$.

Let us evaluate, however, how much the population density C_I decays if a control u is implemented.

We have that

$$Z(x,t) \leq \overline{Z}(x,t), \quad \text{a.e. } x \in \Omega, \text{ for all } t \geq 0.$$

The comparison result for parabolic equations implies that

$$C_I(x,t) \geq C^I(x,t), \quad \text{a.e. } x \in \Omega, \text{ for all } t \geq 0,$$

where C^I is the unique solution to

$$\begin{cases} \dfrac{\partial C^I}{\partial t} = d\Delta C^I + rC^I(M - \kappa C^I) - \delta C^I - p\gamma_1\|\overline{Z}\|_{L^\infty(Q)}(C^I)^2, & (x,t) \in Q, \\ \dfrac{\partial C^I}{\partial n}(x,t) = 0, & (x,t) \in \Sigma, \\ C^I(x,0) = C_{I0}(x), & x \in \Omega. \end{cases}$$

Hence, C_I cannot decay below C^I.

On the other hand, arguing as in Chapter 5, we get that

$$C^I(t) \longrightarrow \tilde{C}^* \quad \text{in } L^\infty(\Omega),$$

as $t \to +\infty$, where $\tilde{C}^*$ is the positive solution to

$$\begin{cases} d\Delta C + rC\left(M - (\kappa + \dfrac{p\gamma_1\|\overline{Z}\|_{L^\infty(Q)}}{r})C\right) - \delta C = 0, & x \in \Omega, \\ \dfrac{\partial C}{\partial n}(x) = 0, & x \in \partial\Omega. \end{cases}$$

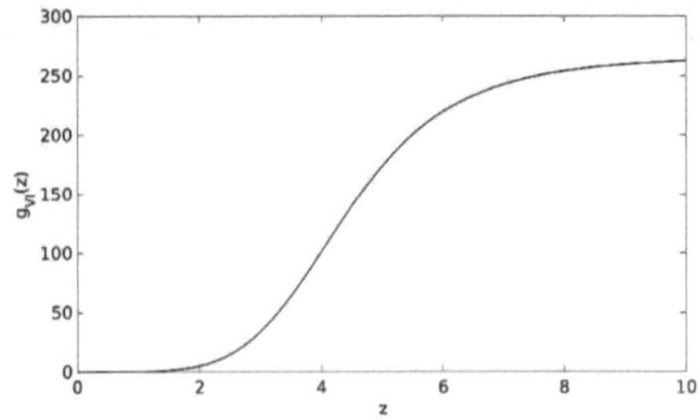

Fig. 6.3 Holling type VI functional response

Hence, we may conclude that for any $\varepsilon > 0$, there exists a $T(\varepsilon) \geq 0$ such that

$$C_I(x,t) \geq \tilde{C}^*(x) - \varepsilon, \quad \text{a.e. } x \in \Omega, \text{ for all } t \geq T(\varepsilon).$$

Remark 6.7 It might be of interest to notice that in the family of Holling type functional responses we might also include the following one, illustrated in Figure 6.3,

$$g_{VI}(z) = \frac{\gamma_1 z^5}{1 + \gamma_2 z^5},$$

with γ_1, γ_2 positive constants.

It can be easily conjectured that in this case the impact of predation on the eradicability of the epidemic would be even less effective.

The above results in particular Theorem 6.2 have confirmed the peculiar role of the parameter M, representing the weed biomass existing in the relevant habitat: Independently of the introduction of predators, cutting weeds has been confirmed as a cost-effective practice for controlling a *Xylella* epidemic in olive orchards (see [25, 48, 62]).

It would be of great interest, from an agronomical point of view, to relate the mathematical threshold parameters, $\tilde{\lambda}_1$ and $\lambda_1(0)$, to some relevant parameters of intervention, such as $M(x)$ (the spatial distribution of the weed biomass) and $u(x)$ (the spatial distribution of the predators inflow). Unfortunately, this is a nontrivial mathematical problem, from a purely analytical point of view. A practical response to this issue can be obtained via adequate computational experiments, as it has been reported in Section 6.4.

On the other hand, the values of $\tilde{\lambda}_1$ and $\lambda_1(0)$ depend, in addition to the possible parameters of intervention $M(x)$ and $u(x)$, upon intrinsic parameters of the relevant agro-ecosystem, which are not yet known numerically. As mentioned in the concluding remarks, it is hopeful that our results may stimulate additional field experiments aimed at the identification of real values of all relevant parameters.

6.4 Numerical Simulations

The numerical strategy adopted to approximate the controlled system, coupled with boundary and initial conditions (including the equations for the trees, see Chapter 5), consists of the finite element method for space discretization and the finite difference method for time discretization. This procedure is the state of the art for the solution of parabolic partial differential equations (PDEs), see, e.g., [168]. See also the computational section of the previous chapter and Appendix B.

Space Discretization We first apply a standard Galerkin procedure to the weak formulations of the controlled system (6.5). As computational domain Ω, we have taken a two-dimensional slab of size $400 \times 80\,\mathrm{Km}^2$, which mimics, e.g., the whole region of Apulia in Southern Italy, from South (right-hand side of the domain) to North (left-hand side of the domain). The rectangular domain has been discretized by a uniform grid of 200×40 bilinear finite elements ($Q1$), yielding a total amount of 8241 discretization nodes. The stiffness matrix is computed exactly, whereas the mass matrix is obtained by applying the mass lumping technique.

Time Discretization After space discretization by finite elements, we obtain the semi-discrete problem that consists of a system of ordinary differential equations (ODEs). We solve these ODEs by employing a first order semi-implicit finite difference scheme, where the linear diffusion terms are approximated by Backward Euler, whereas the nonlinear reaction terms are approximated by Forward Euler. As a result, at each time step, it is required the solution of a linear system of algebraic equations of dimension $5 \times 8241 = 41{,}205$ degrees of freedom. The linear system is solved by Gaussian elimination with the built-in function of MATLAB. The time step size is $\Delta t = 0.0002$ years.

For further details on the numerical discretization of parabolic problems, we refer, e.g., to [168].

6.4.1 *Parameter Calibration*

The values of the simulation parameters are listed in Table 6.2. The diffusion coefficients in the PDEs are set to $d = 1e-4\ \mathrm{Km}^2/\mathrm{year}$ and $d_2 = 0.1d$. Moreover, we have assumed $r_2 = 2/3r,\ c = 0.3,\ K(x) \equiv 1,\ p = 1$. The total simulation time is $T = 20$ years. The initial distributions, used in all next numerical simulations, are the following:

$$\begin{cases} s_1(x,0) = 100 - 20\exp(-100(x_1-380)^2 - 100(x_2-40)^2), \\ i_1(x,0) = 20\exp(-100(x_1-380)^2 - 100(x_2-40)^2), \\ s_2(x,0) = 50, \\ i_2(x,0) = 0, \\ Z(x,0) = 0. \end{cases}$$

Table 6.2 Values of the model parameters

Symbol	Unit	Value range	Simulation value	References
r	Year^{-1}	[37,400]	200	[181, 197]
κ	–	[0.001, 0.05]	0.03	[62]
δ	Year^{-1}	[0.95, 0.99]	0.98	[62]
q	Year^{-1}	[0.2, 0.7]	0.5	[191]
C^0	–	100	100	[62]
ℓ	Year^{-1}	0.01	0.01	[62]
b	–	0.05	0.05	[62]
μ	Year^{-1}	[0.8, 1]	0.9	[174, 175]
α	Year^{-1}	[0.1, 0.5]	0.3	[62]
β	Year^{-1}	[0.5, 1]	1	[92]
ζ	Year^{-1}	[0.2, 0.8]	0.8	[50, 106]

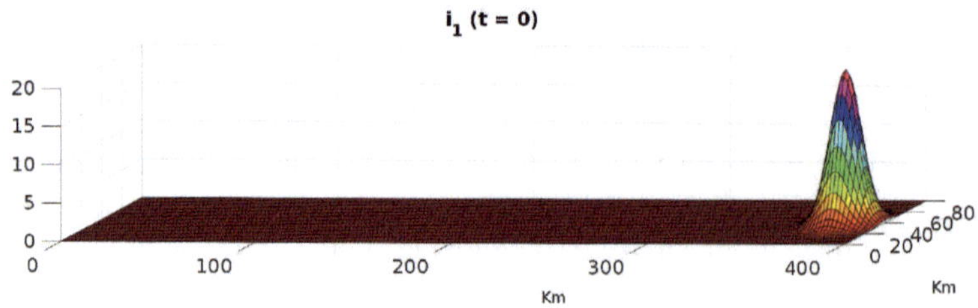

Fig. 6.4 Spatial distribution of the initial infective insect population i_1

Therefore, we assume constant initial distributions for healthy insects and trees, and we localize the initial presence of infected insects on the right-hand side of the domain (Figure 6.4).

6.4.2 Results

We report in this section the results obtained with the MATLAB code

```
run_xylella_pred.m,
```

available on the Birkhäuser website provided in the Preface and also at the link

```
http://www.mat.unimi.it/users/scacchi/ACS_codes/
run_xylella_pred.m
```

In Figures 6.5, 6.6, 6.7, 6.8, 6.9, the outcomes of the computational experiments corresponding to the following scenarios have been reported:

1. Case A1: $M = 0.4$, $u = 2e + 5$, Holling II.
2. Case A2: $M = 0.4$, $u = 2e + 5$, Holling III.
3. Case B11: $M = 1$, $u = 2e + 4$, Holling II.

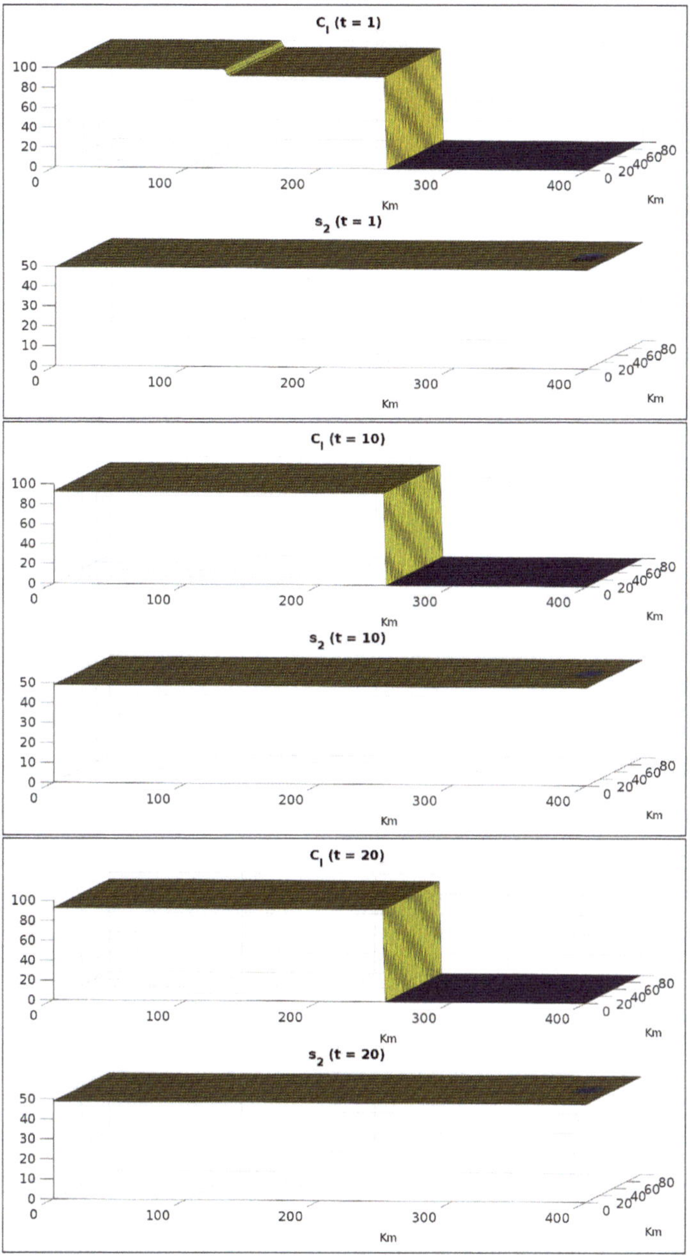

Fig. 6.5 Case A1 ($M = 0.4$, $u = 2e+5$, Holling II). We solve Equations (6.2)–(6.2)–(5.8)–(5.9)–(6.4) with the algorithms and discretization parameters described in Section 6.4 and the model parameters given in Table 6.2. Spatial distribution of the total insect population C_I, and of the uninfected trees s_2, at three time frames. The epidemic is stopped

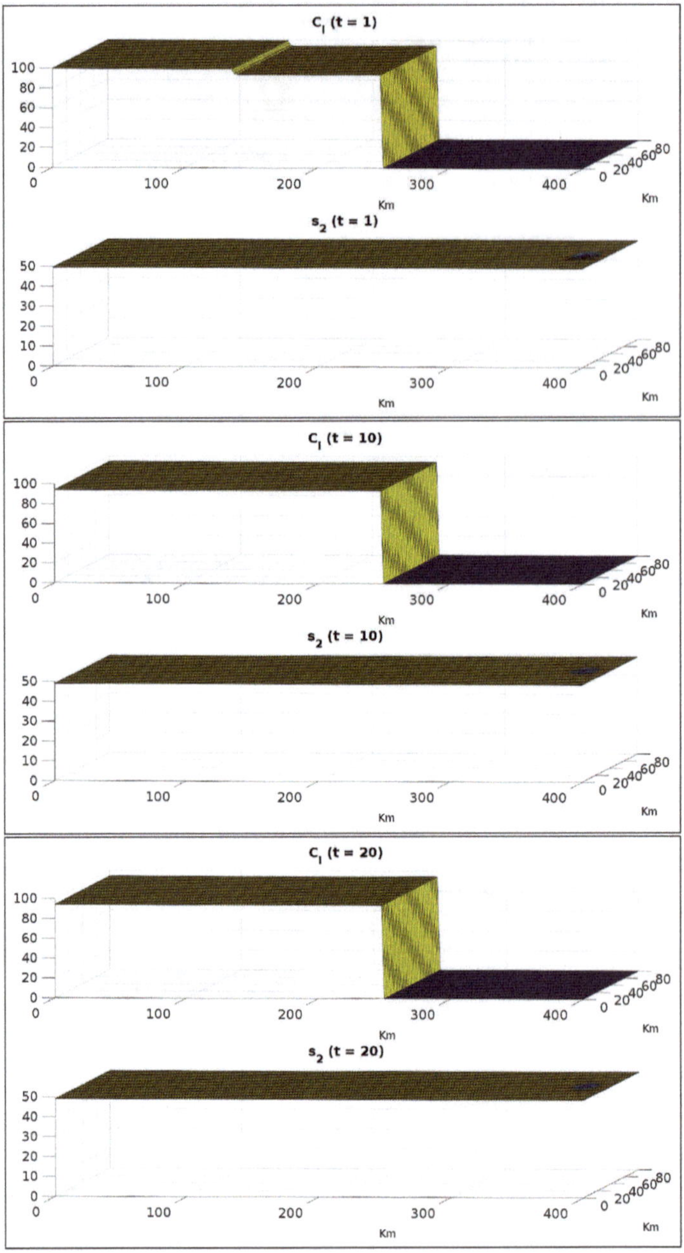

Fig. 6.6 Case A2 ($M = 0.4$, $u = 2e{+}5$, Holling III). We solve Equations (6.2)–(6.2)–(5.8)–(5.9)–(6.4) with the algorithms and discretization parameters described in Section 6.4 and the model parameters given in Table 6.2. Spatial distribution of the total insect population C_I, and of the uninfected trees s_2, at three time frames. The epidemic is stopped

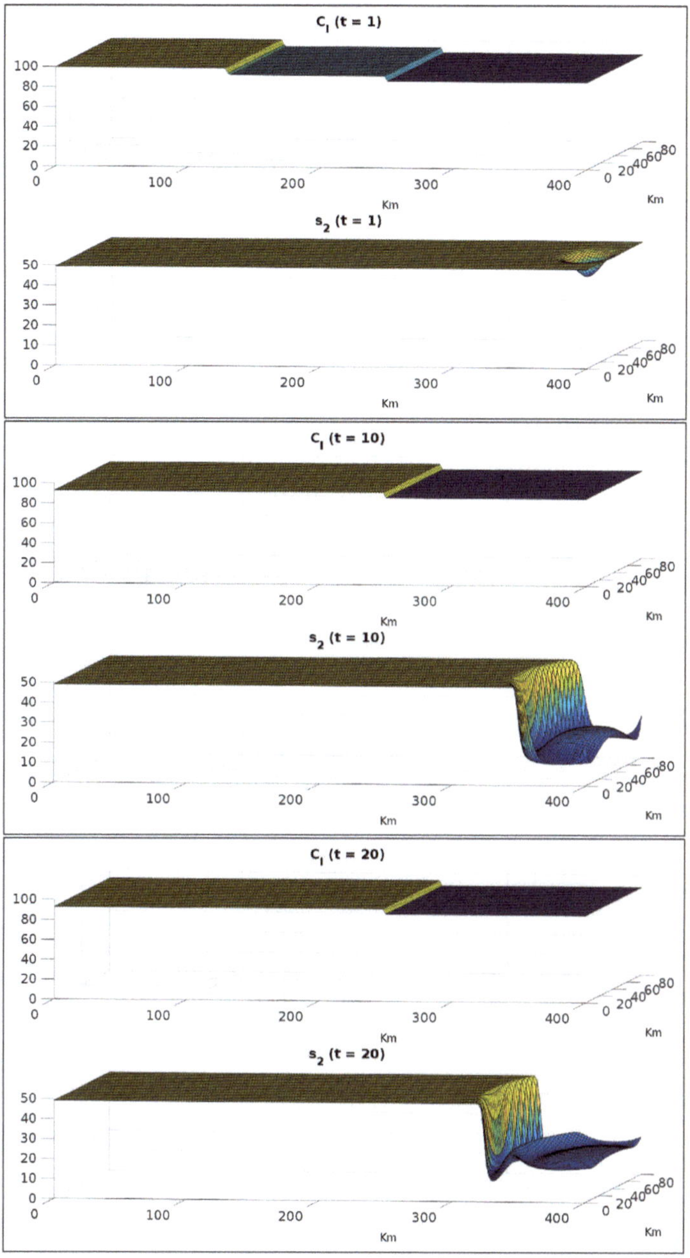

Fig. 6.7 Case B11 ($M = 1, u = 2e + 4$, Holling II). We solve Equations (6.2)–(6.2)–(5.8)–(5.9)–(6.4) with the algorithms and discretization parameters described in Section 6.4 and the model parameters given in Table 6.2. Spatial distribution of the total insect population C_I, and of the uninfected trees s_2, at three time frames. The epidemic propagates

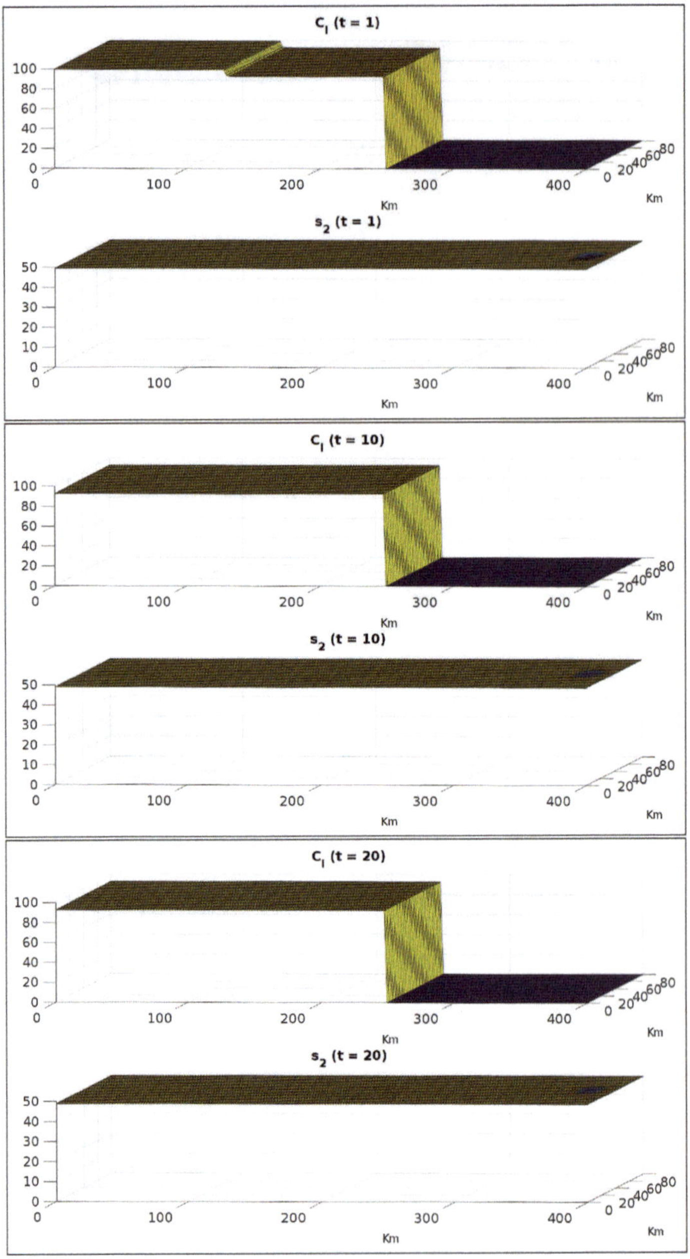

Fig. 6.8 Case B12 ($M = 1, u = 2e + 5$, Holling II). We solve Equations (6.2)–(6.2)–(5.8)–(5.9)–(6.4) with the algorithms and discretization parameters described in Section 6.4 and the model parameters given in Table 6.2. Spatial distribution of the total insect population C_I, and of the uninfected trees s_2, at three time frames. The epidemic is stopped

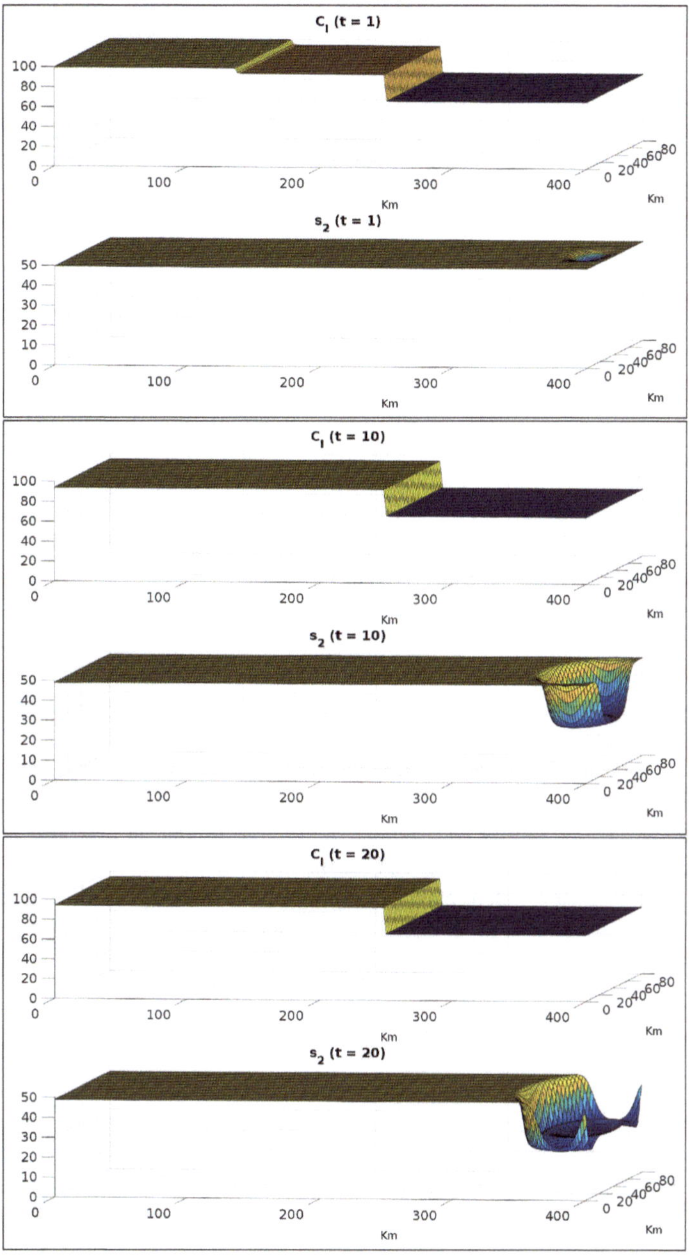

Fig. 6.9 Case B22 ($M = 1, u = 2e+5$, Holling III). We solve Equations (6.2)–(6.2)–(5.8)–(5.9)–(6.4) with the algorithms and discretization parameters described in Section 6.4 and the model parameters given in Table 6.2. Spatial distribution of the total insect population C_I, and of the uninfected trees s_2, at three time frames. The epidemic propagates

4. Case B12: $M = 1, u = 2e + 5$, Holling II.
5. Case B22: $M = 1, u = 2e + 5$, Holling III.

The predation inflow u has been applied only in the region $\omega = \{x > 250\}$.

In the Holling II functional response to predation, we have set $\gamma_1 = 1500,\ \gamma_2 = 5$, whereas in the Holling III functional response to predation we have set $\gamma_1 = 80,\ \gamma_2 = 0.3$.

Coherently with our mathematical analysis, in both cases A1 and A2, in which the level of weed biomass has been kept sufficiently small (M = 0.4), the epidemic is stopped, independently of the specificity of the predators, either for Holling II or Holling III functional response to predation, respectively.

In cases B11 and B12, Holling II has been adopted as predation response functional. They differ only by the level of predation inflow u : smaller in case B11 with respect to case B12. In case B11, the epidemic propagates; in case B12, the epidemic is stopped.

This result highlights the fact that even in the case of predators with a narrow range of possible preys, or in the case of parasitoids, the amount of released control agents represents a crucial aspect that has to be taken into account when planning a control strategy based on their release in the environment. Here it is confirmed that a release of control agents below a certain threshold frustrates the possibility of success of the control strategy.

Given accurate experimental values of all parameters of the mathematical model of the relevant ecosystem, a suitable iteration of the computational experiments B11 and B12 may lead to the identification of the above-mentioned threshold level for u; it is clear that such threshold value will depend upon the selected region of intervention ω, too.

In last case B22, Holling III has been adopted as predation response functional: Even for a "large" predator inflow u, the epidemic persists.

This result represents clear evidence, based on the model presented here, that the use of a H. III predator is not effective in the control of a *Xylella* vector.

6.5 Concluding Remarks

In this chapter, for a *Xylella* epidemic as a case study, we have shown that the use of a H. III predator cannot be considered as a viable solution for the control of a specific pest; it will not able to reach the expected results either in containing or in eradicating the epidemic. The control and the eventual eradication of the epidemic can be reached in the case that specialist predators, possessing the specific pest of interest within its narrow range of preys.

Part IV
Controlling Environmental Pollution in Geographical Economics

Chapter 7
Economic Growth and Pollution Diffusion

7.1 Introduction

As an additional application of the methods presented in this volume, in this chapter, we introduce an environmental issue concerning the control of the pollution produced by human activities.

Standard macroeconomics and environmental economics have been two completely independent research areas for a long time. The seminal paper by Grossman and Kruger [120] has stimulated considerable interest in the relationship between economic development and environmental pollution. The role of taxation for controlling pollution is the key issue of the report [161] (see also references therein) issued by the International Monetary Fund. In view of these policy developments, some recent works tend to develop a global theory combining the above-mentioned two branches of literature (see [57] and [49]).

Here we will present our own results concerning the impact of pollution diffusion on the economic and environmental joint dynamics and consider the optimal control problem of the above -mentioned system subject to capital consumption and the rate of taxation based on the amount of produced pollution.

In [18] and [19], the authors introduced a spatial economic growth model subject to pollution diffusion. In particular, in [18], the authors analyzed the large-time behavior of a spatially structured economic growth model coupling physical capital accumulation and pollution diffusion: This model extends other results in the literature along different directions. Furthermore, a negative feedback to the production function was added, in order to describe the (negative) influence of pollution on capital accumulation. In [19], the authors modified the model presented in [18] by introducing an optimal harvesting control problem by an objective function composed of three terms, namely the intertemporal utility of

S. Aniţa et al., *Mathematical Modeling and Control in Life and Environmental Sciences*, Modeling and Simulation in Science, Engineering and Technology,
https://doi.org/10.1007/978-3-031-49971-5_7

the decision maker, the space-time average of the level of pollution in the habitat, and the disutility due to the imposition of taxation. In [21], two controls have been introduced, namely the level of consumption and, in addition, the level of taxation on physical capital. This includes taxation on machinery, buildings, computers, and any equipment that help to turn the raw materials into finished products or services, and because of these activities, they have a huge impact on pollution emissions.

In the literature, other contributions that explore the spatial dimension in environmental and resource economics can be found in [58, 59, 195], and [34].

The inclusion of a spatial dimension in economic analysis has recently increased relevance and interest. The first studies in economic geography go back to Beckerman [45] and Puu [166], who study regional problems based simply on flow equations. Starting from these works, a new economic geography arises adopting general equilibrium models to analyze the peculiarities of local and global markets and the mobility of production factors (see [114, 141]). More recently, this geographical approach has been introduced in economic growth models to study the implications between accumulation and diffusion of capital on economic dynamics (see [51, 56, 65]). The Solow model [185] with a continuous spatial dimension has been extensively studied. In [65], the authors analyze problems of convergence across regions when capital is mobile, while [56] considers the case in which both capital and labor are mobile. The Ramsey model [169] has been extended to a spatial dimension in [56] and [51], respectively, in average and total utilitarianism version. In [71], alongside the classical Cobb–Douglas production function, the authors have proposed a larger class of convex–concave production functions, thus allowing a larger variety of dynamical scenarios, including the existence of poverty traps.

In this chapter, we report about the key issues concerning the impact of taxation on the production as emerged in [21]. In Section 7.3, an optimal control problem is presented, by acting on both consumption and taxation. Since the localization of the optimal control is not an easy analytical task, we have proposed a computational algorithm for the search by the gradient method. To this aim, in the same Section 7.3, the evaluation of the directional derivative of the cost functional with respect to the two controls has been presented. In Section 7.4, a numerical algorithm that allows to solve the system of four backward–forward reaction–diffusion equations has been reported, together with numerical experiments.

By considering the main scope of this volume, at the end of this chapter, as a challenging open problem, the interested reader may find a regional control problem by reducing the action of the taxation only in an optimally chosen subregion of the whole domain of interest.

7.2 A Spatially Structured Model

Let $k(x, t)$ denote the capital stock held by the representative household located at $x \in \Omega$ (a nonempty bounded domain in $\mathbb{R}^n$, $n = 1, 2$ with a sufficiently smooth boundary), at time $t \geq 0$. The time evolution of the physical capital is modelled by

$$\frac{\partial k}{\partial t}(x,t) = d_1 \Delta k(x,t) + A(x,t) f\big(k(x,t)\big) - \delta_1 k(x,t) - c(x,t)k(x,t), \tag{7.1}$$

for $(x,t) \in \Omega \times (0,+\infty)$.

Here $A(x,t)$ denotes the technological progress available at $x \in \Omega$ and time $t \geq 0$. The parameter δ_1 takes into account the depreciation of the capital. $c(x,t)$ describes the level of consumption per unit of physical capital $k(x,t)$; the quantity $(1 - c(x,t))k(x,t)$ describes the saved capital after consumption.

Equation (7.1) is complemented by an initial capital distribution, $k(x,0)$, and suitable boundary conditions. If we assume that there is no capital flow through the boundary $\partial\Omega$, we impose homogeneous Neumann boundary conditions, i.e.,

$$\frac{\partial k}{\partial n}(x,t) = 0, \quad x \in \partial\Omega,\ t > 0.$$

Concerning the technological progress, here we shall reduce the model to a constant $A(x,t) \equiv A > 0$.

The production function f can be taken of the following form (see [71], and references therein):

$$f(r) = \frac{\alpha_1 r^\gamma}{1 + \alpha_2 r^\gamma},$$

where $\alpha_1 \in (0,+\infty),\ \alpha_2 \in [0,+\infty),\ \gamma \in [1,+\infty)$.

As far as pollution diffusion is concerned, if we denote by $p(x,t)$ the pollution stock faced by a representative household located at $x \in \Omega$, at time $t \geq 0$, we assume the following model:

$$\frac{\partial p}{\partial t}(x,t) = d_2 \Delta p(x,t) + \theta \int_\Omega f(k(x',t))\varphi(x',x)dx' - \delta_2 p(x,t). \tag{7.2}$$

Via the integral term in Equation (7.2), we may see the impact of the economy on pollution production. At any point $x \in \Omega$, pollution is generated due to the production $f(k(x,t))$. In our model, we have supposed that any point $x' \in \Omega$ may have an impact on enhancing pollution at any other point $x \in \Omega$, by any other possible mechanism than random dispersal already described by the Laplace operator. Such mechanisms are described by a kernel φ that satisfies the following hypotheses: $\varphi \in L^\infty(\Omega \times \Omega)$, and $\varphi(x',x) \geq 0$ a.e. $(x',x) \in \Omega \times \Omega$. As a limiting case, one may reduce the kernel φ to a Dirac Delta function, to mean the impact of production on pollution at the same point x. The parameter θ takes into account the "efficiency" of the economic system with respect to pollution production. The parameter δ_2 accounts for the natural decay of pollution.

Equation (7.2) has to be complemented by suitable initial and boundary conditions.

Vice versa in order to model the impact of pollution on production, we may modify Equation (7.1) as follows:

$$\frac{\partial k}{\partial t}(x,t) = d_1 \Delta k(x,t) + \frac{A}{g(p(x,t))} f(k(x,t)) - \delta_1 k(x,t) - c(x,t)k(x,t). \tag{7.3}$$

All parameters, $A, d_1, d_2, \delta_1, \delta_2$, that characterize a specific economic/ecological system are assumed strictly positive.

The function $g : [0, +\infty) \to (0, +\infty)$ models the negative impact of pollution on the production; usual assumptions on g are:

(i) g is continuously differentiable and monotonically increasing.
(ii) $g(0) = 1$ and $\lim_{p \to +\infty} g(p) = +\infty$.

As a typical choice for the function g, later we shall adopt the expression

$$g(p) = 1 + p^2, \quad p \geq 0.$$

7.3 An Optimal Control Problem

From now on, we shall study our economic/ecological system on a finite time horizon $[0, T]$. We shall denote by

$$Q_T = \Omega \times (0, T); \quad \Sigma_T = \partial\Omega \times (0, T).$$

In order to reduce pollution emissions by production processes, local authorities may impose taxes proportional to the level of production; if we denote by $\tau(x, t)$ the environmental taxation rate, we shall modify Equations (7.3)–(7.2) as follows: [57]

$$\begin{cases} \dfrac{\partial k}{\partial t}(x,t) = d_1 \Delta k(x,t) + \dfrac{A}{1 + p(x,t)^2} f((1-\tau(x,t))k(x,t)) \\ \qquad -\delta_1 k(x,t) - c(x,t)k(x,t), \\ \dfrac{\partial p}{\partial t}(x,t) = d_2 \Delta p(x,t) + \theta \displaystyle\int_\Omega f((1-\tau(x',t))k(x',t))\varphi(x',x)dx' \\ \qquad -\delta_2 p(x,t), \end{cases} \tag{7.4}$$

for $(x, t) \in Q_T$.

The term $(1 - \tau(x, t))k(x, t)$ accounts for the amount of the physical capital left to be used in the gross domestic production. An increase in the level of taxation allows the implementation of practices leading to an abatement of the level of pollution emissions, so that we may interpret it as a "green" taxation policy.

In this section, we consider the following optimal control problem with a finite horizon time, in which both $c(x, t)$ and $\tau(x, t)$ play the role of control variables:

$$\underset{(c,\tau)\in U}{Maximize} \left\{ \int_0^T \int_\Omega c(x,t)k(x,t)dx\,dt - \beta_0 \int_0^T \int_\Omega p(x,t)dx\,dt \right.$$
$$\left. -\beta_1 \int_0^T \int_\Omega \tau(x,t)k(x,t)dx\,dt \right\},$$

subject to System (7.4). For simplicity, we may consider homogeneous Neumann boundary conditions

$$\frac{\partial k}{\partial n}(x,t) = \frac{\partial p}{\partial n}(x,t) = 0, \qquad (x,t) \in \Sigma_T, \tag{7.5}$$

and initial conditions

$$k(x,0) = k_0(x), \;\; p(x,0) = p_0(x), \qquad x \in \Omega, \tag{7.6}$$

where $T > 0$ is fixed and $k_0, p_0 \in L^\infty(\Omega)$, $k_0(x) \geq 0$, $p_0(x) \geq 0$ a.e. $x \in \Omega$.

We impose the constraints

$$1 - c(x,t) - \tau(x,t) \geq s, \;\; c(x,t), \tau(x,t) \geq 0, \quad \text{a.e. } (x,t) \in Q_T, \tag{7.7}$$

where $s > 0$ is the saving factor.

$U = \{(c, \tau) \in L^\infty(Q_T) \times L^\infty(Q_T) : (c, \tau) \text{ satisfies } (7.7)\}$ is the set of controls, and (k, p) is the solution to (7.4)–(7.6) corresponding to Q_T and Σ_T. Here β_0, β_1 are positive constants.

The objective function is composed of three terms that we hereby describe:

$T1$: The term $\int_0^T \int_\Omega c(x,t)k(x,t)dx\,dt$, to be maximized, describes the intertemporal utility of the decision maker.

$T2$: The second term $\beta_0 \int_0^T \int_\Omega p(x,t)dx\,dt$, to be minimized, represents an average of the total level of pollution across space and time.

$T3$: The third term $\int_0^T \int_\Omega \tau(x,t)k(x,t)dx\,dt$, to be minimized, takes into account the disutility due to imposition of taxation.

The costate equations for the previous optimal control problem are

$$\frac{\partial \lambda_k}{\partial t}(x,t) = -d_1 \Delta \lambda_k(x,t) - c(x,t) + \beta_1 \tau(x,t)$$
$$-\lambda_k(x,t)\left(\frac{A(1-\tau(x,t))}{1+p(x,t)^2} f'((1-\tau(x,t))k(x,t)) - \delta_1 - c(x,t)\right)$$
$$-(1-\tau(x,t))f'((1-\tau(x,t))k(x,t))\theta \int_\Omega \lambda_p(x',t)\varphi(x,x')dx', \tag{7.8}$$

$$\frac{\partial \lambda_p}{\partial t}(x,t) = -d_2 \Delta \lambda_p(x,t) + \beta_0$$
$$-\lambda_k(x,t)\left(-\frac{2Ap(x,t)}{(1+p(x,t)^2)^2} f((1-\tau(x,t))k(x,t))\right)$$
$$+\delta_2 \lambda_p(x,t) \tag{7.9}$$

subject to homogeneous Neumann boundary conditions and zero final data.

In [20], it has been shown the well-posedness of the initial/boundary problem for System (7.4). Moreover, it has been shown that there exists at least one optimal control (c^*, τ^*).

Let us now proceed with the construction of a computational algorithm for the search of an optimal control by the gradient method.

For the sake of clarity, we shall use the notation $k^{(c,\tau)}$, $p^{(c,\tau)}$, $\lambda_k^{(c,\tau)}$, $\lambda_p^{(c,\tau)}$ for $k, p, \lambda_k, \lambda_p$ when some danger of confusion occurs. Denote by

$$I(c,\tau) = \int_0^T \int_\Omega c(x,t)k^{(c,\tau)}(x,t)dx\,dt$$
$$-\beta_0 \int_0^T \int_\Omega p^{(c,\tau)}(x,t)dx\,dt - \beta_1 \int_0^T \int_\Omega \tau(x,t)k^{(c,\tau)}(x,t)dx\,dt$$

the value of the cost functional corresponding to (c, τ).

Consider an arbitrary $(v_1, v_2) \in L^\infty(Q_T) \times L^\infty(Q_T)$ such that $(c,\tau) + \varepsilon(v_1, v_2) \in U$ for any $\varepsilon > 0$ sufficiently small. Let (z_k, z_p) be the solution to

$$\begin{cases} \dfrac{\partial z_k}{\partial t} = d_1 \Delta z_k + \dfrac{A}{1+(p^{(c,\tau)})^2}\Big[f'((1-\tau)k^{(c,\tau)})(1-\tau)z_k \\ \qquad -f'((1-\tau)k^{(c,\tau)})k^{(c,\tau)}v_2 \Big] - \delta_1 z_k - c z_k - v_1 k^{(c,\tau)} \\ \qquad -\dfrac{2Ap^{(c,\tau)}z_p}{(1+(p^{(c,\tau)})^2)^2} f((1-\tau)k^{(c,\tau)}), \\ \dfrac{\partial z_p}{\partial t} = d_2 \Delta z_p + \theta \displaystyle\int_\Omega f'((1-\tau(x',t))k^{(c,\tau)}(x',t)) \\ \qquad \times[(1-\tau(x',t))z_k(x',t) - v_2(x',t)k^{(c,\tau)}(x',t)]\varphi(x',x)dx' \\ \qquad -\delta_2 z_p, \end{cases} \tag{7.10}$$

for $(x,t) \in Q_T$ subject to homogeneous Neumann boundary conditions and zero initial data. One may prove in a standard way the following auxiliary result.

Lemma 7.1 *The following convergences hold for any $(c,\tau) \in U$:*

(i) $k^{(c,\tau)+\varepsilon(v_1,v_2)} \longrightarrow k^{(c,\tau)}, \quad p^{(c,\tau)+\varepsilon(v_1,v_2)} \longrightarrow p^{(c,\tau)}$ *in* $L^2(Q_T)$, *as* $\varepsilon \to 0+$.

(ii) $\dfrac{k^{(c,\tau)+\varepsilon(v_1,v_2)} - k^{(c,\tau)}}{\varepsilon} \longrightarrow z_k, \quad \dfrac{p^{(c,\tau)+\varepsilon(v_1,v_2)} - p^{(c,\tau)}}{\varepsilon} \longrightarrow z_p$ *in* $L^2(Q_T)$, *as* $\varepsilon \to 0+$.

One may compute now the directional derivative of I:

Theorem 7.2

$$dI(c,\tau)(v_1,v_2) = \int_0^T \int_\Omega k^{(c,\tau)}(x,t)[v_1(x,t)(1-\lambda_k(x,t)) - v_2(x,t)\beta^*(x,t)]dx\,dt,$$

where

$$\beta^*(x,t) = \beta_1$$
$$+ f'((1-\tau(x,t))k^{(c,\tau)}(x,t))\left(\frac{A\lambda_k(x,t)}{1+p^{(c,\tau)}(x,t)^2} + \theta \int_\Omega \lambda_p(x',t)\varphi(x,x')dx'\right).$$

Proof (Sketch of the Proof) For any $\varepsilon > 0$ sufficiently small, we have that

$$I(c+\varepsilon v_1, \tau+\varepsilon v_2) - I(c,\tau) = \int_0^T \int_\Omega [(c+\varepsilon v_1)k^{(c+\varepsilon v_1, \tau+\varepsilon v_2)} - ck^{(c,\tau)}]dx\,dt$$

$$-\int_0^T \int_\Omega \{\beta_0[p^{(c+\varepsilon v_1,\tau+\varepsilon v_2)} - p^{(c,\tau)}] + \beta_1[(\tau+\varepsilon v_2)k^{(c+\varepsilon v_1,\tau+\varepsilon v_2)} - \tau k^{(c,\tau)}]\}dx\,dt.$$

By dividing the above by ε, passing to the limit ($\varepsilon \to 0+$), and then using (i), (ii) in the lemma, we get that

$$\begin{aligned} dI(c,\tau)(v_1,v_2) = & \int_0^T \int_\Omega [v_1 k^{(c,\tau)} + cz_k]dx\,dt - \beta_0 \int_0^T \int_\Omega z_p\,dx\,dt \\ & -\beta_1 \int_0^T \int_\Omega [v_2 k^{(c,\tau)} + \tau z_k]dx\,dt. \end{aligned} \tag{7.11}$$

If we multiply Equation (7.8) by z_k, integrate over Q_T, and use (7.10), we get that

$$-\int_0^T \int_\Omega \Big\{ -d_1 \nabla\lambda_k(x,t)\cdot\nabla z_k(x,t)$$

$$+\frac{A\lambda_k(x,t)}{1+p^2(x,t)}[f'((1-\tau(x,t))k(x,t))(1-\tau(x,t))z_k(x,t)$$

$$-f'((1-\tau(x,t))k(x,t))k(x,t)v_2(x,t)] - \lambda_k(x,t)(\delta_1 z_k(x,t) + cz_k(x,t) + v_1(x,t)k(x,t))$$

$$-\frac{2Ap(x,t)z_p(x,t)\lambda_k(x,t)}{(1+p^2(x,t))^2} f((1-\tau(x,t))k(x,t))\Big\}dx\,dt$$

$$= \int_0^T \int_\Omega \Big\{ d_1 \nabla\lambda_k(x,t)\cdot\nabla z_k(x,t) - cz_k(x,t) + \beta_1\tau(x,t)z_k(x,t)$$

$$-\lambda_k(x,t)z_k(x,t)\left[\frac{A(1-\tau(x,t))}{1+p^2(x,t)} f'((1-\tau(x,t))k(x,t)) - \delta_1 - c\right] - z_k(x,t)(1-\tau(x,t))$$

$$\times f'((1-\tau(x,t))k(x,t))\theta \int_\Omega \lambda_p(x',t)\varphi(x,x')dx' \Big\}\,dx\,dt.$$

It follows that

$$
\begin{aligned}
&\int_0^T \int_\Omega (cz_k(x,t) - \beta_1 \tau(x,t) z_k(x,t)) dx\, dt \\
&= -\int_0^T \int_\Omega \int_\Omega z_k(x,t)(1-\tau(x,t)) f'((1-\tau(x,t))k(x,t)) \theta \lambda_p(x',t) \varphi(x,x') dx'\, dx\, dt \\
&-\int_0^T \int_\Omega \Bigg[\frac{A\lambda_k(x,t)}{(1+p^2(x,t))^2} f'((1-\tau(x,t))k(x,t)) k(x,t) v_2(x,t) + \lambda_k(x,t) v_1(x,t) k(x,t) \\
&+ \frac{2Ap(x,t) z_p(x,t) \lambda_k(x,t)}{(1+p^2(x,t))^2} f((1-\tau(x,t))k(x,t)) \Bigg] dx\, dt.
\end{aligned}
\tag{7.12}
$$

If we multiply Equation (7.9) by z_p, integrate over Q_T, and use (7.10), then we get

$$
\begin{aligned}
&-\int_0^T \int_\Omega \Bigg\{ -d_2 \nabla \lambda_p(x,t) \cdot \nabla z_p(x,t) + \lambda_p(x,t) \theta \int_\Omega f'((1-\tau(x',t))k(x',t)) \\
&\times [(1-\tau(x',t)) z_k(x',t) - k(x',t) v_2(x',t)] \varphi(x',x) dx' - \delta_2 \lambda_p(x,t) z_p(x,t) \Bigg\} dx\, dt \\
&= \int_0^T \int_\Omega \Bigg\{ d_2 \nabla \lambda_p(x,t) \cdot \nabla z_p(x,t) + \beta_0 z_p(x,t) \\
&-\lambda_k(x,t) z_p(x,t) \left[\frac{-2Ap(x,t)}{(1+p^2(x,t))^2} f((1-\tau(x,t))k(x,t)) \right] + \delta_2 \lambda_p(x,t) z_p(x,t) \Bigg\} dx\, dt.
\end{aligned}
$$

It follows that

$$
\begin{aligned}
&-\int_0^T \int_\Omega \beta_0 z_p(x,t) dx\, dt \\
&= \int_0^T \int_\Omega \lambda_k(x,t) z_p(x,t) \frac{2Ap(x,t)}{(1+p^2(x,t))^2} f((1-\tau(x,t))k(x,t)) dx\, dt \\
&+ \theta \int_0^T \int_\Omega \int_\Omega \lambda_p(x,t) f'((1-\tau(x',t))k(x',t)) \\
&\times [(1-\tau(x',t)) z_k(x',t) - k(x',t) v_2(x',t)] \varphi(x',x) dx'\, dx\, dt.
\end{aligned}
\tag{7.13}
$$

By (7.12) and (7.13), we get

$$\int_0^T \int_\Omega (cz_k(x,t) - \beta_1 \tau(x,t) z_k(x,t) - \beta_0 z_p(x,t)) dx\, dt$$

$$= \int_0^T \int_\Omega \Big[- \frac{A\lambda_k(x,t)}{1+p^2(x,t)} f'((1-\tau(x,t))k(x,t))k(x,t)v_2(x,t)$$

$$-v_1(x,t)k(x,t)\lambda_k(x,t) \Big] dx\, dt$$

$$-\theta \int_0^T \int_\Omega \int_\Omega \lambda_p(x,t) f'((1-\tau(x',t))k(x',t))k(x',t)$$

$$v_2(x',t)\varphi(x',x)dx'dx\, dt. \tag{7.14}$$

By (7.11) and (7.14), we get

$$dI(c,\tau)(v_1,v_2) = \int_0^T \int_\Omega k(x,t)[v_1(x,t)(1-\lambda_k(x,t)) - v_2(x,t)\beta^*(x,t)] dx\, dt,$$

and so we get the conclusion of the theorem.

7.4 Computational Issues

In the first paragraph, we describe the conceptual gradient-type algorithm for the optimal control procedure. In the second paragraph, we describe the numerical methods adopted for the solutions of equations (7.4) and (7.8)–(7.9) with the above-mentioned boundary and initial/final conditions. In the last paragraph, we report and discuss the computational results.

7.4.1 Conceptual Algorithm

STEP 0: Set $iter := 0$. Choose maxiter, $\delta > 0$.
STEP 1: Initialize $(c^{(iter)}, \tau^{(iter)}) \in U$.
STEP 2: Compute $k^{(iter)}, p^{(iter)}, \lambda_k^{(iter)}, \lambda_p^{(iter)}, (\beta^*)^{(iter)}$ corresponding to $(c^{(iter)}, \tau^{(iter)})$, and
$I^{(iter)} = I(c^{(iter)}, \tau^{(iter)})$.

STEP 3: Choose $(v_1, v_2) \in L^\infty(Q_T) \times L^\infty(Q_T)$ such that $(c^{(iter)}, \tau^{(iter)}) + \varepsilon_0(v_1, v_2) \in U$ ($\varepsilon_0 \geq 0$) and that $v_1(1 - \lambda_k^{(iter)}) \geq 0$, $-v_2(\beta^*)^{(iter)} \geq 0$, a.e. $(x, t) \in Q_T$.
STEP 4: Compute $(c^{(new)}, \tau^{(new)}) := (c^{(iter)}, \tau^{(iter)}) + \varepsilon_0(v_1, v_2)$.
STEP 5: Compute ξ_0 that maximizes
$I(\xi(c^{(iter)}, \tau^{(iter)}) + (1 - \xi)(c^{(new)}, \tau^{(new)}))$, for $\xi \in [0, 1]$.
Set $(c^{(iter+1)}, \tau^{(iter+1)}) = \xi_0(c^{(iter)}, \tau^{(iter)}) + (1 - \xi_0)(c^{(new)}, \tau^{(new)})$.
STEP 6: If $I^{(iter+1)} - I^{(iter)} < \delta$, then
$(c^*, \tau^*) := (c^{(iter)}, \tau^{(iter)})$
elseif $iter > maxiter$, then
$(c^*, \tau^*) := (c^{(iter)}, \tau^{(iter)})$
else
$iter := iter + 1$ and GO TO STEP 3.

In the numerical tests, we will use:

$$v_1(x, t) \in sgn(1 - \lambda_k^{(iter)}), \quad v_2(x, t) \in -sgn(\beta^*)^{(iter)}, \quad \text{a.e. } (x, t) \in Q_T.$$

7.4.2 *Numerical Methods*

Equations (7.4) and (7.8)–(7.9) are discretized by finite elements in space and finite differences in time.

Space Discretization
We first apply a standard Galerkin procedure to the weak formulations of systems (7.4) and (7.8)–(7.9). The one-dimensional set $\overline{\Omega} = [-1, 1]$ is discretized by a uniform grid of 100 linear finite elements ($\mathbf{P}^1$), yielding a total amount of $N = 101$ discretization nodes. The stiffness matrix is computed exactly, whereas the mass matrix is obtained by applying the mass lumping technique.

In the following, we describe in detail this procedure applied to system (7.4), being the treatment of equations (7.8)–(7.9) analogous.

Let us first denote by V the Sobolev space $H^1(\Omega)$, by $(\cdot, \cdot)$ the L^2- inner product, and by $a_1(\cdot, \cdot)$, $a_2(\cdot, \cdot) : V \times V \to \mathbb{R}$ the bilinear forms

$$a_1(u, v) := \int_\Omega d_1 \nabla u \cdot \nabla v dx, \quad \text{for all } u, v \in V,$$

$$a_2(u, v) := \int_\Omega d_2 \nabla u \cdot \nabla v dx, \quad \text{for all } u, v \in V.$$

Then, using Green's formula, the variational formulation of problem (7.4) reads:

Given $(c, \tau) \in U$, find k, p such that

$$\begin{cases} \dfrac{\partial}{\partial t}(k(t), \phi) + a_1(k(t), \phi) = (G_1(k(t), p(t), c(t), \tau(t)), \phi), & \text{for all } \phi \in V, \\ \dfrac{\partial}{\partial t}(p(t), \phi) + a_2(p(t), \phi) = (G_2(k(t), p(t), c(t), \tau(t)), \phi), & \text{for all } \phi \in V, \end{cases} \tag{7.15}$$

subject to the initial conditions on k and p, and where G_1 and G_2 are the reaction terms in (7.4).

Let us now introduce the $\mathbf{P}^1$ finite element space V^h, associated to a partition of the one-dimensional domain Ω into $N-1$ subintervals of size h. For a detailed description of this space, see Appendix B.

The space semi-discretization of problem (7.15) then reads:

Find $k_h, p_h : (0, T] \to V^h$ such that

$$\begin{cases} \dfrac{\partial}{\partial t}(k_h(t), \phi_h) + a_1(k_h(t), \phi_h) = (G_1(k_h(t), p_h(t), c_h(t), \tau_h(t)), \phi_h), & \text{for all } \phi_h \in V^h, \\ \dfrac{\partial}{\partial t}(p_h(t), \phi_h) + a_2(p_h(t), \phi_h) = (G_2(k_h(t), p_h(t), c_h(t), \tau_h(t)), \phi_h), & \text{for all } \phi_h \in V^h, \end{cases}$$

where $c_h, \tau_h : (0, T] \to V^h$ are convenient finite element approximations of c and τ. The previous variational equation is equivalent to

$$\begin{cases} \dfrac{\partial}{\partial t}(k_h(t), \phi_i) + a_1(k_h(t), \phi_i) = (G_1(k_h(t), p_h(t), c_h(t), \tau_h(t)), \phi_i), \\ \dfrac{\partial}{\partial t}(p_h(t), \phi_i) + a_2(p_h(t), \phi_i) = (G_2(k_h(t), p_h(t), c_h(t), \tau_h(t)), \phi_i), \end{cases} \tag{7.16}$$

for all $i = 1, \ldots, N$, where ϕ_i, $i = 1, \ldots, N$, are the *hat basis functions* of the $\mathbf{P}^1$ finite element space V^h (see Appendix B for the definition of this basis).

Expanding the unknowns $k_h(x,t)$, $p_h(x,t)$ and $c_h(x,t)$, $\tau_h(x,t)$ with respect to the basis $\{\phi_i\}_{i=1}^N$, we have

$$\begin{aligned} k_h(x,t) &= \sum_{j=1}^N K_j(t)\phi_j(x), \quad & p_h(x,t) &= \sum_{j=1}^N P_j(t), \\ c_h(x,t) &= \sum_{j=1}^N C_j(t)\phi_j(x), \quad & \tau_h(x,t) &= \sum_{j=1}^N T_j(t), \end{aligned} \tag{7.17}$$

where $K_j, P_j : [0, T] \to \mathbb{R}$ are unknown scalar functions to be determined, and we assume that $C_j, T_j : [0, T] \to \mathbb{R}$ are given.

In the next steps, we follow closely the approximation of the nonlinear problem presented in Appendix B. Hence, denoting by A_1, A_2 the stiffness matrices associated with the bilinear forms $a_1(\cdot,\cdot)$, $a_2(\cdot,\cdot)$, by M the lumped mass matrix (see Appendix B for the construction of such matrices), and substituting the expressions (7.17) into system (7.16), the semi-discrete problem then reads:

Find $\mathbf{K}, \mathbf{P} : (0, T] \rightarrow \mathbb{R}^N$ such that

$$\begin{cases} \mathrm{M}\dfrac{d\mathbf{K}}{dt}(t) + \mathrm{A}_1\mathbf{K}(t) = \mathrm{M}\mathbf{G}_1(\mathbf{K}(t), \mathbf{P}(t), \mathbf{C}(t), \mathbf{T}(t)), \\ \mathrm{M}\dfrac{d\mathbf{P}}{dt}(t) + \mathrm{A}_2\mathbf{P}(t) = \mathrm{M}\mathbf{G}_2(\mathbf{K}(t), \mathbf{P}(t), \mathbf{C}(t), \mathbf{T}(t)), \end{cases} \tag{7.18}$$

where

$$\begin{aligned} \mathbf{K}(t) &= [\ldots, K_j(t), \ldots]^T, \\ \mathbf{P}(t) &= [\ldots, P_j(t), \ldots]^T, \\ \mathbf{C}(t) &= [\ldots, C_j(t), \ldots]^T, \\ \mathbf{T}(t) &= [\ldots, T_j(t), \ldots]^T, \\ \mathbf{G}_1(\mathbf{K}(t), \mathbf{P}(t), \mathbf{C}(t), \mathbf{T}(t)) &= [\ldots, G_1(K_j(t), P_j(t), C_j(t), T_j(t)), \ldots]^T, \\ \mathbf{G}_2(\mathbf{K}(t), \mathbf{P}(t), \mathbf{C}(t), \mathbf{T}(t)) &= [\ldots, G_2(K_j(t), P_j(t), C_j(t), T_j(t)), \ldots]^T. \end{aligned}$$

Time Discretization

After space discretization by finite elements, we obtain the semi-discrete problem (7.18) that consists of a system of ODEs. We solve this system by employing a first order semi-implicit finite difference scheme (see Example B.1.4 in Appendix B for details), where the linear diffusion and reaction terms are approximated by Backward Euler, whereas the nonlinear reaction terms are approximated by Forward Euler.

In formulae, denoting by Δt the time step size, at the generic time step n, given $\mathbf{K}^n$, $\mathbf{P}^n$, we have to find $\mathbf{K}^{n+1}$, $\mathbf{P}^{n+1}$ such that

$$\begin{cases} \mathrm{M}\dfrac{\mathbf{K}^{n+1} - \mathbf{K}^n}{\Delta t} + \mathrm{A}_1\mathbf{K}^{n+1} = \mathrm{M}\mathbf{G}_1(\mathbf{K}^n, \mathbf{P}^n, \mathbf{C}^n, \mathbf{T}^n), \\ \mathrm{M}\dfrac{\mathbf{P}^{n+1} - \mathbf{P}^n}{\Delta t} + \mathrm{A}_2\mathbf{P}^{n+1} = \mathrm{M}\mathbf{G}_2(\mathbf{K}^n, \mathbf{P}^n, \mathbf{C}^n, \mathbf{T}^n). \end{cases}$$

As a result, at each time step, it is required the solution of the linear system of algebraic equations

$$\begin{bmatrix} \dfrac{1}{\Delta t}\mathrm{M} + \mathrm{A}_1 & 0 \\ 0 & \dfrac{1}{\Delta t}\mathrm{M} + \mathrm{A}_2 \end{bmatrix} \begin{bmatrix} \mathbf{K}^{n+1} \\ \mathbf{P}^{n+1} \end{bmatrix} = \begin{bmatrix} \dfrac{1}{\Delta t}\mathrm{M}\mathbf{K}^n + \mathrm{M}\mathbf{G}_1(\mathbf{K}^n, \mathbf{P}^n, \mathbf{C}^n, \mathbf{T}^n) \\ \dfrac{1}{\Delta t}\mathrm{M}\mathbf{P}^n + \mathrm{M}\mathbf{G}_2(\mathbf{K}^n, \mathbf{P}^n, \mathbf{C}^n, \mathbf{T}^n) \end{bmatrix},$$

which in our specific case has the dimension of 202 degrees of freedom. The linear system is solved by the Gaussian elimination with the built-in function of MATLAB. The time step size is $\Delta t = 0.05$.

For further details on the numerical approximation of parabolic equations, we refer, e.g., to [168].

7.4.3 *Numerical Simulations*

We report in this section the results obtained with the MATLAB code

```
run_opt_tax.m,
```

available on the Birkhäuser web site provided in the Preface and also at the link `http://www.mat.unimi.it/users/scacchi/ACS_codes/run_opt_tax.m`

Let us use the above algorithm with the following parameters' values $d_1 = 1$, $d_2 = 1, \delta_1 = 0.05, \delta_2 = 0.03, A = 1, s = 0.6, \theta = 2, \alpha_1 = 0.7, \alpha_2 = 1, \gamma = 4, \varphi = 0.3$, and initial conditions $k_0(x) = 0.1 \exp\left\{\dfrac{-(x-0.5)^2}{0.1}\right\}$ and $p_0(x) = \exp\{x\}$. We have taken $\beta_0 = \beta_1 = 1$.

Figures 7.1 and 7.2 show the evolution over time of the state variables k and p and of the control variables c and τ, respectively. One can provide an intuitive interpretation of the results expressed in the figures for values of the locations x near to 1: The presence of a high value of pollution at $t = 0$ forces to invest into abatement activities and less in consumptions. This also produces a reduction of the level of physical capital. Instead, when the level of pollution decreases over time and reaches a sustainable level, the model suggests to change the optimal policy and to invest in consumptions (to sustain the economy in the long run) rather than in pollution abatement (τ becomes 0).

7.5 Open Problems

In the above, we have reported about an optimal control problem concerning both the consumption of the physical capital and the level of taxation, by implementing the control actions in the whole domain of interest. We leave it open, as a challenging problem for the interested reader, the case of implementation of the control actions only on an optimally chosen subregion of the whole domain of interest. Of particular interest would be the action of a regional taxation policy.

As an example to be analyzed, one may consider the following regional control by acting on the taxation:

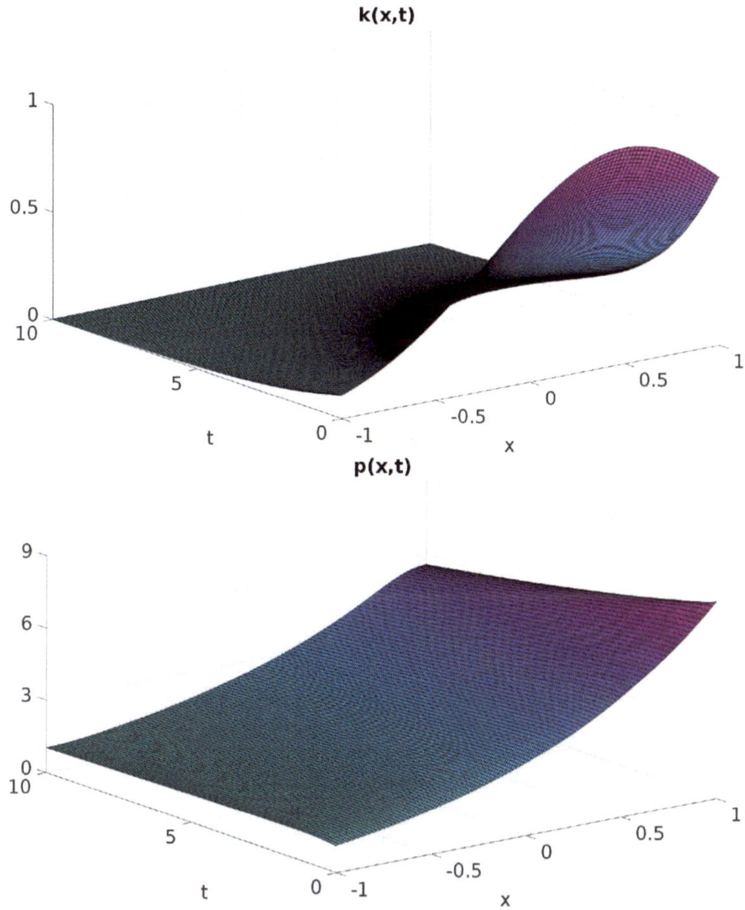

Fig. 7.1 We solve the optimal control problem described in Section 7.3 with the algorithms and discretization parameters reported in Section 7.4. Spatio-temporal evolution of the state variables $k(x,t)$, $p(x,t)$, at the optimal configuration

$$\begin{cases} \dfrac{\partial k}{\partial t}(x,t) = d_1 \Delta k(x,t) + \dfrac{A(x,t)}{1+p(x,t)^2} f((1-\mathbb{I}_\omega(x)\tau(x,t))k(x,t)), \\ \qquad\qquad -\delta_1 k(x,t) - c(x,t)k(x,t), \\ \dfrac{\partial p}{\partial t}(x,t) = d_2 \Delta p(x,t) + \theta \displaystyle\int_\Omega f((1-\mathbb{I}_\omega(x')\tau(x',t))k(x',t))\varphi(x',x)dx' \\ \qquad\qquad -\delta_2 p(x,t), \end{cases} \tag{7.19}$$

for $(x,t) \in Q_T$, where ω is a nontrivial open subset of Ω.

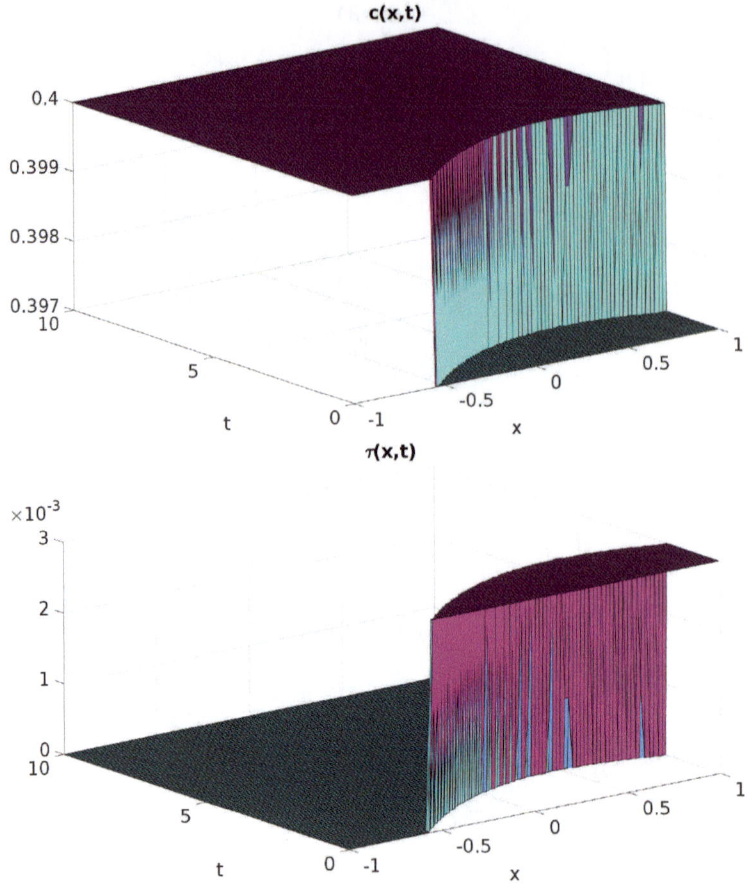

Fig. 7.2 We solve the optimal control problem described in Section 7.3 with the algorithms and discretization parameters reported in Section 7.4. Spatio-temporal evolution of the controls $c(x, t)$, $\tau(x, t)$, at the optimal configuration

In this case, the optimal control problem would be

$$\underset{\substack{(c,\tau)\in U\\ \omega\subset\Omega}}{Maximize}\left\{\int_0^T\int_\Omega c(x,t)k(x,t)dx\,dt-\beta_0\int_0^T\int_\Omega p(x,t)dx\,dt\right.$$

$$\left.-\beta_1\int_0^T\int_\omega \tau(x,t)k(x,t)dx\,dt\right\},$$

subject to System (7.19).

Another possible open problem would consider the costs due to a regional pollution treatment, as follows:

$$\begin{cases} \dfrac{\partial k}{\partial t}(x,t) = d_1 \Delta k(x,t) + \dfrac{A(x,t)}{1+p(x,t)^2} f((1-\mathbb{I}_\omega(x)\tau(x,t))k(x,t)) \\ \qquad -\delta_1 k(x,t) - c(x,t)k(x,t), \\ \dfrac{\partial p}{\partial t}(x,t) = d_2 \Delta p(x,t) + \theta \displaystyle\int_\Omega f((1-\mathbb{I}_\omega(x')\tau(x',t))k(x',t))\varphi(x',x)dx' \\ \qquad -\delta_2 p(x,t) - \mathbb{I}_\omega(x)\xi(x,t)p(x,t), \end{cases} \tag{7.20}$$

for $(x,t) \in Q_T$, where ω is a nontrivial open subset of Ω.

About the control function, one may assume that $\xi \in L^\infty(\omega \times (0,T))$, such that $0 \le \xi(x,t) \le L$, a.e. $(x,t) \in \omega \times (0,T)$, for a given constant L.

In this case, the optimal control problem would be

$$\underset{\substack{(c,\tau,\xi)\in\widetilde{U},\\ \omega\subset\Omega}}{Maximize} \left\{ \int_0^T \int_\Omega c(x,t)k(x,t)dx\,dt - \beta_0 \int_0^T \int_\Omega p(x,t)dx\,dt \right.$$
$$- \beta_1 \int_0^T \int_\omega \tau(x,t)k(x,t)dx\,dt$$
$$\left. -\beta_2 \int_0^T \int_\omega \xi(x,t)p(x,t)dx\,dt \right\},$$

subject to System (7.20). Here $\widetilde{U} := U \times \{\xi \in L^\infty(\omega \times (0,T)) : 0 \le \xi(x,t) \le L,\ \text{a.e. } (x,t) \in \omega \times (0,T)\}$ and β_2 is a positive constant.

It is clear that the solutions of the above problems would depend upon the geographical distribution of the potentially polluting production activities; in this respect, we may expect a crucial role played by the functional parameter $A(x,t)$. This fact would confirm the relevance of spatially structured models for geographical economics.

Appendix A
Special Topics for Integro-Differential Systems

In this appendix main results and methods concerning principal eigenvalues and comparison results for the solutions to integro-differential systems have been presented.

They have been used throughout the present book either in their exact form or slightly adapted to the context.

A.1 Principal Eigenvalue for Self-Adjoint Elliptic Operators. Rayleigh's Formula

Let $\Omega \subset \mathbb{R}^N$ ($N \in \mathbb{N}^*$) be a bounded domain with a C^2-boundary $\partial\Omega$. We denote by $\|\cdot\|_p$ the norm of $L^p(\Omega)$.

Consider the following eigenvalue problem

$$\begin{cases} -d\Delta\varphi(x) + a(x)\varphi(x) = \lambda\varphi(x), \ x \in \Omega, \\ \dfrac{\partial\varphi}{\partial n}(x) = 0, \qquad\qquad\qquad\qquad x \in \partial\Omega, \end{cases} \tag{A.1}$$

where d is a positive constant, $a \in L^\infty(\Omega)$.

Since the linear operator $\mathcal{T} : L^2(\Omega) \longrightarrow L^2(\Omega)$, defined by $\mathcal{T}f = \varphi$, where $\varphi \in H^1(\Omega)$ is the unique weak solution to

$$\begin{cases} -d\Delta\varphi(x) + (a(x) + \|a\|_\infty + 1)\varphi(x) = f(x), \ x \in \Omega, \\ \dfrac{\partial\varphi}{\partial n}(x) = 0, \qquad\qquad\qquad\qquad\qquad\qquad\quad x \in \partial\Omega, \end{cases}$$

S. Aniţa et al., *Mathematical Modeling and Control in Life and Environmental Sciences*, Modeling and Simulation in Science, Engineering and Technology,
https://doi.org/10.1007/978-3-031-49971-5

is self-adjoint and compact, then we know (see, e.g., [55, 103]) that the eigenvalues for (A.1) are such that

$$\lambda_1 < \lambda_2 \le \lambda_3 \le \cdots ;$$

λ_1 is a simple eigenvalue (i.e., the space of its corresponding eigenvectors has the dimension one). λ_1 is called the *principal eigenvalue* of (A.1).

The following Rayleigh's formula holds

$$\lambda_1 = \min_{\varphi \in H^1(\Omega)\setminus\{0\}} \frac{d\int_\Omega |\nabla\varphi(x)|^2 dx + \int_\Omega a(x)|\varphi(x)|^2 dx}{\int_\Omega |\varphi(x)|^2 dx}$$

(this quotient is called Rayleigh's quotient)

$$= \min_{\varphi \in H^1(\Omega), \|\varphi\|_2 = 1} \left\{ d\int_\Omega |\nabla\varphi(x)|^2 dx + \int_\Omega a(x)|\varphi(x)|^2 dx \right\}. \tag{A.2}$$

The minimum in (A.2) is attained for and only for the eigenvectors corresponding to (A.1) and λ_1. By Rayleigh's formula we get that there exists a nonnegative eigenvector φ_1 corresponding to (A.1) and λ_1.

It is possible to prove that, for a sufficiently smooth boundary $\partial\Omega$ (in our present case it is sufficient to take it in C^3), φ_1 is in $C^1(\overline{\Omega})$ and strictly positive in $\overline{\Omega}$. A more detailed discussion on this argument will be presented in the next section.

In what follows we shall deduce three consequences of Rayleigh's formula that are used throughout the present book.

Corollary A.1 *Consider the following eigenvalue problems*

$$\begin{cases} -d\Delta\varphi(x) + a_k(x)\varphi(x) = \lambda\varphi(x), & x \in \Omega, \\ \dfrac{\partial\varphi}{\partial n}(x) = 0, & x \in \partial\Omega, \end{cases} \tag{A.3}$$

where $a_k \in L^\infty(\Omega)$, $k \in \{1, 2\}$, $a_1(x) \le a_2(x)$, *a.e.* $x \in \Omega$.

If we denote by λ_1^k *the principal eigenvalue of (A.3), then*

$$\lambda_1^1 \le \lambda_1^2.$$

This follows immediately by (A.2).

Corollary A.2 *If* $\{a_k\}_{k\in\mathbb{N}^*} \subset L^\infty(\Omega)$ *satisfy*

$$a_k(x) \longrightarrow a^*(x), \quad |a_k(x)| \le b(x), \quad \text{a.e. } x \in \Omega,$$

where $b \in L^\infty(\Omega)$, then

$$\lambda_1^k \longrightarrow \lambda_1^*,$$

where λ_1^k is the principal eigenvalue of (A.1) corresponding to $a := a_k$ and λ_1^* is the principal eigenvalue of (A.1) corresponding to $a := a^*$.

Proof Indeed, let $\varphi_1^k \in H^1(\Omega)$ such that $\|\varphi_1^k\|_2 = 1$ and

$$\lambda_1^k = d \int_\Omega |\nabla \varphi_1^k(x)|^2 dx + \int_\Omega a_k(x) |\varphi_1^k(x)|^2 dx, \tag{A.4}$$

and $\varphi_1^* \in H^1(\Omega)$ such that $\|\varphi_1^*\|_2 = 1$ and

$$\lambda_1^* = d \int_\Omega |\nabla \varphi_1^*(x)|^2 dx + \int_\Omega a^*(x) |\varphi_1^*(x)|^2 dx. \tag{A.5}$$

By (A.4) and (A.2) we have that

$$\lambda_1^k \le d \int_\Omega |\nabla \varphi_1^*(x)|^2 dx + \int_\Omega a_k(x) |\varphi_1^*(x)|^2 dx \tag{A.6}$$

and so

$$|\lambda_1^k| \le d \int_\Omega |\nabla \varphi_1^*|^2 dx + \int_\Omega b |\varphi_1^*|^2 dx,$$

and

$$d \int_\Omega |\nabla \varphi_1^k|^2 dx \le d \int_\Omega |\nabla \varphi_1^*|^2 dx + \int_\Omega b |\varphi_1^*|^2 dx + \|b\|_\infty.$$

We conclude that $\{\varphi_1^k\}_{k \in \mathbb{N}^*}$ is a bounded sequence in $H^1(\Omega)$ and so, there exists a subsequence such that

$$\varphi_1^{k_j} \longrightarrow \varphi_1, \quad \text{weakly in } H^1(\Omega),$$

$$\varphi_1^{k_j} \longrightarrow \varphi_1, \quad \text{in } L^2(\Omega) \text{ and a.e. } x \in \Omega.$$

We get that $\|\varphi_1\|_2 = 1$, and by (A.4) we deduce that

$$d\int_\Omega |\nabla\varphi_1|^2dx + \int_\Omega a^*|\varphi_1|^2dx \le \liminf_{j\to+\infty} \lambda_1^{k_j} \le \limsup_{j\to+\infty} \lambda_1^{k_j}$$

$$\le d\int_\Omega |\nabla\varphi_1^*|^2dx + \int_\Omega a^*|\varphi_1^*|^2dx = \lambda_1^*$$

(by (A.6) and (A.5)). That means $\varphi_1 = c\varphi_1^*$ ($c \in \mathbb{R}$) and since $\|\varphi_1\|_2 = \|\varphi_1^*\|_2 = 1$, we get that $\varphi_1 = \varphi_1^*$ or $\varphi_1 = -\varphi_1^*$. We conclude that $\lim_{j\to+\infty} \lambda_1^{k_j} = \lambda_1^*$. Arguing by contradiction we get that actually $\lim_{k\to+\infty} \lambda_1^k = \lambda_1^*$. □

Corollary A.3 *Let $\lambda_{1\gamma}(\omega)$ be the principal eigenvalue for*

$$\begin{cases} -d\Delta\varphi(x) + a(x)\varphi(x) + \gamma\mathbb{I}_\omega(x)\varphi(x) = \lambda\varphi(x), & x \in \Omega, \\ \dfrac{\partial\varphi}{\partial n}(x) = 0, & x \in \partial\Omega, \end{cases} \tag{A.7}$$

where γ is a nonnegative constant and ω is an open subset of Ω with a C^2-boundary $\partial\omega$, such that $\overline{\omega} \subset \Omega$ and $\Omega \setminus \overline{\omega}$ is a domain.

By Rayleigh's formula we get that

$$\lambda_{1\gamma}(\omega) = \min_{\varphi\in H^1(\Omega), \|\varphi\|_2=1} \left\{ d\int_\Omega |\nabla\varphi|^2dx + \int_\Omega a|\varphi|^2dx + \gamma\int_\omega |\varphi|^2dx \right\}.$$

It is obvious that $\gamma \mapsto \lambda_{1\gamma}(\omega)$ is an increasing function and so there exists $\lim_{\gamma\to+\infty} \lambda_{1\gamma}(\omega) = \tilde{\lambda}$, and $\omega \mapsto \lambda_{1\gamma}(\omega)$ is increasing as well.

Let us prove that, for a fixed ω,

$$\lim_{\gamma\to+\infty} \lambda_{1\gamma}(\omega) = \lambda_1(\omega),$$

where $\lambda_1(\omega)$ is the principal eigenvalue to

$$\begin{cases} -d\Delta\varphi(x) + a(x)\varphi(x) = \lambda\varphi(x), & x \in \Omega \setminus \overline{\omega}, \\ \varphi(x) = 0, & x \in \partial\omega, \\ \dfrac{\partial\varphi}{\partial n}(x) = 0, & x \in \partial\Omega. \end{cases} \tag{A.8}$$

Recall that the eigenvalues of (A.8) are

$$\lambda_1(\omega) < \lambda_2(\omega) \le \lambda_2(\omega) \le \cdots,$$

$\lambda_1(\omega)$ being simple (see [55, 103]) and that

$$\lambda_1 = \min_{V\setminus\{0\}} \frac{d\int_{\Omega\setminus\overline{\omega}} |\nabla\varphi(x)|^2 dx + \int_{\Omega\setminus\overline{\omega}} a(x)|\varphi(x)|^2 dx}{\int_{\Omega\setminus\overline{\omega}} |\varphi(x)|^2 dx}$$

$$= \min_{\varphi\in V, \|\varphi\|_{L^2(\Omega\setminus\overline{\omega})}=1} \left\{ d\int_{\Omega\setminus\overline{\omega}} |\nabla\varphi(x)|^2 dx + \int_{\Omega\setminus\overline{\omega}} a(x)|\varphi(x)|^2 dx \right\},$$

where $V = \{v \in H^1(\Omega \setminus \overline{\omega}) : v(x) = 0, \text{ a.e. } x \in \partial\omega\}$.

Proof Let φ_1 be an eigenvector corresponding to (A.8) and $\lambda_1(\omega)$, such that $\|\varphi_1\|_{L^2(\Omega\setminus\overline{\omega})} = 1$; let $\varphi_{1\gamma}$ be an eigenvector corresponding to (A.7) and $\lambda_{1\gamma}(\omega)$, such that $\|\varphi_{1\gamma}\|_2 = 1$.

This yields

$$\begin{aligned} \lambda_{1\gamma}(\omega) &= d\int_\Omega |\nabla\varphi_{1\gamma}|^2 dx + \int_\Omega a|\varphi_{1\gamma}|^2 dx + \gamma\int_\omega |\varphi_{1\gamma}|^2 dx \\ &\le d\int_\Omega |\nabla\bar{\varphi}_1|^2 dx + \int_\Omega a|\bar{\varphi}_1|^2 dx + \gamma\int_\omega |\bar{\varphi}_1|^2 dx = \lambda_1(\omega), \end{aligned} \tag{A.9}$$

where $\bar{\varphi}_1$ is the extension by 0 of φ_1 on $\overline{\omega}$. Hence,

$$\lim_{\gamma\to+\infty} \lambda_{1\gamma}(\omega) = \tilde{\lambda} \le \lambda_1(\omega).$$

On the other hand we get that $\{\gamma\int_\omega |\varphi_{1\gamma}|^2 dx\}_{\gamma>0}$ is bounded and so

$$\varphi_{1\gamma} \longrightarrow 0 \quad \text{in } L^2(\omega), \text{ as } \gamma \to +\infty$$

and $\{\varphi_{1\gamma}\}_{\gamma>0}$ is bounded in $H^1(\Omega)$. We may infer that on a subsequence (also denoted by $\{\varphi_{1\gamma}\}$) we get that

$$\varphi_{1\gamma} \longrightarrow \tilde{\varphi}, \quad \text{weakly in } H^1(\Omega), \text{ and in } H^1(\Omega\setminus\overline{\omega}),$$

$$\varphi_{1\gamma} \longrightarrow \tilde{\varphi}, \quad \text{in } L^2(\Omega),$$

as $\gamma \to +\infty$, and so, $\tilde{\varphi}(x) = 0$, a.e. $x \in \omega$, $\tilde{\varphi} \in V$ and $\|\tilde{\varphi}\|_{L^2(\Omega\setminus\overline{\omega})} = 1$.

By (A.9) we may infer that

$$\tilde{\lambda} \ge d\int_{\Omega\setminus\overline{\omega}} |\nabla\tilde{\varphi}|^2 dx + \int_{\Omega\setminus\overline{\omega}} a|\tilde{\varphi}|^2 dx \ge \lambda_1(\omega)$$

(by Rayleigh's formula). So $\tilde{\lambda} = \lambda_1(\omega)$, and the conclusion follows ($\lim_{\gamma\to+\infty} \lambda_{1\gamma}(\omega) = \tilde{\lambda} = \lambda_1(\omega)$). □

A.2 Principal Eigenvalues for Some Non Self-Adjoint Operators

Let X be a real Banach space and $\mathcal{C}$ be a *cone* in X, i.e., $\mathcal{C}$ is a closed and convex subset of X and satisfies $\lambda\mathcal{C} \subset \mathcal{C}$, for any $\lambda \geq 0$ and $\mathcal{C} \cap (-\mathcal{C}) = \{0\}$. The cone provides an ordering "$\leq$" in X:

$$x \leq y \text{ iff } y - x \in \mathcal{C}$$

and a strict ordering "$<$" in X:

$$x < y \text{ iff } y - x \in \mathcal{C} \setminus \{0\}.$$

We assume that $\mathcal{C}$ is a *total cone* ($\overline{\mathcal{C} - \mathcal{C}} = X$).

We say that the linear operator $\mathcal{T} \in L(X)$ is *strongly positive* if for any $0 < x$ (positive vector) we have that $\mathcal{T}x \in int(\mathcal{C})$.

Let us recall a version of the Krein–Rutman theorem (see Theorem 1.2 in [81, 95, 102]):

Theorem A.4 (Krein–Rutman) *If $int(\mathcal{C}) \neq \emptyset$ and if $\mathcal{T} \in L(X)$ is compact and strongly positive, then*

(1) The spectral radius $r(\mathcal{T}) > 0$ is a geometrically simple eigenvalue of $\mathcal{T}$ pertaining to an eigenvector $x_0 \in int(\mathcal{C})$.
(2) For any eigenvalue $\lambda \neq r(\mathcal{T})$, we have $|\lambda| < r(\mathcal{T})$.
(3) If a real eigenvalue of $\mathcal{T}$ has a positive eigenvector, then the eigenvalue is $r(\mathcal{T})$.
(4) $r(\mathcal{T})$ is an eigenvalue of $\mathcal{T}^*$ with an eigenvector x_0^* which is a strictly positive functional (i.e., $\langle x_0^*, x\rangle_{X^*,X} > 0$, for any $x > 0$). Moreover, any eigenvector of $\mathcal{T}^*$ in $\mathcal{C}^* \setminus \{0\}$ corresponds to $r(\mathcal{T})$ (here $\mathcal{C}^* = \{x^* \in X^* \ :\ \langle x^*, x\rangle_{X^*,X} \geq 0,\ for\ any\ x \in \mathcal{C}\}$).

The eigenvalue $r(\mathcal{T})$ is called the *principal eigenvalue* of $\mathcal{T}$.

Let us discuss now some applications of the Krein–Rutman theorem.

Let $\Omega \subset \mathbb{R}^N$ ($N \in \mathbb{N}^*$) be a bounded domain with a C^2-boundary $\partial\Omega$ and ω be an open subset of Ω. Consider the following eigenvalue problem

$$\begin{cases} -d\Delta\varphi(x) + a(x)\varphi(x) - \theta\displaystyle\int_{\Omega} k(x,x')\varphi(x')dx' \\ \qquad\qquad + \mathbb{I}_{\omega}(x)\gamma\varphi(x) = \lambda\varphi(x), & x \in \Omega, \\ \dfrac{\partial\varphi}{\partial n}(x) + \alpha\varphi(x) = 0, & x \in \partial\Omega, \end{cases} \tag{A.10}$$

where $a \in L^\infty(\Omega)$, θ, α are nonnegative constants, $k \in L^\infty(\Omega \times \Omega)$, $k(x,x') \geq 0$, a.e. $(x,x') \in \Omega \times \Omega$, d, γ are positive constants.

There is no diminishment of the generality if we assume that $a(x) \geq a_0$, a.e. $x \in \Omega$ and a_0 is a sufficiently large positive constant.

For any $\zeta \in L^2(\Omega)$, the following problem

$$\begin{cases} -d\Delta\varphi(x) + a(x)\varphi(x) - \theta \int_\Omega k(x, x')\varphi(x')dx' \\ \qquad\qquad + \mathbb{I}_\omega(x)\gamma\varphi(x) = \zeta(x), & x \in \Omega, \\ \frac{\partial\varphi}{\partial n}(x) + \alpha\varphi(x) = 0, & x \in \partial\Omega \end{cases} \tag{A.11}$$

admits a unique weak solution $\varphi = \mathcal{T}\zeta \in H^1(\Omega)$ (this follows via Lax–Milgram's lemma for a_0 sufficiently large).

Let the Banach space $X = C^1(\overline{\Omega})$ and the total cone

$$\mathcal{C} = \{u \in X \, : \, u(x) \geq 0, \; x \in \overline{\Omega}\}.$$

It is obvious that

$$int(\mathcal{C}) = \{u \in X \, : \, u(x) > 0, \; x \in \overline{\Omega}\}$$

(see also [81]). Since $a, k, \mathbb{I}_\omega$ are essentially bounded we get that $\int_\Omega k(\cdot, x')\varphi(x')dx' \in L^\infty(\Omega)$ and via Theorem 2.4.2.7 in [119] that for any $p > N$ and $\zeta \in C^1(\overline{\Omega})$, $\mathcal{T}\zeta \in W^{2,p}(\Omega)$. On the other hand by Theorem 2.7 in [180] we get that for any $p > N$ there exists a positive constant c_p such that $\|\mathcal{T}\zeta\|_{W^{2,p}(\Omega)} \leq c_p\|\zeta\|_{C^1(\overline{\Omega})}$, for any $\zeta \in C^1(\overline{\Omega})$.

Since $W^{2,p}(\Omega)$ is compactly embedded in $C^1(\overline{\Omega})$, for any $p > N$, we may conclude that $\mathcal{T} \in L(X)$ is compact.

On the other hand, it follows in a standard manner (by multiplying the equation (A.11) by φ^- and integrating over Ω) that, for any $\zeta \in \mathcal{C} \setminus \{0\}$, $\varphi(x) \geq 0$, a.e. $x \in \Omega$ (for a_0 sufficiently large). Moreover, in the same way we get that

$$\tilde{\varphi}(x) \leq \varphi(x), \quad \text{for any } x \in \overline{\Omega},$$

where $\tilde{\varphi}$ is the solution to

$$\begin{cases} -d\Delta\varphi(x) + (\|a\|_{L^\infty(\Omega)} + \gamma)\varphi(x) = \zeta(x), \; x \in \Omega, \\ \frac{\partial\varphi}{\partial n}(x) + \alpha\varphi(x) = 0, & x \in \partial\Omega. \end{cases}$$

A result in [3] amounts to saying that, if $\partial\Omega$ is C^3, then there exists a positive constant θ_0 such that

$$\theta_0 \leq \tilde{\varphi}(x), \quad \text{for any } x \in \overline{\Omega},$$

and consequently $\varphi \in int(\mathcal{C})$. That means $\mathcal{T} : X \to X$ is strongly positive.

We get now all the conclusions in the above theorem. Actually λ is an eigenvalue to (A.10) if and only if $\frac{1}{\lambda}$ is an eigenvalue for $\mathcal{T}$ and the corresponding eigenvectors are the same. The conclusions of the Krein–Rutman's theorem can be transferred to the eigenvalue problem (A.10).

The eigenvalue $\lambda_{1\gamma} = \frac{1}{r(\mathcal{T})}$ is called the *principal eigenvalue* for (A.10). There exists a corresponding eigenvector (eigenfunction) $\varphi_{1\gamma} \in int(\mathcal{C})$, i.e., $\varphi_{1\gamma} \in C^1(\overline{\Omega})$, $\varphi_{1\gamma}(x) > 0$, for any $x \in \overline{\Omega}$.

Assume that for the rest of this section Ω has a C^3-boundary $\partial\Omega$.

Lemma A.5 *If* $0 < \gamma_1 < \gamma_2$, *then* $\lambda_{1\gamma_1} < \lambda_{1\gamma_2}$.

Proof Let γ be a positive real constant; φ_1 be an eigenvector corresponding to the principal eigenvalue $\lambda_{1\gamma}$ to (A.10) such that $\varphi_1 > 0$ for $x \in \overline{\Omega}$. Let φ_1^* be a positive eigenvector corresponding to the principal eigenvalue $\tilde{\lambda}_{1\gamma}$ of the adjoint problem (which has the same form as (A.10))

$$\begin{cases} -d\Delta\varphi^*(x) + a(x)\varphi^*(x) - \theta \displaystyle\int_{\Omega} k(x', x)\varphi^*(x')dx' \\ \qquad\qquad + \mathbb{I}_\omega(x)\gamma\varphi^*(x) = \lambda\varphi^*(x), & x \in \Omega, \\ \dfrac{\partial\varphi^*}{\partial n}(x) + \alpha\varphi^*(x) = 0, & x \in \partial\Omega. \end{cases} \tag{A.12}$$

By (A.10) and taking into account (A.12) we get after an easy calculation that

$$\int_{\Omega} \varphi_1(x)\Big[- d\Delta\varphi_1^*(x) + a(x)\varphi_1^*(x) - \theta \int_{\Omega} k(x', x)\varphi_1^*(x')dx' + \mathbb{I}_\omega(x)\gamma\varphi_1^*(x)\Big]dx$$

$$= \lambda_{1\gamma} \int_{\Omega} \varphi_1(x)\varphi_1^*(x)dx.$$

By (A.12) we obtain

$$\tilde{\lambda}_{1\gamma} \int_{\Omega} \varphi_1\varphi_1^* dx = \lambda_{1\gamma} \int_{\Omega} \varphi_1\varphi_1^* dx$$

and so $\tilde{\lambda}_{1\gamma} = \lambda_{1\gamma}$.

That means $\lambda_{1\gamma}$ is the principal eigenvalue for the adjoint problem (A.12) as well.

Let $\varphi_{1\gamma_1}$ be a positive eigenvector corresponding to $\gamma := \gamma_1$ and to the principal eigenvalue $\lambda_{1\gamma_1}$ to (A.10). Let $\varphi_{1\gamma_2}^*$ be a positive eigenvector corresponding to $\gamma := \gamma_2$ and to the principal eigenvalue $\lambda_{1\gamma_2}$ of the adjoint problem (A.12).

If we multiply (A.12) (with $\gamma := \gamma_2$, $\lambda := \lambda_{1\gamma_2}$, $\varphi^* := \varphi^*_{1\gamma_2}$) by $\varphi_{1\gamma_1}$, integrate over Ω and take into account (A.10), we get that

$$\lambda_{1\gamma_1} \int_\Omega \varphi_{1\gamma_1}\varphi^*_{1\gamma_2} dx + (\gamma_2 - \gamma_1) \int_\omega \varphi_{1\gamma_1}\varphi^*_{1\gamma_2} dx = \lambda_{1\gamma_2} \int_\Omega \varphi_{1\gamma_1}\varphi^*_{1\gamma_2} dx$$

and so

$$\lambda_{1\gamma_2} - \lambda_{1\gamma_1} = (\gamma_2 - \gamma_1)\frac{\int_\omega \varphi_{1\gamma_1}\varphi^*_{1\gamma_2} dx}{\int_\Omega \varphi_{1\gamma_1}\varphi^*_{1\gamma_2} dx} > 0$$

(because $\int_\omega \varphi_{1\gamma_1}\varphi^*_{1\gamma_2} dx > 0$). The conclusion follows. □

Remark A.6 By the last part of the proof we also get that $\gamma \mapsto \lambda_{1\gamma}$ is a Lipschitz continuous function (with Lipschitz constant 1).

In the same manner as in the proof of Lemma A.5 it follows that the principal eigenvalue of (A.10) depends increasingly on a and ω and decreasingly on k.

Let ω be a nonempty open subset of Ω, with a C^2−boundary $\partial\omega$, such that $\overline{\omega} \subset \Omega$, and $\Omega \setminus \overline{\omega}$ is a domain. Consider the following eigenvalue problem

$$\begin{cases} -d\Delta\varphi(x) + a(x)\varphi(x) - \theta \int_{\Omega\setminus\overline{\omega}} k(x,x')\varphi(x')dx' = \lambda\varphi(x), & x \in \Omega \setminus \overline{\omega}, \\ \varphi(x) = 0, & x \in \partial\omega, \\ \dfrac{\partial\varphi}{\partial n}(x) + \alpha\varphi(x) = 0, & x \in \partial\Omega. \end{cases} \tag{A.13}$$

By Krein–Rutman's theorem we get similar results for (A.13) as those for problem (A.10). In this case we take the Banach space $X = \{u \in C^1(\overline{\Omega} \setminus \omega) \ : \ u(x) = 0,\ x \in \partial\omega\}$ and the total cone

$$\mathcal{C} = \{u \in X \ : \ u(x) \geq 0 \text{ in } \overline{\Omega}\}.$$

It is obvious that

$$int(\mathcal{C}) = \{u \in X \ : \ u(x) > 0 \text{ in } \overline{\Omega} \setminus \overline{\omega},\ \frac{\partial u}{\partial n}(x) < 0,\ x \in \partial\omega\},$$

where $n(x)$ is the outward (with respect to $\Omega \setminus \overline{\omega}$) normal versor at $x \in \omega$ (see also [81]). Moreover, since $a, k, \mathbb{I}_\omega$ are essentially bounded we may infer as before that $\tilde{\mathcal{T}} \in L(X)$ is compact and strongly positive. Here $\tilde{\mathcal{T}}\zeta = \varphi$ is the weak solution to

$$\begin{cases} -d\Delta\varphi(x) + a(x)\varphi(x) - \theta \displaystyle\int_{\Omega\setminus\overline{\omega}} k(x,x')\varphi(x')dx' = \zeta(x), & x \in \Omega \setminus \overline{\omega}, \\ \varphi(x) = 0, & x \in \partial\omega, \\ \dfrac{\partial\varphi}{\partial n}(x) + \alpha\varphi(x) = 0, & x \in \partial\Omega. \end{cases}$$

Let $\lambda_1(\omega)$ be the principal eigenvalue to (A.13) and to its dual problem

$$\begin{cases} -d\Delta\varphi(x) + a(x)\varphi(x) - \theta \displaystyle\int_{\Omega\setminus\overline{\omega}} k(x',x)\varphi(x')dx' = \lambda\varphi(x), & x \in \Omega \setminus \overline{\omega}, \\ \varphi(x) = 0, & x \in \partial\omega, \\ \dfrac{\partial\varphi}{\partial n}(x) + \alpha\varphi(x) = 0, & x \in \partial\Omega \end{cases} \tag{A.14}$$

and φ_1^* be an eigenvector corresponding to the eigenvalue $\lambda_1(\omega)$ and to (A.14), which satisfies $\varphi_1^* > 0$, for $x \in \Omega \setminus \overline{\omega}$.

Theorem A.7 *The following convergence holds*

$$\lambda_{1\gamma} \longrightarrow \lambda_1(\omega), \quad \textit{as } \gamma \to +\infty.$$

Proof Let us prove firstly that for any $\gamma > 0$ we have that $\lambda_{1\gamma} < \lambda_1(\omega)$.

Let $\varphi_{1\gamma}$ be an eigenvector corresponding to $\gamma, \lambda := \lambda_{1\gamma}$ and (A.10) and such that $\varphi_{1\gamma} > 0$ for $x \in \Omega$, $\|\varphi_{1\gamma}\|_{L^2(\Omega)} = 1$.

If we multiply (A.10) (corresponding to $\gamma, \lambda := \lambda_{1\gamma}$ and $\varphi_{1\gamma}$) by φ_1^* and integrate over $\Omega \setminus \overline{\omega}$ we get after an easy calculation and taking into account (A.14) that

$$\int_{\Omega\setminus\overline{\omega}} \varphi_{1\gamma}(x)\Big[- d\Delta\varphi_1^*(x) + a(x)\varphi_1^*(x) - \theta \int_{\Omega\setminus\overline{\omega}} k(x',x)\varphi_1^*(x')dx'\Big]dx$$

$$+ d \int_{\partial\omega} \varphi_{1\gamma}(x)\frac{\partial\varphi_1^*}{\partial n}(x)d\sigma - \theta \int_{\Omega\setminus\overline{\omega}} \varphi_1^*(x) \int_{\omega} k(x,x')\varphi_{1\gamma}(x')dx'\,dx$$

$$= \lambda_{1\gamma} \int_{\Omega\setminus\overline{\omega}} \varphi_{1\gamma}(x)\varphi_1^*(x)dx.$$

It follows that

$$\lambda_1(\omega) \int_{\Omega\setminus\overline{\omega}} \varphi_{1\gamma}\varphi_1^* dx + d \int_{\partial\omega} \varphi_{1\gamma}\frac{\partial\varphi_1^*}{\partial n}d\sigma - \theta \int_{\Omega\setminus\overline{\omega}} \varphi_1^*(x) \int_{\omega} k(x,x')\varphi_{1\gamma}(x')dx'\,dx$$

$$= \lambda_{1\gamma} \int_{\Omega\setminus\overline{\omega}} \varphi_{1\gamma}\varphi_1^* dx.$$

Since $\int_{\partial\omega} \varphi_{1\gamma} \frac{\partial \varphi_1^*}{\partial n} d\sigma < 0$, $\theta \int_{\Omega\setminus\overline{\omega}} \varphi_1^*(x) \int_{\omega} k(x,x')\varphi_{1\gamma}(x')dx'\,dx \geq 0$ and $\int_{\Omega\setminus\overline{\omega}} \varphi_{1\gamma}\varphi_1^* dx > 0$, we may infer that $\lambda_{1\gamma} < \lambda_1(\omega)$.

Since $\gamma \mapsto \lambda_{1\gamma}$ is an increasing function, it follows that there exists $\lim_{\gamma\to+\infty} \lambda_{1\gamma} = \tilde{\lambda} \leq \lambda_1(\omega)$.

Let us sketch now the proof of the fact that actually $\tilde{\lambda} = \lambda_1(\omega)$.

If we multiply (A.10) (corresponding to γ, $\lambda := \lambda_{1\gamma}$ and $\varphi_{1\gamma}$) by $\varphi_{1\gamma}$ and integrate over Ω we obtain, after an easy evaluation, that $\{\varphi_{1\gamma}\}_{\gamma>0}$ is bounded in $H^1(\Omega)$, $\left\{\gamma \int_{\omega} |\varphi_{1\gamma}|^2 dx\right\}_{\gamma>0}$ is bounded and so we get that there exists a subsequence (denoted by the same index γ) and a $\tilde{\varphi} \in H^1(\Omega)$ such that

$$\varphi_{1\gamma} \longrightarrow \tilde{\varphi}, \quad \text{weakly in } H^1(\Omega) \text{ as in } H^1(\omega),$$

$$\varphi_{1\gamma} \longrightarrow \tilde{\varphi}, \quad \text{in } L^2(\Omega),$$

$$\varphi_{1\gamma} \longrightarrow 0, \quad \text{in } L^2(\omega),$$

as $\gamma \to +\infty$. We may infer that in fact $\varphi_{1\gamma} \longrightarrow 0$ weakly in $H^1(\omega)$ and weakly in $L^2(\partial\omega)$ and that $\varphi_{1\gamma} \longrightarrow \tilde{\varphi}$ weakly in $L^2(\partial\Omega)$.

If we multiply (A.10) (corresponding to γ, $\lambda := \lambda_{1\gamma}$ and $\varphi_{1\gamma}$) by $\varphi \in \{u \in H^1(\Omega\setminus\overline{\omega}) : u(x) = 0,\ x \in \partial\omega\}$ and integrate over $\Omega\setminus\overline{\omega}$, then passing to the limit ($\gamma \to +\infty$) we obtain that $\tilde{\varphi}$ is a weak solution to (A.13) corresponding to $\lambda := \tilde{\lambda}$; $\|\tilde{\varphi}\|_{L^2(\Omega\setminus\overline{\omega})} = 1$, $\tilde{\varphi}(x) \geq 0$, a.e. $x \in \Omega\setminus\overline{\omega}$ and $\tilde{\varphi}(x) = 0$, a.e. $x \in \partial\omega$. In fact $\tilde{\varphi}$ is a positive eigenvector, and so $\tilde{\lambda} = \lambda_1(\omega)$ and we get the conclusion. □

Other properties of the principal eigenvalues and corresponding eigenvectors used throughout this book follow using similar arguments.

A.3 The Diffusion Equation with General Boundary Conditions

We start by reminding the derivation of the usual diffusion equation, as reported in [9, p.185].

Consider the diffusion of a population or any other substance, whose spatial density at spatial position x and time t is denoted by $y(x,t)$ (here $x \in \overline{\Omega} \subset \mathbb{R}^N$ is the habitat, $N \in \mathbb{N}^*$ and $t \geq 0$); we assume that Ω is a bounded domain with a sufficiently smooth boundary $\partial\Omega$.

The function y is such that the population at time $t \geq 0$, in any region $V \subset \Omega$ (we take open and bounded subsets V with C^1-class boundary ∂V) is given by

$$Y_V(t) = \int_V y(x,t)dx.$$

Let V be an arbitrary subregion of Ω with the above mentioned properties.

If we denote by $f(x,t)$ any contribution to the dynamics of population at $x \in \Omega$, $t \geq 0$, a natural conservation law for the change of the population in region V is given by

$$\int_V \left[\frac{\partial y}{\partial t}(x,t) - div_x(\gamma(x,t)\nabla_x y(x,t)) - f(x,t)\right] dx = 0. \tag{A.15}$$

Since the subregion V is arbitrary, the integrand must be zero, so that

$$\frac{\partial y}{\partial t}(x,t) - div_x(\gamma(x,t)\nabla_x y(x,t)) - f(x,t) = 0. \tag{A.16}$$

Equation (A.16) has been obtained by imposing the balance of the relevant spatial densities in the interior of the domain Ω, while the boundary conditions acting on $\partial\Omega$ are appended later.

Typical boundary conditions are the following ones.

- The *Dirichlet condition* has the form:

$$y(x,t) = \varphi(x,t), \qquad x \in \partial\Omega,\ t \geq 0,$$

  where φ is a given function. When $\varphi \equiv 0$, this indicates (for biological populations) a completely inhospitable boundary $\partial\Omega$.
- The *Neumann condition* has the form:

$$\frac{\partial y}{\partial n}(x,t) = \varphi(x,t), \qquad x \in \partial\Omega,\ t \geq 0,$$

  where φ is a given function. For biological populations, for instance, this condition means that the population flow through the boundary is φ. If $\varphi \equiv 0$, then there is no population flow through the boundary, and the population is isolated in the habitat Ω.
- The *Robin condition* has the form

$$\frac{\partial y}{\partial n}(x,t) + \alpha(x,t)y(x,t) = \varphi(x,t), \qquad x \in \partial\Omega,\ t \geq 0,$$

  where α and φ are given functions. This condition follows usually (for populations), from the fact that the population flow through the boundary is proportional to the difference between the population density $y(x,t)$ inside Ω, and the population density $\tilde{y}(x,t)$ outside Ω:

$$\frac{\partial y}{\partial n}(x,t) = -\alpha(x,t)(y(x,t) - \tilde{y}(x,t)), \qquad x \in \partial\Omega,\ t \geq 0,$$

so that

$$\frac{\partial y}{\partial n}(x,t) + \alpha(x,t)y(x,t) = \alpha(x,t)\tilde{y}(x,t), \qquad x \in \partial\Omega,\ t \geq 0.$$

This perspective misses essential connections between the diffusion equation and the boundary conditions. This aspect has been overcome in a recent paper [117] by generalizing the standard balance equation (A.15)—applied to Ω—as follows

$$\begin{aligned}
&\int_\Omega \frac{\partial y}{\partial t}(x,t)dx + \int_{\partial\Omega} \frac{\partial y}{\partial t}(x,t)d\sigma(x) \\
&\quad = \int_\Omega div_x(\gamma(x,t)\nabla_x y(x,t))dx + \int_\Omega f(x,t)dx \\
&\quad + \int_{\partial\Omega} \phi(x,t,Op(y)(x,t))d\sigma(x),
\end{aligned} \tag{A.17}$$

where $\phi(x,t,Op(y)(x,t))$ denotes a possible source on the boundary ($Op(y)(x,t)$ denotes any integro-differential operator acting on y evaluated at x and t); $\int_{\partial\Omega} \ldots d\sigma(x)$ denotes a surface integral.

It is clear that, if we impose the balance in Ω as given by (A.15), (A.17) reduces to

$$\int_{\partial\Omega} \left[\frac{\partial y}{\partial t}(x,t) - \phi(x,t,Op(y)(x,t))\right] d\sigma(x). \tag{A.18}$$

Clearly a sufficient condition for (A.18) to hold is that

$$\frac{\partial y}{\partial t}(x,t) = \phi(x,t,Op(y)(x,t)), \quad \text{for} \quad x \in \partial\Omega,\ t > 0. \tag{A.19}$$

It is now possible to derive all usual boundary conditions by specializing the term $\phi(x,t,Op(y)(x,t))$, as shown in [117].

For example let us impose

$$\phi(x,t,Op(y)(x,t)) = g(t), \quad \text{for} \quad x \in \partial\Omega,\ t > 0,$$

for g a nontrivial function of time only.

From (A.19) we obtain

$$\frac{\partial y}{\partial t}(x,t) = g(t), \quad \text{for} \quad x \in \partial\Omega,\ t > 0.$$

If y is sufficiently regular, we may state that

$$\frac{\partial}{\partial n}\frac{\partial y}{\partial t}(x,t) = \frac{\partial}{\partial t}\frac{\partial y}{\partial n}(x,t) = 0,$$

for $x \in \partial\Omega$, $t > 0$.

This implies

$$\frac{\partial y}{\partial n}(x,t) = F(x), \quad \text{for} \quad x \in \partial\Omega,\ t > 0,$$

which is a usual inhomogeneous Neumann boundary condition.

A.4 Comparison Theorems

Here $\Omega \subset \mathbb{R}^N$ ($N \geq 2$) is a bounded domain with a C^1-boundary $\partial\Omega$. Denote $Q = \Omega \times (0, +\infty)$, $\Sigma = \partial\Omega \times (0, +\infty)$.

In this volume we deal with several reaction–diffusion systems of the following form

$$\begin{cases} \dfrac{\partial v_1}{\partial t}(x,t) = c_1 \Delta v_1(x,t) + f_1(x, v_1(x,t), v_2(x,t)) \\ \qquad\qquad + \bar{f}_1(\cdot, v_1(\cdot,t), v_2(\cdot,t))(x) + \tilde{f}_1(x,t), \ (x,t) \in Q, \\ \dfrac{\partial v_2}{\partial t}(x,t) = c_2 \Delta v_2(x,t) + f_2(x, v_1(x,t), v_2(x,t)) \\ \qquad\qquad + \bar{f}_2(\cdot, v_1(\cdot,t), v_2(\cdot,t))(x) + \tilde{f}_2(x,t), \ (x,t) \in Q, \\ \dfrac{\partial v_1}{\partial n}(x,t) + \alpha_1(x) v_1(x,t) = \varphi_1(x,t), \qquad (x,t) \in \Sigma, \\ \dfrac{\partial v_2}{\partial n}(x,t) + \alpha_2(x) v_2(x,t) = \varphi_2(x,t), \qquad (x,t) \in \Sigma, \\ v_1(x,0) = v_{10}(x), \ v_2(x,0) = v_{20}(x), \qquad x \in \Omega, \end{cases} \tag{A.20}$$

where c_1, c_2 are positive constants, $f_1, f_2 : \Omega \times \mathbb{R} \times \mathbb{R} \longrightarrow \mathbb{R}$, $\tilde{f}_1, \tilde{f}_2 : Q \longrightarrow \mathbb{R}$, $\alpha_1, \alpha_2 : \partial\Omega \to \mathbb{R}$, $\varphi_1, \varphi_2 : \Sigma \to \mathbb{R}$, $v_{10}, v_{20} : \Omega \to \mathbb{R}$, are measurable functions. Here, $\bar{f}_1, \bar{f}_2 : \Omega \times L^2(\Omega) \times L^2(\Omega) \to \mathbb{R}$ describe some reaction terms containing integrals.

Definition A.8 We say that (v_1, v_2) is a (weak) solution to (A.20) if $v_1, v_2 \in C([0,T]; L^2(\Omega)) \cap AC((0,T]; L^2(\Omega)) \cap L^2(0,T; H^1(\Omega)) \cap L^2_{loc}((0,T]; H^2(\Omega))$, for any $T \in (0, +\infty)$ and if

$$\begin{cases} \dfrac{\partial v_1}{\partial t}(x,t) = c_1 \Delta v_1(x,t) + f_1(x, v_1(x,t), v_2(x,t)) \\ \qquad\qquad + \bar{f}_1(\cdot, v_1(\cdot,t), v_2(\cdot,t))(x) + \tilde{f}_1(x,t), \text{ a.e. } (x,t) \in Q, \\ \dfrac{\partial v_2}{\partial t}(x,t) = c_2 \Delta v_2(x,t) + f_2(x, v_1(x,t), v_2(x,t)) \\ \qquad\qquad + \bar{f}_2(\cdot, v_1(\cdot,t), v_2(\cdot,t))(x) + \tilde{f}_2(x,t), \text{ a.e. } (x,t) \in Q, \\ \dfrac{\partial v_1}{\partial n}(x,t) + \alpha_1(x) v_1(x,t) = \varphi_1(x,t), \quad \text{a.e. } (x,t) \in \Sigma, \\ \dfrac{\partial v_2}{\partial n}(x,t) + \alpha_2(x) v_2(x,t) = \varphi_2(x,t), \quad \text{a.e. } (x,t) \in \Sigma, \\ v_1(x,0) = v_{10}(x), \ v_2(x,0) = v_{20}(x), \quad \text{a.e. } x \in \Omega. \end{cases}$$

The existence and uniqueness for all the reaction–diffusion systems in this volume follow by using Banach's fixed point theorem and the comparison result for the solutions to the linear parabolic equations.

In general we deal with nonnegative solutions.

For any $i \in \{1,2\}$, let Γ_{i1}, Γ_{i2} be C^1-$(N-1)$-dimensional submanifolds of $\partial\Omega$ such that $\Gamma_{i1} \cap \Gamma_{i2} = \emptyset$, $\overline{\Gamma}_{i1} \cup \overline{\Gamma}_{i2} = \partial\Omega$ and $\overline{\Gamma}_{i1} \cap \overline{\Gamma}_{i2}$ is a C^1-$(N-2)$-dimensional submanifold.

If we consider homogeneous Dirichlet boundary conditions on the measurable subsets Γ_{11}, Γ_{21}, respectively, and the above-mentioned Robin boundary conditions on Γ_{12}, Γ_{22}, respectively, then a solution (v_1, v_2), by definition, is such that for any $i \in \{1,2\}$: $v_i \in C([0,T]; L^2(\Omega)) \cap AC((0,T]; L^2(\Omega)) \cap L^2(0,T; V_i) \cap L^2_{loc}((0,T]; H^2(\Omega))$, for any $T \in (0,+\infty)$, where $V_i = \{h \in H^1(\Omega) : v(x) = 0, \text{ a.e. } x \in \Gamma_{i1}\}$. On the other hand it satisfies the equations, the boundary conditions and initial conditions, almost everywhere, respectively.

A.4.1 Comparison Results for Parabolic Equations

From now on we shall take as ω an open subset of Ω, with a C^1-boundary $\partial\omega$ (note that ω could be an empty set), such that $\overline{\omega} \subset \Omega$ and $\Omega \setminus \overline{\omega}$ is a domain.

Denote $\widetilde{\Omega} = \Omega \setminus \overline{\omega}$, $\Gamma_1 = \partial\omega$, $\Gamma_2 = \partial\Omega$, $\widetilde{Q} = \widetilde{\Omega} \times (0,+\infty)$, $\Sigma_1 = \Gamma_1 \times (0,+\infty)$, $\Sigma_2 = \Gamma_2 \times (0,+\infty)$; $\|\cdot\|_p$ denotes the norm of $L^p(\widetilde{\Omega})$ or of $L^p(\widetilde{\Omega} \times \widetilde{\Omega})$.

Theorem A.9 *Let w_i ($i \in \{1,2\}$) be nonnegative solutions to*

$$\begin{cases} \dfrac{\partial w_i}{\partial t}(x,t) = d\Delta w_i(x,t) + F(x, w_i(x,t)) + G_i(x,t), & (x,t) \in \widetilde{Q}, \\ w_i(x,t) = 0, & (x,t) \in \Sigma_1, \\ \dfrac{\partial w_i}{\partial n}(x,t) + \alpha_i(x) w_i(x,t) = \varphi_i(x,t), & (x,t) \in \Sigma_2, \\ w_i(x,0) = w_{0i}(x), & x \in \widetilde{\Omega}, \end{cases}$$

respectively, where $d \in (0, +\infty)$, $w_{01}, w_{02} \in L^2(\widetilde{\Omega})$,

$$w_{01}(x) \le w_{02}(x), \quad a.e.\ x \in \widetilde{\Omega},$$

$F \in L^2_{loc}(\overline{\widetilde{Q}})$, *and for almost any* $x \in \widetilde{\Omega}$ *we have that*

$$F(x, \cdot) \in C^1([0, +\infty)), \quad \frac{\partial F}{\partial w}(x, w) \le \theta, \quad a.e.\ (x, w) \in \widetilde{\Omega} \times [0, +\infty)$$

($\theta \in \mathbb{R}$), $G_1, G_2 \in L^2_{loc}(\overline{\widetilde{Q}})$,

$$G_1(x, t) \le G_2(x, t), \quad a.e.\ (x, t) \in \widetilde{Q},$$

$\alpha_1, \alpha_2 \in L^\infty(\Gamma_2)$, $\alpha_1(x) \ge \alpha_2(x) \ge 0$, *a.e.* $x \in \Gamma_2$, *and* $\varphi_1, \varphi_2 \in L^2_{loc}(\Sigma_2)$, $\varphi_1(x, t) \le \varphi_2(x, t)$, *a.e.* $(x, t) \in \Sigma_2$.

Then

$$w_1(x, t) \le w_2(x, t), \quad a.e.\ x \in \widetilde{\Omega},\ for\ any\ t \ge 0.$$

Proof Function $w = w_2 - w_1$ is a solution to

$$\begin{cases} \dfrac{\partial w}{\partial t}(x, t) = d\Delta w(x, t) + (F(x, w_2(x, t)) - F(x, w_1(x, t)) \\ \qquad +G_2(x, t) - G_1(x, t), & (x, t) \in \widetilde{Q}, \\ w(x, t) = 0, & (x, t) \in \Sigma_1, \\ \dfrac{\partial w}{\partial n}(x, t) + \alpha_2(x) w(x, t) = \varphi_2(x, t) - \varphi_1(x, t) \\ \qquad +(\alpha_1(x) - \alpha_2(x)) w_1(x, t), & (x, t) \in \Sigma_2, \\ w(x, 0) = w_{02}(x) - w_{01}(x), & x \in \widetilde{\Omega}. \end{cases} \tag{A.21}$$

If we multiply (A.21) by w^- and integrate over $(0, t) \times \widetilde{\Omega}$ we get that

$$-\frac{1}{2}\int_{\widetilde{\Omega}} |w^-(x, t)|^2 dx \ge d \int_0^t \int_{\widetilde{\Omega}} |\nabla w^-(x, s)|^2 dx\, ds$$

$$+ \int_0^t \int_{\widetilde{\Omega}} \frac{\partial F}{\partial w}(x, \xi(x, s)) w(x, s) w^-(x, s) dx\, ds$$

$$+ d \int_0^t \int_{\Gamma_2} [-\alpha_2 w + \varphi_2 - \varphi_1 + (\alpha_1 - \alpha_2) w_1](x, s) w^-(x, s) d\sigma\, ds$$

$$\ge -\theta \int_0^t \int_{\widetilde{\Omega}} |w^-(x, s)|^2 dx\, ds,$$

for any $t \geq 0$. By Gronwall's inequality we get that

$$w^-(x,t) = 0, \quad \text{a.e. } x \in \widetilde{\Omega}, \text{ for any } t \geq 0,$$

and the conclusion. □

A.4.2 Comparison Results for Integro-Parabolic Equations

Consider the following model for population dynamics with logistic term and migration:

$$\begin{cases} \dfrac{\partial w}{\partial t}(x,t) = d\Delta w(x,t) + a(x)w(x,t) \\ \qquad -c(x)w(x,t)^2 + \displaystyle\int_{\widetilde{\Omega}} k(x,x')w(x',t)dx', & (x,t) \in \widetilde{Q}, \\ w(x,t) = 0, & (x,t) \in \Sigma_1, \\ \dfrac{\partial w}{\partial n}(x,t) + \alpha(x)w(x,t) = 0, & (x,t) \in \Sigma_2, \\ w(x,0) = w_0(x), & x \in \widetilde{\Omega}, \end{cases} \tag{A.22}$$

where $d > 0$, and $a,\ c \in L^\infty(\widetilde{\Omega})$, $c(x) \geq 0$, a.e. $x \in \widetilde{\Omega}$, $k \in L^\infty(\widetilde{\Omega} \times \widetilde{\Omega})$, $k(x,x') \geq 0$, a.e. $(x,x') \in \widetilde{\Omega} \times \widetilde{\Omega}$, $\alpha \in L^\infty(\Gamma_2)$, $\alpha(x) \geq 0$, a.e. $x \in \Gamma_2$, $w_0 \in L^\infty(\widetilde{\Omega})$, $w_0(x) \geq 0$, a.e. $x \in \widetilde{\Omega}$; $\widetilde{\Omega}$, Γ_1, Γ_2, Σ_1, Σ_2 were defined in the previous subsection.

Theorem A.10 *There exists a unique nonnegative (weak) solution to (A.22).*

Proof (Sketch) If w is a nonnegative solution to (A.22), then by multiplying by w and integrating over $\widetilde{\Omega}$ we get that

$$\frac{d}{dt}\left(\int_{\widetilde{\Omega}} |w(x,t)|^2\,dx\right) \leq (\|a\|_\infty + \|k\|_2)\int_{\widetilde{\Omega}} |w(x,t)|^2 dx, \quad \text{a.e. } t > 0,$$

and so

$$\int_{\widetilde{\Omega}} |w(x,t)|^2 dx \leq e^{(\|a\|_\infty + \|k\|_2)t}\|w_0\|_2^2, \quad \text{for any } t \geq 0.$$

It follows that

$$G(x,t) = \int_{\widetilde{\Omega}} k(x,x')w(x',t)dx' \leq \|k\|_\infty \operatorname{meas}(\widetilde{\Omega})^{\frac{1}{2}} e^{\frac{1}{2}(\|a\|_\infty + \|k\|_2)t}\|w_0\|_2 = h(t),$$

a.e. $x \in \widetilde{\Omega}$, for any $t \geq 0$.

Using Theorem A.9 we get that

$$w(x,t) \leq \bar{w}(x,t), \quad \text{a.e. } x \in \widetilde{\Omega}, \text{ for any } t \geq 0,$$

where $\bar{w}$ is the unique weak solution to the linear problem

$$\begin{cases} \dfrac{\partial \bar{w}}{\partial t}(x,t) = d\Delta \bar{w}(x,t) + \|a\|_\infty \bar{w}(x,t) + h(t), \ (x,t) \in \widetilde{Q}, \\ \dfrac{\partial \bar{w}}{\partial n}(x,t) = 0, & (x,t) \in \widetilde{\Sigma}, \\ \bar{w}(x,0) = \|w_0\|_\infty, & x \in \widetilde{\Omega} \end{cases}$$

($\widetilde{\Sigma} = \partial\widetilde{\Omega} \times (0,+\infty)$). The existence and uniqueness of the solution follows as in [38]. Actually, $\bar{w}$ is independent of x and is the solution to the following ODE:

$$\begin{cases} \bar{w}'(t) = \|a\|_\infty \bar{w}(t) + h(t), \ t > 0, \\ \bar{w}(0) = \|w_0\|_\infty . \end{cases}$$

We may infer that $\bar{w} \in L^\infty_{loc}([0,+\infty))$.

The existence and uniqueness of a unique weak solution to (A.22) follows if for an arbitrary but fixed $T > 0$ we apply Banach's fixed point theorem in the space $X = \{y \in C([0,T]; L^2(\widetilde{\Omega})) \ : \ 0 \leq y(x,t) \leq M, \ \text{a.e. } x \in \widetilde{\Omega}, \text{ for any } t \in [0,T]\}$, where M is a sufficiently large constant. □

Remark A.11 If we consider the following problem

$$\begin{cases} \dfrac{\partial w}{\partial t}(x,t) = d\Delta w(x,t) + a(x)w(x,t) \\ \qquad -c(x)w(x,t)^2 + \displaystyle\int_{\widetilde{\Omega}} k(x,x')w(x',t)dx' \\ \qquad +\tilde{G}(x,t), & (x,t) \in \widetilde{Q}, \\ w(x,t) = 0, & (x,t) \in \Sigma_1, \\ \dfrac{\partial w}{\partial n}(x,t) + \alpha(x)w(x,t) = 0, & (x,t) \in \Sigma_2, \\ w(x,0) = w_0(x), & x \in \widetilde{\Omega}, \end{cases} \tag{A.23}$$

where $\tilde{G} \in L^\infty_{loc}(\overline{\widetilde{Q}})$, $\tilde{G}(x,t) \geq 0$, a.e. $(x,t) \in \widetilde{Q}$, (A.23) admits a unique nonnegative solution. Moreover, this solution belongs to $L^\infty_{loc}(\overline{\widetilde{Q}})$.

The proof follows via a slight adaption of the one of Theorem A.10.

Theorem A.12 *Let w_i ($i \in \{1, 2\}$) be nonnegative solutions to*

$$\begin{cases} \dfrac{\partial w_i}{\partial t}(x,t) = d\Delta w_i(x,t) + a_i(x)w_i(x,t) \\ \qquad -c_i(x)w(x,t)^2 + \displaystyle\int_{\widetilde{\Omega}} k_i(x,x')w(x',t)dx' \\ \qquad +\tilde{G}_i(x,t), & (x,t) \in \widetilde{Q}, \\ w_i(x,t) = 0, & (x,t) \in \Sigma_1, \\ \dfrac{\partial w_i}{\partial n}(x,t) + \alpha_i(x)w_i(x,t) = 0, & (x,t) \in \Sigma_2, \\ w_i(x,0) = w_{0i}(x), & x \in \widetilde{\Omega}, \end{cases}$$

where $a_i,\ c_i \in L^\infty(\widetilde{\Omega})$, $i \in \{1, 2\}$, $a_1(x) \le a_2(x)$, a.e. $x \in \widetilde{\Omega}$, $c_1(x) \ge c_2(x) \ge 0$, a.e. $x \in \widetilde{\Omega}$, $k_i \in L^\infty(\widetilde{\Omega} \times \widetilde{\Omega})$, $k_2(x,x') \ge k_1(x,x') \ge 0$, a.e. $(x,x') \in \widetilde{\Omega} \times \widetilde{\Omega}$, $\tilde{G}_i \in L^\infty_{loc}(\overline{\widetilde{Q}})$, $\tilde{G}_2(x,t) \ge \tilde{G}_1(x,t) \ge 0$, a.e. $(x,t) \in \widetilde{Q}$, $\alpha_i \in L^\infty(\Gamma_2)$, $\alpha_1(x) \ge \alpha_2(x) \ge 0$, a.e. $x \in \Gamma_2$, $w_{0i} \in L^\infty(\widetilde{\Omega})$, $w_{02}(x) \ge w_{01}(x) \ge 0$, a.e. $x \in \widetilde{\Omega}$.

Then

$$w_1(x,t) \le w_2(x,t), \quad a.e.\ x \in \widetilde{\Omega},\ for\ any\ t \ge 0.$$

Proof Let $z = w_2 - w_1$; z is a weak solution to

$$\begin{cases} \dfrac{\partial z}{\partial t} = d\Delta z + a_1 z - c_1(w_1 + w_2)z + \displaystyle\int_{\widetilde{\Omega}} k_1(x,x')z(x',t)dx' \\ \quad +(a_2 - a_1)w_2 + (c_1 - c_2)w_2^2 \\ \quad + \displaystyle\int_{\widetilde{\Omega}} (k_2 - k_1)(x,x')w_2(x',t)dx' \\ \quad +\tilde{G}_2 - \tilde{G}_1, & (x,t) \in \widetilde{Q}, \\ z(x,t) = 0, & (x,t) \in \Sigma_1, \\ \dfrac{\partial z}{\partial n}(x,t) + \alpha_1(x)z(x,t) = (\alpha_1(x) - \alpha_2(x))w_2(x,t), & (x,t) \in \Sigma_2, \\ z(x,0) = w_{02}(x) - w_{01}(x), & x \in \widetilde{\Omega}. \end{cases}$$

Let T be an arbitrary, but fixed positive constant. We have that $w_1, w_2 \in L^\infty(\widetilde{\Omega} \times (0,T))$.

If we multiply the equation in z by z^- and integrate over $(0,t) \times \widetilde{\Omega}$ ($t \in (0,T]$), then we get after an easy calculation and taking into account that $(a_2-a_1)w_2+(c_1-c_2)w_2^2 + \int_{\widetilde{\Omega}}(k_2 - k_1)(x,x')w_2(x',t)dx' + \tilde{G}_2 - \tilde{G}_1 \ge 0$, a.e. $(x,t) \in \widetilde{\Omega} \times (0,T)$ and $(\alpha_1 - \alpha_2)w_2 \ge 0$, a.e. $(x,t) \in \Gamma_2 \times (0,T)$, that

$$-\frac{1}{2}\|z^-(t)\|_2^2 \geq -\int_0^t \int_{\widetilde{\Omega}} (a_1(x) - c_1(x)(w_1(x,x) + w_2(x,s)))|z^-(x,s)|^2 dx\, ds$$

$$-\int_0^t \int_{\widetilde{\Omega}} z^-(x,s) \int_{\widetilde{\Omega}} k_1(x,x')z^-(x',s)dx'\, dx\, ds.$$

It follows that there exists a constant $\bar{M} \geq 0$ such that

$$\|z^-(t)\|_2^2 \leq \bar{M} \int_0^t \|z^-(s)\|_2^2 ds, \quad \text{for any } t \in [0,T],$$

and via Gronwall's lemma we conclude that $z^-(x,t) = 0$, a.e. $x \in \widetilde{\Omega}$, for any $t \in [0,T]$, and since $T > 0$ is arbitrary, the conclusion of the theorem. □

Appendix B
Essentials of Numerical Methods

The aim of this appendix is to introduce a non-expert reader to the field of numerical approximation of partial differential equations with low order finite element methods and to their implementation. We do not enter into the theoretical details, since our focus is to provide a simple tutorial for the implementation of the algorithms adopted throughout the book. We assume that the reader has some basic notions of numerical linear algebra.

B.1 $\mathbf{P}^1$ Finite Elements in 1D

B.1.1 Example 1: The Stationary Diffusion Equation

Consider as a model problem the following second order differential equation, posed on the domain $\Omega = (0, 1)$ and augmented with Dirichlet boundary conditions:

Problem B.1 Given $f : \Omega \to \mathbb{R}$ and $\alpha, \beta \in \mathbb{R}$, find $u : \overline{\Omega} \to \mathbb{R}$ such that

$$\begin{cases} -\dfrac{d^2u}{dx^2}(x) = f(x), \quad x \in \Omega, \\ u(0) = \alpha, \\ u(1) = \beta. \end{cases} \tag{B.1}$$

We refer to Problem B.1 as *strong problem* or *strong formulation*. Unfortunately, such formulation is not adequate to deal with general applications, where the load term f might be not sufficiently regular to guarantee that the solution u admits a continuous second derivative as required by equation (B.1).

S. Aniţa et al., *Mathematical Modeling and Control in Life and Environmental Sciences*, Modeling and Simulation in Science, Engineering and Technology,
https://doi.org/10.1007/978-3-031-49971-5

Therefore, we need an alternative formulation of the differential problem (B.1), called *weak formulation*, that allows reducing the order of derivation of the solution u.

B.1.1.1 Weak Formulation

For sake of simplifying the discussion, we suppose now that *homogeneous Dirichlet boundary conditions* hold, i.e.,

$$\alpha = \beta = 0.$$

In order to derive the *weak formulation* of Problem B.1, we first multiply equation (B.1) by an arbitrary *test function* v, such that

$$v(0) = v(1) = 0,$$

and then we integrate on Ω, thus obtaining

$$-\int_0^1 \frac{d^2u}{dx^2}(x)v(x)\,dx = \int_0^1 f(x)\,v(x)\,dx.$$

After integrating by parts the left-hand side, we finally get the integral equation

$$\int_0^1 \frac{du}{dx}(x)\frac{dv}{dx}(x)\,dx = \int_0^1 f(x)\,v(x)\,dx. \tag{B.2}$$

Note that, in this derivation, we have proceeded formally, without worrying whether the operations were allowed. In order to provide a well posed formulation of equation (B.2), even for not regular data f, we need to introduce the following function spaces:

$$\begin{aligned}
L^2(\Omega) &:= \left\{ v : \Omega \to \mathbb{R} : \ v \text{ is Lebesgue measurable, } \int_0^1 |v(x)|^2\,dx < +\infty \right\}, \\
H^1(\Omega) &:= \left\{ v \in L^2(\Omega) : \ \frac{dv}{dx} \in L^2(\Omega) \right\}, \\
H_0^1(\Omega) &:= \left\{ v \in H^1(\Omega) : \ v(0) = v(1) = 0 \right\}, \\
H^2(\Omega) &:= \left\{ v \in H^1(\Omega) : \ \frac{d^2v}{dx^2} \in L^2(\Omega) \right\},
\end{aligned}$$

where the derivatives $\frac{dv}{dx}$ and $\frac{d^2v}{dx^2}$ must be intended in the sense of distributions and the equal signs in the definition of $H_0^1(\Omega)$ must be intended in the sense of traces; see, e.g., [167] for further details. In the following, we will set $V = H_0^1(\Omega)$.

Let us then define:

- The $L^2(\Omega)$ inner product

$$(u, v) := \int_0^1 u(x)\, v(x)\, dx, \qquad \text{for all } u, v \in L^2(\Omega).$$

- The $L^2(\Omega)$ norm

$$||u||_0 := \sqrt{(u, u)}, \qquad \text{for all } u \in L^2(\Omega).$$

- The $H^1(\Omega)$ norm

$$||u||_1 := \sqrt{(u, u) + \int_0^1 \left|\frac{du}{dx}(x)\right|^2 dx}, \qquad \text{for all } u \in H^1(\Omega).$$

- The $H^1(\Omega)$ semi-norm

$$|u|_1 := \sqrt{\int_0^1 \left|\frac{du}{dx}(x)\right|^2 dx}, \qquad \text{for all } u \in H^1(\Omega).$$

- The $H^2(\Omega)$ semi-norm

$$|u|_2 := \sqrt{\int_0^1 \left|\frac{d^2u}{dx^2}(x)\right|^2 dx}, \qquad \text{for all } u \in H^2(\Omega).$$

Recall that, thanks to the Poincaré inequality (see [167]), the $|\cdot|_1$ semi-norm is a norm in $H_0^1(\Omega)$, equivalent to the $||\cdot||_1$ norm.

Let us now define the bilinear form $a(\cdot,\cdot) : V \times V \to \mathbb{R}$ such that

$$a(u, v) := \int_0^1 \frac{du}{dx}(x) \frac{dv}{dx}(x)\, dx, \quad \text{for all } u, v \in V.$$

It is easy to verify that the bilinear form $a(\cdot,\cdot)$ is

- *Continuous*, i.e.,

$$\text{there exists } M > 0 \text{ such that } a(u, v) \leq M\, ||u||_1 ||v||_1, \quad \text{for all } u, v \in V.$$

- *Coercive*, i.e.,

$$\text{there exists } \alpha > 0 \text{ such that } a(u, u) \geq \alpha\, ||u||_1^2, \quad \text{for all } u \in V.$$

We can now enunciate the *weak formulation* of Problem B.1:

Problem B.2 Given $f \in L^2(\Omega)$, find $u \in V$ such that

$$a(u, v) = (f, v), \quad \text{for all } v \in V.$$

The well-posedness of Problem B.2 follows from the *Lax–Milgram lemma*, see [167]. The finite element method is a particular case of Galerkin method that constructs an approximation of the solution u of Problem B.2 in a finite dimensional subspace of V.

B.1.1.2 Discretization with $\mathbf{P}^1$ Finite Elements

Let us introduce a partition $\mathcal{T}^h$ of the unit interval domain $\Omega = (0, 1)$ into $N - 1$ subintervals $E_j = [x_j, x_{j+1}]$, for $j = 1, \ldots, N - 1$, of length $h = \dfrac{1}{N-1}$, with

$$0 = x_1 < x_2 < \ldots < x_{N-1} < x_N = 1.$$

The resulting N equispaced points x_j, for $j = 1, \ldots, N$, are called *discretization nodes*, whereas the parameter h is called *mesh size*. For sake of simplifying the discussion, we have supposed that the subintervals E_j have the same length h. However, the generalization to variable length h is straightforward.

Definition B.1 We call one-dimensional $\mathbf{P}^1$ finite element space the finite dimensional function space

$$V^h := \{v_h \in C(\overline{\Omega}) : v_h|_{E_j} \in \mathbb{P}_1, \text{ for all } E_j \in \mathcal{T}^h\},$$

where $\mathbb{P}_1$ denotes the space of polynomials of degree 1 in one variable.

We first observe that $V^h \subset H^1(\Omega)$. The dimension of the space V^h coincides with the number of discretization nodes N. A basis for V^h consists of the so-called *hat functions* $\{\phi_i\}_{i=1}^N$, such that $\phi_i(x_j) = \delta_{ij}$, for all $i, j = 1, \ldots, N$, where δ_{ij} is the Kronecker delta; see Figure B.1.

We also define the following subset of V^h,

$$V_0^h := \{v_h \in V^h : v_h(0) = v_h(1) = 0\},$$

where we will look for the approximate solution. Clearly $V_0^h \subset V$ and $\{\phi_i\}_{i=2}^{N-1}$ is a basis for V_0^h, whose dimension is $N - 2$, i.e., the number of interior discretization nodes. It is now possible to enunciate our discrete problem:

Problem B.3 Given $f \in L^2(\Omega)$, find $u_h \in V_0^h$ such that

$$a(u_h, v_h) = (f, v_h), \quad \text{for all } v_h \in V_0^h.$$

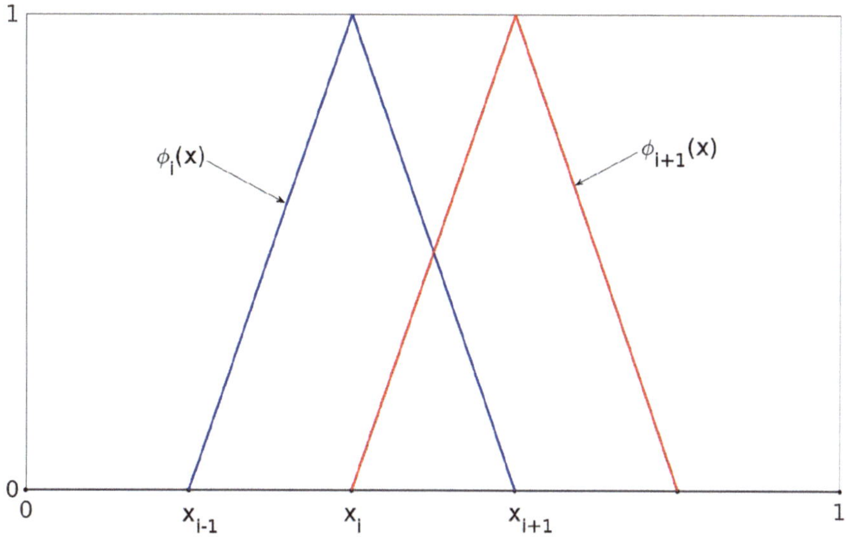

Fig. B.1 Example of *hat functions* in 1D

It is reasonable to expect that, the smaller is the mesh size h, the better u_h approximates the exact solution u. In fact, it holds the following theorem [167, 168]:

Theorem B.4 *Let $u \in V$ and $u_h \in V_0^h$ be the solutions of problems B.2 and B.3, respectively. Suppose also that $u \in H^2(\Omega)$. Then the following error estimates hold:*

$$||u_h - u||_1 \leq C_1 h|u|_2,$$
$$||u_h - u||_0 \leq C_0 h^2|u|_2,$$

being C_1 and C_0 constants independent of h and u.

B.1.1.3 Implementation

Since $\{\phi_i\}_{i=2}^{N-1}$ is a basis for V_0^h, Problem B.3 is equivalent to finding $u_h \in V_0^h$ such that

$$a(u_h, \phi_i) = (f, \phi_i), \quad \text{for all } i = 2, \dots, N-1. \tag{B.3}$$

Expanding the unknown u_h with respect to the basis $\{\phi_i\}_{i=2}^{N-1}$, we have

$$u_h(x) = \sum_{j=2}^{N-1} U_j \phi_j(x), \tag{B.4}$$

where U_j, for $j = 2, \ldots, N-1$, are $N-2$ unknown scalar values to be determined. Set then $U_1 = U_N = 0$, because of the homogeneous Dirichlet boundary conditions. Substituting now the expression (B.4) into equation (B.3), we have

$$a(u_h, \phi_i) = a\left(\sum_{j=2}^{N-1} U_j\phi_j, \phi_i\right) = \sum_{j=2}^{N-1} U_j a(\phi_j, \phi_i) = (f, \phi_i),$$

for all $i = 2, \ldots, N-1$. Hence, the discrete problem B.3 reduces to the following problem:

Problem B.4 Given $f \in L^2(\Omega)$, find U_j, for $j = 2, \ldots, N-1$, such that

$$\sum_{j=2}^{N-1} U_j a(\phi_j, \phi_i) = (f, \phi_i), \qquad \text{for all } i = 2, \ldots, N-1. \tag{B.5}$$

Thus, in order to find u_h, we have to solve the linear system (B.5). Let us now define the vectors:

$$\begin{aligned} &\mathbf{I} = [2, 3, \ldots, N-1], \\ &\mathbf{U} = [U_1, \ldots, U_N]^T, \\ &\hat{\mathbf{U}} = \mathbf{U}(\mathbf{I}) = [U_2, \ldots, U_{N-1}]^T, \\ &\mathbf{F} = [F_1, \ldots, F_N]^T, \\ &\hat{\mathbf{F}} = \mathbf{F}(\mathbf{I}) = [F_2, \ldots, F_{N-1}]^T, \end{aligned} \tag{B.6}$$

with $F_i = (f, \phi_i)$, for $i = 1, \ldots, N$. Then define the matrix:

$$\hat{\mathrm{K}} = \mathrm{K}(\mathbf{I}, \mathbf{I}),$$

being K the $N \times N$ matrix such that

$$\mathrm{K} = \{k_{ij}\}_{i,j=1}^N, \quad k_{ij} = a(\phi_j, \phi_i), \quad \text{for all } i, j = 1, \ldots, N. \tag{B.7}$$

The matrix K is called *stiffness matrix*. Note that matrix K is symmetric positive semi-definite, with the kernel constituted by the constant vector. With the notation

$$\mathrm{K}(\mathbf{I}, \mathbf{I}) = \mathrm{K}(2:N-1, 2:N-1),$$

we indicate the submatrix of K constituted by the rows $i = 2, \ldots, N-1$ and the columns $j = 2, \ldots, N-1$.

With this notation, the linear system (B.5) in compact form becomes:

$$\hat{\mathrm{K}}\hat{\mathbf{U}} = \hat{\mathbf{F}}.$$

Note that matrix $\hat{\mathrm{K}}$ is symmetric positive definite.

The crucial points of the implementation are therefore the assembling of the stiffness matrix K and of the right-hand side vector **F**.

Assembling the Stiffness Matrix Observe that the generic element of matrix K is

$$k_{ij} = a(\phi_j, \phi_i) = \int_0^1 \frac{d\phi_j}{dx}\frac{d\phi_i}{dx}\,dx = \sum_{l=1}^{N-1}\int_{x_l}^{x_{l+1}} \frac{d\phi_j}{dx}\frac{d\phi_i}{dx}\,dx.$$

As a consequence, to assemble K, we can proceed with the following double loop algorithm:

$$\begin{aligned}
&\text{for } i = 1, \ldots, N\\
&\quad \text{for } j = 1, \ldots, N\\
&\qquad \text{for } l = 1, \ldots, N-1\\
&\qquad\quad k_{ij} = k_{ij} + \int_{x_l}^{x_{l+1}} \frac{d\phi_j}{dx}\frac{d\phi_i}{dx}\,dx\\
&\qquad \text{end for}\\
&\quad \text{end for}\\
&\text{end for}
\end{aligned} \tag{B.8}$$

By exchanging the order of the for loops, we can re-write algorithm (B.8) as follows:

$$\begin{aligned}
&\text{for } l = 1, \ldots, N-1\\
&\quad \text{for } i = 1, \ldots, N\\
&\qquad \text{for } j = 1, \ldots, N\\
&\qquad\quad k_{ij} = k_{ij} + \int_{x_l}^{x_{l+1}} \frac{d\phi_j}{dx}\frac{d\phi_i}{dx}\,dx\\
&\qquad \text{end for}\\
&\quad \text{end for}\\
&\text{end for}
\end{aligned} \tag{B.9}$$

Thanks to the tight support of the *hat basis functions* $\{\phi_i\}_{i=1}^N$, only two basis functions do not vanish on the generic subinterval $[x_l, x_{l+1}]$, i.e., ϕ_l and ϕ_{l+1}, thus algorithm (B.9) becomes:

$$\begin{aligned}
&\text{for } l = 1, \ldots, N-1\\
&\quad \text{for } i = l, l+1\\
&\qquad \text{for } j = l, l+1\\
&\qquad\quad k_{ij} = k_{ij} + \int_{x_l}^{x_{l+1}} \frac{d\phi_j}{dx}\frac{d\phi_i}{dx}\,dx\\
&\qquad \text{end for}\\
&\quad \text{end for}\\
&\text{end for}
\end{aligned} \tag{B.10}$$

It remains to compute the integrals

$$\int_{x_l}^{x_{l+1}} \frac{d\phi_j}{dx}\frac{d\phi_i}{dx}\,dx.$$

To this end, we assume that, up to a shift, the generic subinterval $E_l = [x_l, x_{l+1}]$ is $\hat{E} = [0, h]$. The two *hat basis functions* living on $\hat{E}$ are

$$\hat{\phi}_1(x) = 1 - \frac{x}{h} \quad \text{and} \quad \hat{\phi}_2(x) = \frac{x}{h}.$$

Define the 2×2 matrix $\mathrm{K}^{loc} = \{k_{ij}^{loc}\}_{i,j=1}^{2}$ by

$$k_{ij}^{loc} = \int_0^h \frac{d\hat{\phi}_j}{dx}\frac{d\hat{\phi}_i}{dx}\,dx, \quad i, j = 1, 2.$$

After some basic computations, we have:

$$\mathrm{K}^{loc} = \frac{1}{h}\begin{bmatrix} 1 & -1 \\ -1 & 1 \end{bmatrix}.$$

Finally, the algorithm to assemble the *stiffness matrix* K can be written in compact form as follows:

for $l = 1, \ldots, N-1$
  $\mathrm{K}(l : l+1, l : l+1) = \mathrm{K}(l : l+1, l : l+1) + \mathrm{K}^{loc}$
end for

Assembling the Mass Matrix We call *mass matrix* M the $N \times N$ matrix such that

$$\mathrm{M} = \{m_{ij}\}_{i,j=1}^{N}, \quad m_{ij} = (\phi_j, \phi_i), \quad \text{for all } i, j = 1, \ldots, N. \tag{B.11}$$

This matrix is used to assemble the right-hand side vector **F** in this example and it will be used in the next examples to approximate reaction–diffusion and time-dependent problems.

Proceeding as in the assembling of the *stiffness matrix*, the algorithm to assemble the *mass matrix* reads:

for $l = 1, \ldots, N-1$
  for $i = l, l+1$
    for $j = l, l+1$
      $k_{ij} = k_{ij} + \int_{x_l}^{x_{l+1}} \phi_j \phi_i\,dx$
    end for
  end for
end for

To compute the integrals

$$\int_{x_l}^{x_{l+1}} \phi_j \phi_i \, dx,$$

we focus again on the interval $\hat{E} = [0, h]$. Define the 2×2 matrix $\mathrm{M}^{loc} = \{m_{ij}^{loc}\}_{i,j=1}^2$ by

$$m_{ij}^{loc} = \int_0^h \hat{\phi}_j \hat{\phi}_i \, dx, \quad i, j = 1, 2.$$

If the integrals above are approximated using the trapezoidal rule, the matrix M^{loc} becomes

$$\mathrm{M}^{loc} = \frac{h}{2} \begin{bmatrix} 1 & 0 \\ 0 & 1 \end{bmatrix}.$$

This technique that yields a local diagonal *mass matrix*, and consequently the resulting global diagonal *mass matrix*, is known in the literature as *mass lumping*; see [168]. The algorithm to assemble M is then:

for $l = 1, \ldots, N - 1$
  $\mathrm{M}(l : l + 1, l : l + 1) = \mathrm{M}(l : l + 1, l : l + 1) + \mathrm{M}^{loc}$
end for

Assembling the Right-Hand Side Vector The generic element F_i, for $i = 2, \ldots, N - 1$, of the right-hand side vector **F** can be computed as follows:

$$\begin{aligned} F_i &= (f, \phi_i) \\ &\approx \left(\sum_{j=1}^{N} f(x_j) \, \phi_j, \phi_i \right) \\ &= \sum_{j=1}^{N} f(x_j)(\phi_j, \phi_i) \\ &= \sum_{j=1}^{N} f(x_j) \, m_{ij}, \end{aligned} \tag{B.12}$$

where we have approximated f with its finite element interpolant, i.e.,

$$f(x) \approx \sum_{j=1}^{N} f(x_j) \, \phi_j(x).$$

Handling the Case of General Dirichlet Boundary Conditions In case of general Dirichlet boundary conditions, i.e., $\alpha, \beta \neq 0$, Problem B.3 is equivalent to finding $u_h \in V^h$ such that

$$a(u_h, \phi_i) = (f, \phi_i), \quad \text{for all } i = 2, \ldots, N-1, \tag{B.13}$$

with $u_h(0) = \alpha$ and $u_h(1) = \beta$. Expanding the unknown u_h with respect to the basis $\{\phi_i\}_{i=1}^N$, we have

$$u_h(x) = \alpha\phi_1(x) + \sum_{j=2}^{N-1} U_j\phi_j(x) + \beta\phi_N(x), \tag{B.14}$$

where the coefficients U_j, $j = 2, \ldots, N-1$ have to be determined. Setting $U_1 = \alpha$, $U_N = \beta$ and substituting the expression (B.14) into equation (B.13), we obtain

$$a(u_h, \phi_i) = U_1 a(\phi_1, \phi_i) + \sum_{j=2}^{N-1} U_j a(\phi_j, \phi_i) + U_N a(\phi_N, \phi_i) = (f, \phi_i),$$

for all $i = 2, \ldots, N-1$. Hence, the discrete problem B.3, in case of general Dirichlet boundary conditions, reduces to the following linear system:

Problem B.5 Given $f \in L^2(\Omega)$, find U_j, for $j = 2, \ldots, N-1$, such that

$$\sum_{j=2}^{N-1} U_j a(\phi_j, \phi_i) = (f, \phi_i) - U_1 a(\phi_1, \phi_i) - U_N a(\phi_N, \phi_i), \tag{B.15}$$

for all $i = 2, \ldots, N-1$. Let us now define the vector $\mathbf{U}_b = [U_1, 0, \ldots, 0, U_N]^T \in \mathbb{R}^N$. Then, recalling the previous definitions, the linear system (B.15) in compact form becomes:

$$\hat{\mathrm{K}}\hat{\mathbf{U}} = \hat{\mathbf{F}} - \mathrm{K}(\mathbf{I}, 1:N)\mathbf{U}_b.$$

B.1.1.4 Numerical Tests

We report in this section the results obtained with the MATLAB code

```
run_stat_diff_1d.m,
```

available on the Birkhäuser website provided in the Preface and also at the link

```
http://www.mat.unimi.it/users/scacchi/...
ACS_codes/run_stat_diff_1d.m
```

Table B.1 Example 1, one-dimensional stationary diffusion equation. L^2 relative error $err = \frac{||u_h - u||_0}{||u||_0}$ as a function of the mesh size h, thus of the number of nodes N

$1/h$	N	err
8	9	5.34e-4
16	17	1.38e-4
32	33	3.51e-5
64	65	8.86e-6
128	129	2.23e-6
256	257	5.58e-7

As a numerical example, we consider the solution of equation (B.1), imposing as exact solution

$$u(x) = -\frac{1}{2}x^3 - \frac{1}{4}\sin(2x) + \left(1 + \frac{1}{2} + \frac{1}{4}\sin(2)\right)x.$$

We perform a convergence test by reducing the mesh size h, thus increasing the number of nodes N, as detailed in Table B.1. The behavior of the L^2 relative error,

$$err := \frac{||u_h - u||_0}{||u||_0},$$

reported in Table B.1 and Figure B.2 shows that the method converges with order 2, since halving the mesh size h yields a reduction of the error of a factor 4.

B.1.2 Example 2: The Stationary Reaction–Diffusion Equation

Consider as a model problem the following linear reaction–diffusion equation in strong form, posed on the domain $\Omega = (0, 1)$ and augmented with homogeneous Neumann boundary conditions:

Problem B.6 Given $f : \Omega \to \mathbb{R}$, find $u : \overline{\Omega} \to \mathbb{R}$ such that

$$\begin{cases} -\dfrac{d^2u}{dx^2}(x) + u(x) = f(x), & x \in \Omega, \\ \dfrac{du}{dx}(0) = \dfrac{du}{dx}(1) = 0. \end{cases} \tag{B.16}$$

B.1.2.1 Weak Formulation

In order to derive the *weak formulation* of Problem B.6, we first multiply equation (B.16) by an arbitrary *test function* v and then we integrate on Ω, thus obtaining

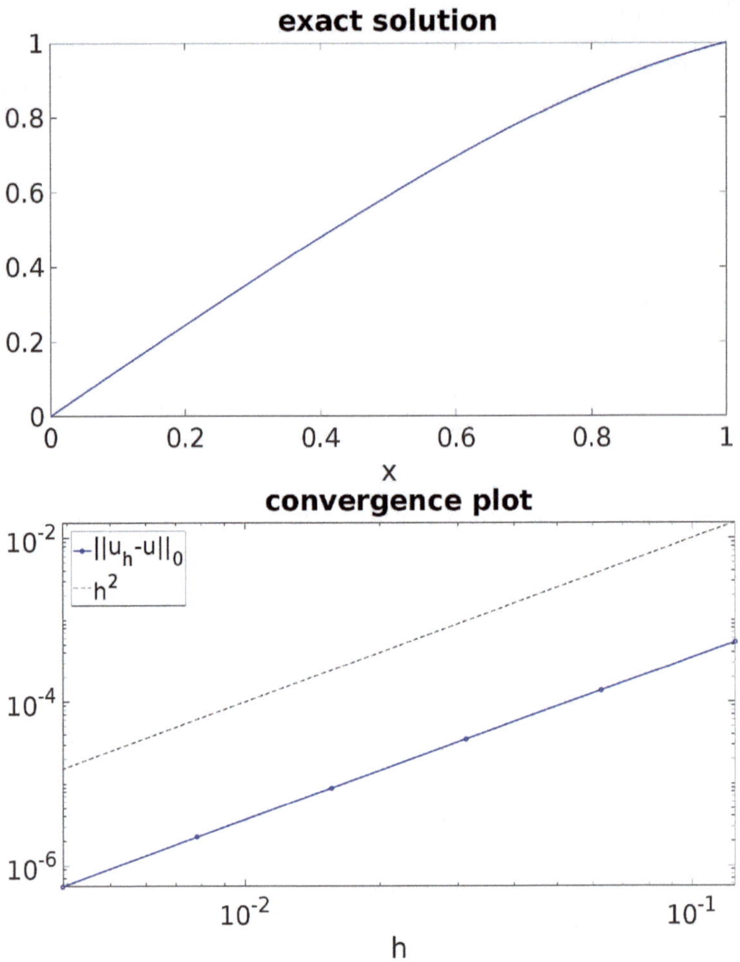

Fig. B.2 Example 1, one-dimensional stationary diffusion equation. Exact solution (top) and convergence (bottom) plots

$$-\int_0^1 \frac{d^2u}{dx^2}(x)\, v(x)\, dx + \int_0^1 u(x)\, v(x)\, dx = \int_0^1 f(x)\, v(x)\, dx.$$

After integrating by parts the first term of the left-hand side, we finally get the integral equation

$$\int_0^1 \frac{du}{dx}(x)\frac{dv}{dx}(x)\, dx + \int_0^1 u(x)\, v(x)\, dx = \int_0^1 f(x)\, v(x)\, dx.$$

In the following, we will set $V = H^1(\Omega)$. Let us then define the bilinear form $a(\cdot,\cdot) : V \times V \to \mathbb{R}$ such that

$$a(u, v) := \int_0^1 \frac{du}{dx}(x)\frac{dv}{dx}(x)\,dx + \int_0^1 u(x)\,v(x)\,dx, \qquad \text{for all } u, v \in V. \tag{B.17}$$

It is easy to verify that the bilinear form $a(\cdot,\cdot)$ is *continuous* and *coercive*.

We can now enunciate the *weak formulation* of Problem B.6:

Problem B.7 Given $f \in L^2(\Omega)$, find $u \in V$ such that

$$a(u, v) = (f, v), \qquad \text{for all } v \in V.$$

B.1.2.2 Discretization with $\mathbf{P}^1$ Finite Elements

Let $\mathcal{T}^h$ be a partition of the domain $\Omega = (0, 1)$ into $N - 1$ subintervals $E_j = [x_j, x_{j+1}]$, for $j = 1, \ldots, N-1$, of length $h = \frac{1}{N-1}$, and let V^h be the associated $\mathbf{P}^1$ finite element space, defined in Definition B.1. It is now possible to enunciate the discrete problem:

Problem B.8 Given $f \in L^2(\Omega)$, find $u_h \in V^h$ such that

$$a(u_h, v_h) = (f, v_h), \qquad \text{for all } v_h \in V^h. \tag{B.18}$$

Note that convergence estimates analogous to those given in Theorem B.4 can be derived in this case, too.

B.1.2.3 Implementation

Denoting by $\{\phi_i\}_{i=1}^N$ the *hat functions* basis of V^h, Problem B.8 is equivalent to finding $u_h \in V^h$ such that

$$a(u_h, \phi_i) = (f, \phi_i), \qquad \text{for all } i = 1, \ldots, N. \tag{B.19}$$

Expanding the unknown u_h with respect to the basis $\{\phi_i\}_{i=1}^N$, we get

$$u_h(x) = \sum_{j=1}^{N} U_j \phi_j(x), \tag{B.20}$$

where U_j, for $j = 1, \ldots, N$, are N unknown scalar values to be determined. Substituting now the expression (B.20) into equation (B.19), we obtain

$$a(u_h, \phi_i) = a\left(\sum_{j=1}^{N} U_j\phi_j, \phi_i\right) = \sum_{j=1}^{N} U_j a(\phi_j, \phi_i) = (f, \phi_i), \qquad \text{for all } i = 1, \ldots, N.$$

Hence, the discrete problem B.8 reduces to the following linear system:

Problem B.9 Given $f \in L^2(\Omega)$, find U_j, for $j = 1, \ldots, N$, such that

$$\sum_{j=1}^{N} U_j a(\phi_j, \phi_i) = (f, \phi_i), \qquad \text{for all } i = 1, \ldots, N. \tag{B.21}$$

Let us now denote by **U** and **F** the vectors defined in (B.6), and by $\mathrm{K} = \{k_{ij}\}$ and $\mathrm{M} = \{m_{ij}\}$ the *stiffness* and *mass* matrices defined in (B.7) and (B.11), respectively. Then, according to the definition of the bilinear form $a(\cdot, \cdot)$ in (B.17), we have

$$a(\phi_j, \phi_i) = k_{ij} + m_{ij}, \qquad \text{for all } i, j = 1, \ldots, N.$$

Thus, defining the matrix

$$\mathrm{A} = \mathrm{K} + \mathrm{M},$$

the linear system (B.21) in compact form becomes:

$$\mathrm{A}\mathbf{U} = \mathbf{F}. \tag{B.22}$$

B.1.2.4 Numerical Tests

We report in this section the results obtained with the MATLAB code

```
run_stat_reac_diff_1d.m,
```

available on the Birkhäuser website provided in the Preface and also at the link

```
http://www.mat.unimi.it/users/scacchi/...
ACS_codes/run_stat_reac_diff_1d.m
```

As a numerical example, we consider the solution of equation (B.16), imposing as exact solution

$$u(x) = \cos(\pi x).$$

We perform a convergence test by reducing the mesh size h, thus increasing the number of nodes N, as detailed in Table B.2. The behavior of the L^2 relative error,

$$err := \frac{||u_h - u||_0}{||u||_0},$$

Table B.2 Example 2, one-dimensional stationary reaction–diffusion equation. L^2 relative error $err = \frac{||u_h - u||_0}{||u||_0}$ as a function of the mesh size h, thus of the number of nodes N

$1/h$	N	err
8	9	1.17e-2
16	17	2.92e-3
32	33	7.30e-4
64	65	1.82e-4
128	129	4.56e-5
256	257	1.14e-5

reported in Table B.2 and Figure B.3 shows that the method converges with order 2, since halving the mesh size h yields a reduction of the error of a factor 4.

B.1.3 Example 3: The Heat Equation

We now move to a nonstationary problem, and in particular we consider the heat equation on the one-dimensional domain $\Omega = (0, 1)$, during a time interval $(0, T)$, being $T > 0$, with homogeneous Neumann boundary conditions:

Problem B.10 Given $f : \Omega \times (0, T) \to \mathbb{R}$ and $u_0 : \Omega \to \mathbb{R}$, find $u : \overline{\Omega} \times [0, T] \to \mathbb{R}$ such that

$$\begin{cases} \dfrac{\partial u}{\partial t}(x,t) - \dfrac{\partial^2 u}{\partial x^2}(x,t) = f(x,t), & (x,t) \in \Omega \times (0,T), \\ \dfrac{\partial u}{\partial x}(0,t) = \dfrac{\partial u}{\partial x}(1,t) = 0, & t \in (0,T), \\ u(x,0) = u_0(x), & x \in \Omega. \end{cases} \tag{B.23}$$

In order to approximate numerically problem (B.23), we will introduce a *weak formulation*, as we did in the two previous examples.

B.1.3.1 Weak Formulation

Proceeding formally, we first multiply equation (B.23), for each $t > 0$, by a test function $v = v(x)$ and then we integrate on Ω. Setting $V = H^1(\Omega)$, denoting by $(\cdot, \cdot)$ the $L^2(\Omega)$ inner product and defining the bilinear form $a(\cdot, \cdot) : V \times V \to \mathbb{R}$,

$$a(u, v) = \int_0^1 \frac{\partial u}{\partial x}(x) \frac{\partial v}{\partial x}(x)\, dx,$$

the *weak formulation* of Problem B.10 reads:

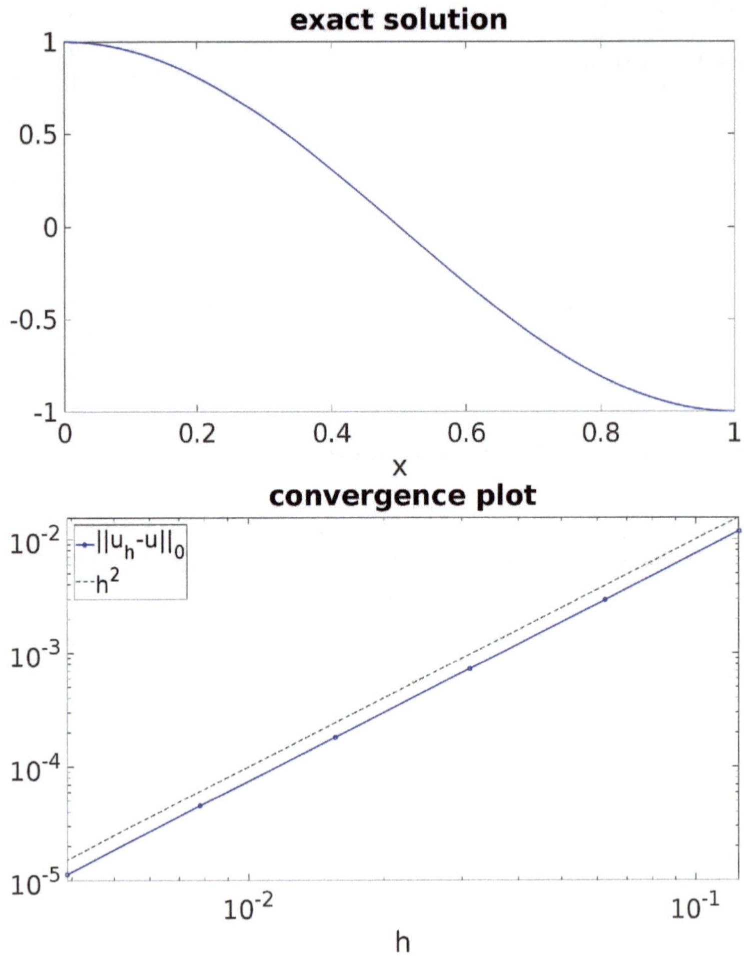

Fig. B.3 Example 2, one-dimensional stationary reaction–diffusion equation. Exact solution (top) and convergence (bottom) plots

Problem B.11 Given $f : (0, T) \to L^2(\Omega)$ and $u_0 \in L^2(\Omega)$, find $u : (0, T] \to V$ such that

$$\left(\frac{\partial u}{\partial t}(t), v\right) + a(u(t), v) = (f(t), v), \quad \text{for all } v \in V, \tag{B.24}$$

with $u(0) = u_0$.

For the mathematical analysis of Problem B.11, we refer, e.g., to [168].

B.1.3.2 Semi-discretization in Space with $\mathbf{P}^1$ Finite Elements

Let $\mathcal{T}^h$ be a partition of the domain $\Omega = (0, 1)$ into $N - 1$ subintervals $E_j = [x_j, x_{j+1}]$, for $j = 1, \dots, N-1$, of length $h = \dfrac{1}{N-1}$, and let V^h be the associated $\mathbf{P}^1$ finite element space, defined in Definition B.1. The semi-discretization in space of Problem B.11 reads:

Problem B.12 Given $f : (0, T) \to L^2(\Omega)$ and $u_0 \in L^2(\Omega)$, find $u_h : (0, T] \to V^h$ such that

$$\left(\frac{\partial u_h}{\partial t}(t), v_h\right) + a(u_h(t), v_h) = (f(t), v_h), \qquad \text{for all } v_h \in V^h, \tag{B.25}$$

with $u_h(0) = u_0^h$, being u_0^h a suitable approximation of u_0 in V^h.

For the convergence analysis of the semi-discrete problem B.12, we refer to [167, 168]; see also [189].

B.1.3.3 Implementation of the Semi-discretization in Space

Denoting by $\{\phi_i\}_{i=1}^N$ the *hat functions* basis of V^h, Problem B.12 is equivalent to finding $u_h : (0, T] \to V^h$ such that

$$\left(\frac{\partial u_h}{\partial t}(t), \phi_i\right) + a(u_h(t), \phi_i) = (f(t), \phi_i), \qquad \text{for all } i = 1, \dots, N, \tag{B.26}$$

with $u_h(0) = u_0^h$. Expanding the unknown $u_h(t)$ with respect to the basis $\{\phi_i\}_{i=1}^N$, we get

$$u_h(x, t) = \sum_{j=1}^N U_j(t)\phi_j(x), \tag{B.27}$$

where $U_j(t)$, for $j = 1, \dots, N$, are N unknown scalar functions to be determined. Substituting now the expression (B.27) into equation (B.26), we obtain

$$\begin{aligned}
&\left(\frac{\partial u_h}{\partial t}(t), \phi_i\right) + a(u_h, \phi_i) \\
&= \left(\frac{\partial\left(\sum_{j=1}^N U_j(t)\phi_j\right)}{\partial t}, \phi_i\right) + a\left(\sum_{j=1}^N U_j(t)\phi_j, \phi_i\right) \\
&= \sum_{j=1}^N \frac{dU_j}{dt}(t)(\phi_j, \phi_i) + \sum_{j=1}^N U_j(t)a(\phi_j, \phi_i) = (f(t), \phi_i),
\end{aligned}$$

for all $i = 1, \ldots, N$. Hence, the semi-discrete problem B.12 reduces to the following system of linear ordinary differential equations (ODEs):

Problem B.13 Given $f : (0, T) \to L^2(\Omega)$ and $u_0 \in L^2(\Omega)$, find $U_j : (0, T] \to \mathbb{R}$, for $j = 1, \ldots, N$, such that

$$\sum_{j=1}^{N} \frac{dU_j}{dt}(t)(\phi_j, \phi_i) + \sum_{j=1}^{N} U_j(t)a(\phi_j, \phi_i) = (f(t), \phi_i), \tag{B.28}$$

for all $i = 1, \ldots, N$, with $U_j(0) = u_0(x_j)$, for all $j = 1, \ldots, N$.

Let us now denote by $\mathbf{U}, \mathbf{F} : (0, T] \to \mathbb{R}^N$ the vector functions

$$\mathbf{U}(t) = [U_1(t), \ldots, U_N(t)]^T,$$
$$\mathbf{F}(t) = [F_1(t), \ldots, F_N(t)]^T,$$

with $F_i(t) = (f(t), \phi_i)$, for all $i = 1, \ldots, N$. Note that $F_i(t)$ can be computed as detailed in (B.12) for the stationary case. Furthermore, let $\mathrm{K} = \{k_{ij}\}$ and $\mathrm{M} = \{m_{ij}\}$ be the *stiffness* and *mass* matrices defined in (B.7) and (B.11), respectively. Then, the ODEs system (B.28) in compact form becomes:

$$\begin{cases} \mathrm{M}\dfrac{d\mathbf{U}}{dt}(t) + \mathrm{A}\mathbf{U}(t) = \mathbf{F}(t), \quad t \in (0, T), \\ \mathbf{U}(0) = \mathbf{U}^0, \end{cases} \tag{B.29}$$

with

$$\mathbf{U}^0 = [u_0(x_1), \ldots, u_0(x_N)]^T. \tag{B.30}$$

B.1.3.4 Time Discretization

System (B.29) can be approximated by any numerical method for ODEs [64]. Here we adopt the Backward Euler method. Let us introduce a partition of the time interval $(0, T)$ into M subintervals $[t_n, t_{n+1}]$, for $n = 0, \ldots, M-1$, of equal size $\tau = \dfrac{T}{M}$. Thus $t_0 = 0$ and $t_M = T$. A variable time step τ could be used as well. Set now $\mathbf{U}^n = \mathbf{U}(t_n)$.

Then, starting from $\mathbf{U}^0$ given in (B.30), the Backward Euler method consists of computing the solution $\mathbf{U}^{n+1}$ at the generic time step t_{n+1}, for $n = 0, \ldots, M-1$, as follows:

$$\mathrm{M}\frac{\mathbf{U}^{n+1} - \mathbf{U}^n}{\tau} + \mathrm{A}\mathbf{U}^{n+1} = \mathbf{F}(t_{n+1}). \tag{B.31}$$

Thus, at the generic time step t_{n+1}, given $\mathbf{U}^n$ computed at the previous time step, we have to find $\mathbf{U}^{n+1}$ by solving the linear system

$$\left(\frac{1}{\tau}\mathrm{M}+\mathrm{A}\right)\mathbf{U}^{n+1}=\frac{1}{\tau}\mathrm{M}\mathbf{U}^n+\mathbf{F}(t_{n+1}).$$

For the convergence analysis of the fully discrete problem (B.31), we refer to [167, 168, 189].

B.1.3.5 Numerical Tests

We report in this section the results obtained with the MATLAB code

```
run_heat_1d.m,
```

available on the Birkhäuser website provided in the Preface and also at the link

```
http://www.mat.unimi.it/users/scacchi/ACS_codes/
run_heat_1d.m
```

As a numerical example, we consider the solution of equation (B.23), with $\Omega = (0, 1)$ and $T = 0.1$, imposing as exact solution

$$u(x,t) = e^{-3t}\cos^2(6\pi x).$$

We first perform a convergence test in space, by reducing the mesh size h, thus increasing the number of nodes N, as detailed in Table B.3, keeping fixed the time step size $\tau = 0.0001$. Then, we perform a convergence test in time, by reducing the time step size τ, as detailed in Table B.4, keeping fixed the mesh size $h = 1/1024$, thus $N = 1025$. The behavior of the error,

$$err := \max_{t_n,\, n=0,\dots,M} \frac{||u_h(\cdot,t_n)-u(\cdot,t_n)||_0}{||u(\cdot,t_n)||_0},$$

reported in Tables B.3 and B.4 and Figure B.4 shows that the convergence of the method is of order 2 in space, since halving the mesh size h yields a reduction of the error of a factor 4, and of order 1 in time, since halving the time step size τ yields a reduction of the error of a factor 2. Finally, we report in Figure B.5 the exact solution and the approximated solution for $h = 1/128$ and $\tau = 0.01$.

B.1.4 Example 4: A Nonlinear Reaction–Diffusion Equation

We conclude the one-dimensional problems with a nonlinear reaction–diffusion equation on $\Omega = (0, 1)$, during a time interval $(0, T)$, being $T > 0$, with homogeneous Neumann boundary conditions:

Table B.3 Example 3, one-dimensional heat equation on $\Omega = (0, 1)$, $T = 0.1$. Error $err = \max_{t_n, n=0,\dots,M} \frac{||u_h(\cdot, t_n) - u(\cdot, t_n)||_0}{||u(\cdot, t_n)||_0}$ as a function of the mesh size h, thus of the number of nodes N, with fixed $\tau = 0.0001$

$1/h$	N	err
16	17	3.56e-1
32	33	7.12e-2
64	65	1.69e-2
128	129	4.19e-3
256	257	1.05e-3
512	513	2.65e-4

Table B.4 Example 3, one-dimensional heat equation on $\Omega = (0, 1)$, $T = 0.1$. Error $err = \max_{t_n, n=0,\dots,M} \frac{||u_h(\cdot, t_n) - u(\cdot, t_n)||_0}{||u(\cdot, t_n)||_0}$ as a function of the time step size τ, with fixed $h = 1/1024$, thus $N = 1025$

τ	err
2.00e-2	8.48e-3
1.00e-2	4.26e-3
5.00e-3	2.14e-3
2.50e-3	1.07e-3
1.25e-3	5.39e-4
6.25e-4	2.76e-4

Problem B.14 Given $u_0 : \Omega \to \mathbb{R}$, find $u : \overline{\Omega} \times [0, T] \to \mathbb{R}$ such that

$$\begin{cases} \dfrac{\partial u}{\partial t}(x,t) - \dfrac{\partial^2 u}{\partial x^2}(x,t) + f(u(x,t)) = 0, & (x,t) \in \Omega \times (0,T), \\ \dfrac{\partial u}{\partial x}(0,t) = \dfrac{\partial u}{\partial x}(1,t) = 0, & t \in (0,T), \\ u(x,0) = u_0(x), & x \in \Omega, \end{cases} \tag{B.32}$$

where $f : \mathbb{R} \to \mathbb{R}$ is a nonlinear function, representing the reaction term.

We introduce now a *weak formulation* of problem (B.32), as we did in the previous examples.

B.1.4.1 Weak Formulation

Proceeding formally, we first multiply equation (B.32), for each $t > 0$, by a test function $v = v(x)$ and then we integrate on Ω. Setting $V = H^1(\Omega)$, denoting by $(\cdot, \cdot)$ the $L^2(\Omega)$ inner product and by $a(\cdot, \cdot)$ the bilinear form introduced in the previous example, the *weak formulation* of Problem B.10 reads:

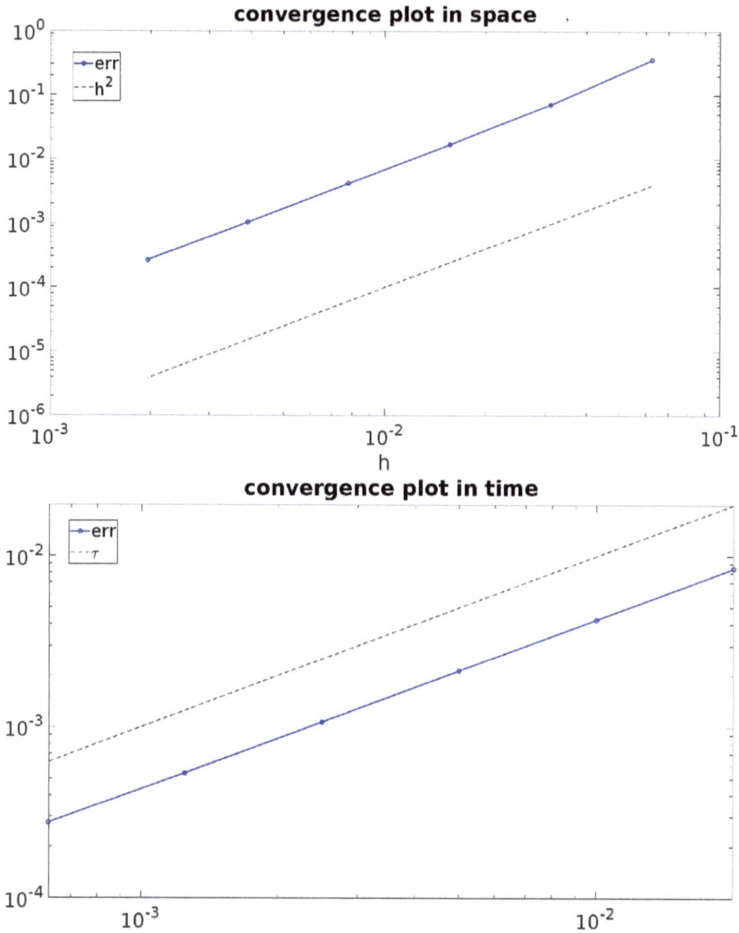

Fig. B.4 Example 3, one-dimensional heat equation on $\Omega = (0, 1)$, $T = 0.1$. Convergence plots in space (top) and time (bottom)

Problem B.15 Given $u_0 \in L^2(\Omega)$, find $u : (0, T] \to V$ such that

$$\left(\frac{\partial u}{\partial t}(t), v\right) + a(u(t), v) + (f(u(t)), v) = 0, \quad \text{for all } v \in V,$$

with $u(0) = u_0$.

Under suitable assumptions on the nonlinear term f, it can be proven that Problem B.15 admits a unique solution.

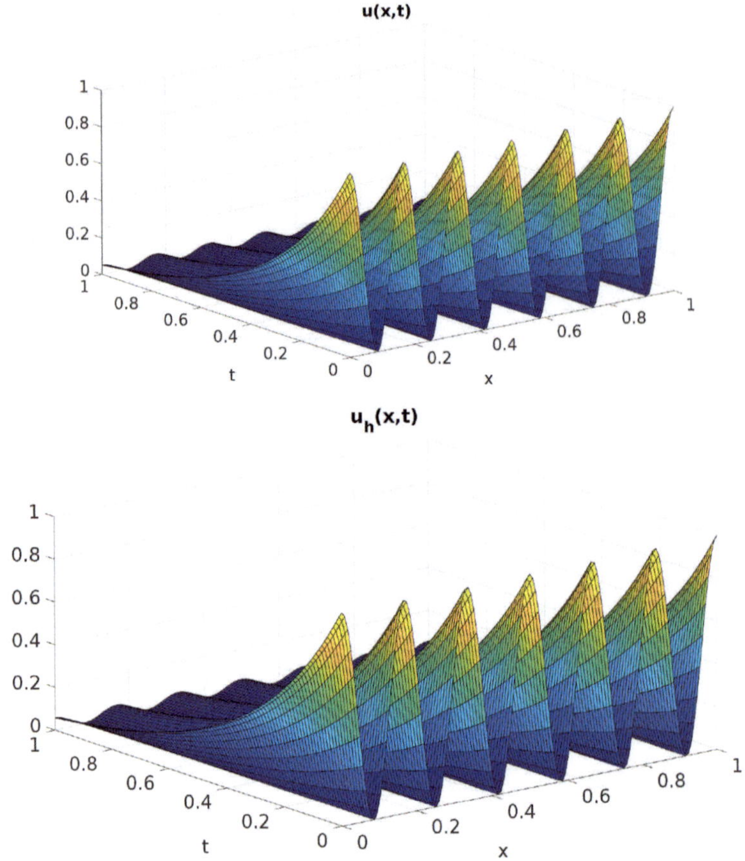

Fig. B.5 Example 3, one-dimensional heat equation on $\Omega = (0, 1)$, $T = 1$. Exact solution $u(x, t)$ (top) and approximated solution $u_h(x, t)$ (bottom) computed with the discretization parameters $h = 1/128$ and $\tau = 0.01$

B.1.4.2 Semi-discretization in Space with $\mathbf{P}^1$ Finite Elements

Let $\mathcal{T}^h$ be a partition of the domain $\Omega = (0, 1)$ into $N - 1$ subintervals $E_j = [x_j, x_{j+1}]$, for $j = 1, \ldots, N-1$, of length $h = \dfrac{1}{N-1}$, and let V^h be the associated $\mathbf{P}^1$ finite element space, defined in Definition B.1. The semi-discretization in space of Problem B.15 reads:

Problem B.16 Given $u_0 \in L^2(\Omega)$, find $u_h : (0, T] \to V^h$ such that

$$\left(\frac{\partial u_h}{\partial t}(t), v_h\right) + a(u_h(t), v_h) + (f(u_h(t)), v_h) = 0, \quad \text{for all } v_h \in V^h, \tag{B.33}$$

with $u_h(0) = u_0^h$, being u_0^h a suitable approximation of u_0 in V^h.

B.1.4.3 Implementation of the Semi-discretization in Space

Denoting by $\{\phi_i\}_{i=1}^N$ the *hat functions* basis of V^h, Problem B.16 is equivalent to finding $u_h : (0, T] \to V^h$ such that

$$\left(\frac{\partial u_h}{\partial t}(t), \phi_i\right) + a(u_h(t), \phi_i) + (f(u_h(t)), \phi_i) = 0, \quad \text{for all } i = 1, \dots, N, \tag{B.34}$$

with $u_h(0) = u_0^h$. Expanding the unknown $u_h(t)$ with respect to the basis $\{\phi_i\}_{i=1}^N$, we get

$$u_h(x, t) = \sum_{j=1}^N U_j(t)\phi_j(x), \tag{B.35}$$

where $U_j(t)$, for $j = 1, \dots, N$, are N unknown scalar functions to be determined. Substituting now the expression (B.35) into equation (B.34), we obtain

$$\begin{aligned}
&\left(\frac{\partial u_h}{\partial t}(t), \phi_i\right) + a(u_h, \phi_i) + (f(u_h), \phi_i) \\
&= \left(\frac{\partial\left(\sum_{j=1}^N U_j(t)\phi_j\right)}{\partial t}, \phi_i\right) + a\left(\sum_{j=1}^N U_j(t)\phi_j, \phi_i\right) + \left(f\left(\sum_{j=1}^N U_j(t)\phi_j\right), \phi_i\right) \\
&= \sum_{j=1}^N \frac{dU_j}{dt}(t)(\phi_j, \phi_i) + \sum_{j=1}^N U_j(t)a(\phi_j, \phi_i) + \left(f\left(\sum_{j=1}^N U_j(t)\phi_j\right), \phi_i\right) = 0,
\end{aligned}$$

for all $i = 1, \dots, N$. We introduce now the following approximation of the nonlinear term:

$$\left(f\left(\sum_{j=1}^N U_j(t)\phi_j\right), \phi_i\right) \approx \left(\sum_{j=1}^N f(U_j(t))\phi_j, \phi_i\right) = \sum_{j=1}^N f(U_j(t))(\phi_j, \phi_i).$$

Then, the semi-discrete problem B.16 reduces to the following system of nonlinear ODEs:

Problem B.17 Given $u_0 \in L^2(\Omega)$, find $U_j : (0, T] \to \mathbb{R}$, for $j = 1, \dots, N$, such that

$$\sum_{j=1}^N \frac{dU_j}{dt}(t)(\phi_j, \phi_i) + \sum_{j=1}^N U_j(t)a(\phi_j, \phi_i) + \sum_{j=1}^N f(U_j(t))(\phi_j, \phi_i) = 0, \tag{B.36}$$

for all $i = 1, \dots, N$, with $U_j(0) = u_0(x_j)$, for all $j = 1, \dots, N$.

Let us now denote by $\mathbf{U}(t), \mathbf{F}(\mathbf{U}(t)) \in \mathbb{R}^N$ the vectors

$$\begin{aligned}\mathbf{U}(t) \quad &= [U_1(t), \ldots, U_N(t)]^T,\\ \mathbf{F}(\mathbf{U}(t)) &= [f(U_1(t)), \ldots, f(U_N(t))]^T,\end{aligned}$$

and by $\mathrm{K} = \{k_{ij}\}$ and $\mathrm{M} = \{m_{ij}\}$ the *stiffness* and *mass* matrices defined in (B.7) and (B.11), respectively. Then, the ODEs system (B.36) in compact form becomes:

$$\begin{cases}\mathrm{M}\dfrac{d\mathbf{U}}{dt}(t) + \mathrm{A}\mathbf{U}(t) + \mathrm{M}\mathbf{F}(\mathbf{U}(t)) = 0, \qquad t \in (0, T),\\ \mathbf{U}(0) = \mathbf{U}^0,\end{cases} \tag{B.37}$$

with

$$\mathbf{U}^0 = [u_0(x_1), \ldots, u_0(x_N)]^T. \tag{B.38}$$

B.1.4.4 Time Discretization

System (B.37) can be approximated by any numerical method for ODEs [64]. Here we adopt a semi-implicit method, where the linear diffusion term is treated implicitly, whereas the nonlinear reaction term is treated explicitly. Let us introduce a partition of the time interval $(0, T)$ into M subintervals $[t_n, t_{n+1}]$, for $n = 0, \ldots, M-1$, of equal size $\tau = \frac{T}{M}$. Thus $t_0 = 0$ and $t_M = T$. A variable time step τ could be used as well. Set now $\mathbf{U}^n = \mathbf{U}(t_n)$.

Then, starting from $\mathbf{U}^0$ given in (B.38), our semi-implicit method consists of computing the solution $\mathbf{U}^{n+1}$ at the generic time step t_{n+1}, for $n = 0, \ldots, M-1$, as follows:

$$\mathrm{M}\frac{\mathbf{U}^{n+1} - \mathbf{U}^n}{\tau} + \mathrm{A}\mathbf{U}^{n+1} + \mathrm{M}\mathbf{F}(\mathbf{U}^n) = 0.$$

Thus, at the generic time step t_{n+1}, given $\mathbf{U}^n$ computed at the previous time step, we have to find $\mathbf{U}^{n+1}$ by solving the linear system

$$\left(\frac{1}{\tau}\mathrm{M} + \mathrm{A}\right)\mathbf{U}^{n+1} = \frac{1}{\tau}\mathrm{M}\mathbf{U}^n - \mathrm{M}\mathbf{F}(\mathbf{U}^n).$$

B.1.4.5 Numerical Tests

We report in this section the results obtained with the MATLAB code

```
run_nagumo_1d.m,
```

available on the Birkhäuser website provided in the Preface and also at the link

```
http://www.mat.unimi.it/users/scacchi/...
ACS_codes/run_nagumo_1d.m
```

As a numerical example, we consider the solution of the one-dimensional Nagumo equation,

$$\begin{cases} \dfrac{\partial u}{\partial t}(x,t) - \sigma \dfrac{\partial^2 u}{\partial x^2}(x,t) = bu(x,t)(u(x,t)-\beta)(\delta - u(x,t)) & \\ \qquad\qquad +I_{app}(x,t), & (x,t) \in \Omega \times (0,T), \\ \dfrac{\partial u}{\partial x}(0,t) = \dfrac{\partial u}{\partial x}(1,t) = 0, & t \in (0,T), \\ u(x,0) = u_0(x), & x \in \Omega, \end{cases}$$

with $\Omega = (0,1)$, $T = 20$, initial datum $u_0(x) = 0$, *applied current*

$$I_{app}(x,t) = \begin{cases} 10, & \text{for } 0 \le x \le 0.04, \ 0 \le t \le 1, \\ 0, & \text{otherwise,} \end{cases}$$

$\sigma = 1e-3$, $b = 5$, $\beta = 0.1$, $\delta = 1$. As numerical parameters we set $N = 65$ and $\tau = 0.1$. Figure B.6 reports the spatio-temporal evolution of the solution, where we observe the propagation of the traveling wave from the left of the interval domain $(0, 1)$, where the current is applied at the beginning of the simulation, toward the right end.

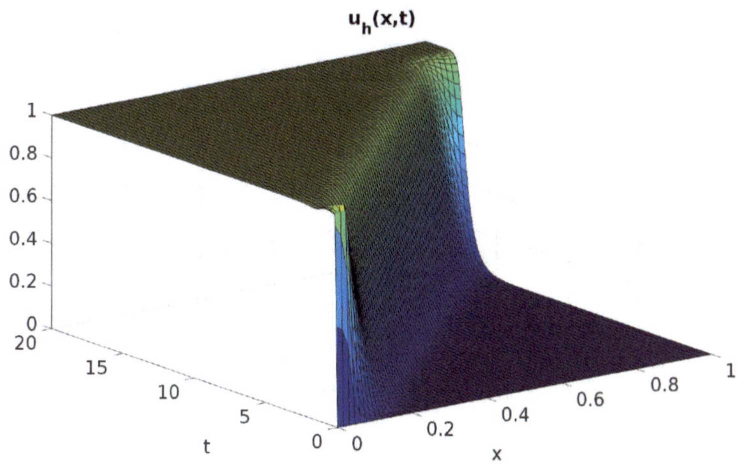

Fig. B.6 Example 4, one-dimensional Nagumo equation on $\Omega = (0, 1)$, $T = 20$. Approximated solution $u_h(x, t)$, computed with the discretization parameters $h = 1/64$ and $\tau = 0.1$

B.2 Q^1 Finite Elements in 2D

B.2.1 Example 5: The Two-Dimensional Stationary Diffusion Equation

Let $\Omega = (0,1)^2$ be the unit square domain and let $\mathbf{x} = (x_1, x_2) \in \overline{\Omega}$. Consider as a model problem the Poisson equation on Ω with homogeneous Dirichlet boundary conditions:

Problem B.18 Given $f : \Omega \to \mathbb{R}$, find $u : \overline{\Omega} \to \mathbb{R}$ such that

$$\begin{cases} -\Delta u(\mathbf{x}) = f(\mathbf{x}), & \mathbf{x} \in \Omega, \\ u(\mathbf{x}) = 0, & \mathbf{x} \in \partial\Omega. \end{cases} \tag{B.39}$$

We refer to Problem B.18 as *strong problem* or *strong formulation.* As discussed in the one-dimensional case, in order to deal with load terms f presenting low regularity, we need an alternative formulation of the problem (B.39), called *weak formulation*, that allows reducing the order of derivation of the solution u.

B.2.1.1 Weak Formulation

In order to derive the *weak formulation* of Problem B.18, we first multiply equation (B.39) by an arbitrary *test function* v vanishing on $\partial\Omega$, and then we integrate on Ω, thus obtaining

$$-\int_\Omega \Delta u(\mathbf{x})\, v(\mathbf{x})\, d\mathbf{x} = \int_\Omega f(\mathbf{x})\, v(\mathbf{x})\, d\mathbf{x}.$$

After applying the Green formula to the left-hand side, we finally get the integral equation

$$\int_\Omega \nabla u(\mathbf{x}) \cdot \nabla v(\mathbf{x})\, d\mathbf{x} = \int_\Omega f(\mathbf{x})\, v(\mathbf{x})\, d\mathbf{x}. \tag{B.40}$$

Note that, in this derivation, we have proceeded formally, without worrying whether the operations were allowed. In order to provide a well posed formulation of equation (B.40), even for not regular data f, we need to introduce suitable function spaces.

Before doing this, we need to introduce the multi-index notation for the derivatives. Let $\alpha = (\alpha_1, \alpha_2, \ldots, \alpha_n)$ be an n-uple of nonnegative integers (called multi-index) and let $g : H \to \mathbb{R}$ be a function defined in $H \subset \mathbb{R}^n$. We will use the following notation:

$$D^\alpha g = \frac{\partial^{|\alpha|} g(\mathbf{x})}{\partial x_1^{\alpha_1} \partial x_2^{\alpha_2} \dots \partial x_n^{\alpha_n}},$$

being $|\alpha| = \alpha_1 + \alpha_2 + \ldots + \alpha_n$ the length of the multi-index coinciding with the order of differentiation of g.

We can now define the following function spaces:

$$\begin{aligned}
L^2(\Omega) &:= \{v : \Omega \to \mathbb{R} : \ v \text{ is Lebesgue measurable}, \int_\Omega |v(\mathbf{x})|^2 \, d\mathbf{x} < +\infty\}, \\
H^1(\Omega) &:= \{v \in L^2(\Omega) : \ \frac{\partial v}{\partial x_1}, \frac{\partial v}{\partial x_2} \in L^2(\Omega)\}, \\
H_0^1(\Omega) &:= \{v \in H^1(\Omega) : \ v = 0 \text{ on } \partial\Omega\}, \\
H^2(\Omega) &:= \{v \in H^1(\Omega) : \ D^\alpha v \in L^2(\Omega), \quad \text{for all } \alpha : \ |\alpha| = 2\},
\end{aligned}$$

where all derivatives must be intended in the sense of distributions and the equal signs in the definition of $H_0^1(\Omega)$ must be intended in the sense of traces; see, e.g., [167] for further details. In the following, we will set $V = H_0^1(\Omega)$.

Let us then define:

- The $L^2(\Omega)$ inner product

$$(u, v) := \int_\Omega u(\mathbf{x}) \, v(\mathbf{x}) \, d\mathbf{x}, \quad \text{for all } u, v \in L^2(\Omega).$$

- The $L^2(\Omega)$ norm

$$||u||_0 := \sqrt{(u, u)}, \quad \text{for all } u \in L^2(\Omega).$$

- The $H^1(\Omega)$ norm

$$||u||_1 := \sqrt{(u, u) + \int_\Omega |\nabla u(\mathbf{x})|^2 \, d\mathbf{x}}, \quad \text{for all } u \in H^1(\Omega).$$

- The $H^1(\Omega)$ semi-norm

$$|u|_1 := \sqrt{\int_\Omega |\nabla u(\mathbf{x})|^2 \, d\mathbf{x}}, \quad \text{for all } u \in H^1(\Omega).$$

- The $H^2(\Omega)$ semi-norm

$$|u|_2 := \sqrt{\sum_{|\alpha|=2} \int_\Omega (D^\alpha u(\mathbf{x}))^2 \, d\mathbf{x}}, \quad \text{for all } u \in H^2(\Omega).$$

Recall that, thanks to the Poincaré inequality (see [167]), the $|\cdot|_1$ semi-norm is a norm in $H_0^1(\Omega)$, equivalent to the $||\cdot||_1$ norm.

Let us now define the bilinear form $a(\cdot,\cdot) : V \times V \to \mathbb{R}$ such that

$$a(u, v) := \int_\Omega \nabla u(\mathbf{x}) \cdot \nabla v(\mathbf{x})\, d\mathbf{x}, \quad \text{for all } u, v \in V.$$

It is easy to verify that the bilinear form $a(\cdot,\cdot)$ is

- *Continuous*, i.e.,

$$\text{there exists } M > 0 \text{ such that } a(u, v) \le M\, ||u||_1 ||v||_1, \quad \text{for all } u, v \in V.$$

- *Coercive*, i.e.,

$$\text{there exists } \alpha > 0 \text{ such that } a(u, u) \ge \alpha\, ||u||_1^2, \quad \text{for all } u \in V.$$

We can now enunciate the *weak formulation* of Problem B.18:

Problem B.19 Given $f \in L^2(\Omega)$, find $u \in V$ such that

$$a(u, v) = (f, v), \quad \text{for all } v \in V.$$

The well-posedness of Problem B.19 follows from the *Lax–Milgram lemma*, see [167]. The finite element method is a particular case of Galerkin method that constructs an approximation of the solution u of Problem B.19 in a finite dimensional subspace of V.

B.2.1.2 Discretization with $\mathbf{Q}^1$ Finite Elements

Let us introduce a partition $\mathcal{T}^h$ of the unit square domain $\Omega = (0, 1)^2$, consisting of a Cartesian grid of squares E_j (called *elements*) with equal edge h. Let N^{el} be the number of elements. The parameter h is called *mesh size*. Let N be the number of all square vertices $\mathbf{x}_j$, which are called *discretization nodes*. For sake of simplifying the discussion, we have supposed that the squares E_j have the same edge h. However, it is possible to consider partitions constituted by general quadrilaterals. This generalization is beyond the scope of this book. We refer the interested readers to [52, 136, 168].

Definition B.3 We call two-dimensional $\mathbf{Q}^1$ finite element space the finite dimensional function space

$$V^h := \{v_h \in C(\overline{\Omega}) : \; v_h|_{E_j} \in \mathbb{Q}_1, \quad \text{for all } E_j \in \mathcal{T}^h\},$$

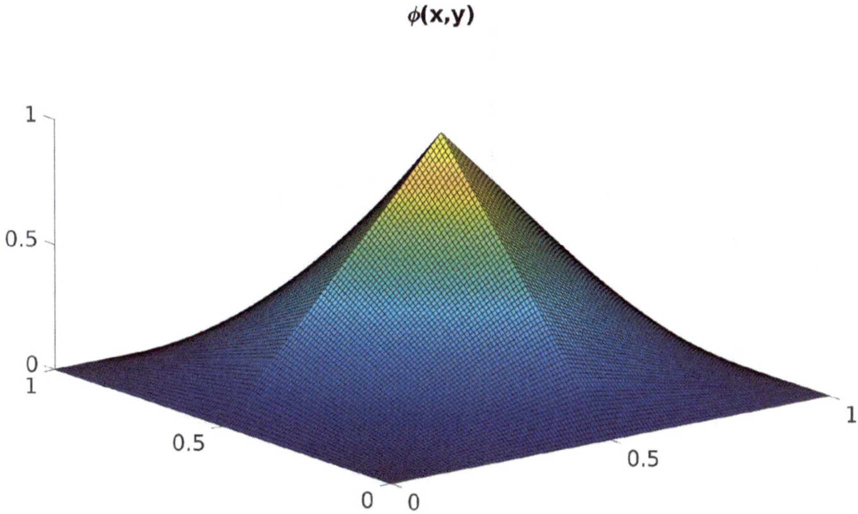

Fig. B.7 Example of $\mathbf{Q}^1$ *hat function* in 2D

where $\mathbb{Q}_1$ denotes the space of bilinear polynomials, i.e., functions of the general form

$$p(x_1, x_2) = a + bx_1 + cx_2 + dx_1x_2.$$

We first observe that $V^h \subset H^1(\Omega)$. The dimension of the space V^h coincides with the number of discretization nodes N. A basis for V^h consists of the so-called *hat functions* $\{\phi_i\}_{i=1}^N$, such that $\phi_i(\mathbf{x}_j) = \delta_{ij}$, for all $i, j = 1, \ldots, N$, where δ_{ij} is the Kronecker delta; see Figure B.7.

We also define the following subset of V^h,

$$V_0^h := \{v_h \in V^h : \ v_h = 0 \text{ on } \partial\Omega\},$$

where we will look for the approximate solution. Clearly $V_0^h \subset V$. Denote by $\mathbf{I}$ and $\mathbf{B}$ the sets of indices associated with the interior and boundary discretization nodes, respectively. Then $\{\phi_i\}_{i\in\mathbf{I}}$ is a basis for V_0^h, whose dimension coincides with the number of interior discretization nodes. It is now possible to enunciate our discrete problem:

Problem B.20 Given $f \in L^2(\Omega)$, find $u_h \in V_0^h$ such that

$$a(u_h, v_h) = (f, v_h), \quad \text{for all } v_h \in V_0^h.$$

It is reasonable to expect that, the smaller is the mesh size h, the better u_h approximates the exact solution u. In fact, it holds the following theorem [167, 168]:

Theorem B.4 *Let $u \in V$ and $u_h \in V_0^h$ be the solutions of problems B.19 and B.20, respectively. Suppose also that $u \in H^2(\Omega)$. Then the following error estimates hold:*

$$||u_h - u||_1 \le C_1 h |u|_2,$$
$$||u_h - u||_0 \le C_0 h^2 |u|_2,$$

being C_1 and C_0 constants independent of h and u.

B.2.1.3 Implementation

Since $\{\phi_i\}_{i \in \mathbf{I}}$ is a basis for V_0^h, Problem B.20 is equivalent to finding $u_h \in V_0^h$ such that

$$a(u_h, \phi_i) = (f, \phi_i), \quad \text{for all } i \in \mathbf{I}. \tag{B.41}$$

Expanding the unknown u_h with respect to the basis $\{\phi_i\}_{i \in \mathbf{I}}$, we have

$$u_h(x) = \sum_{j \in \mathbf{I}} U_j \phi_j(x), \tag{B.42}$$

where U_j, for $j \in \mathbf{I}$, are unknown scalar values to be determined. Set then $U_j = 0$, for all $j \in \mathbf{B}$, because of the homogeneous Dirichlet boundary conditions. Substituting now the expression (B.42) into equation (B.41), we have

$$a(u_h, \phi_i) = a\left(\sum_{j \in \mathbf{I}} U_j \phi_j, \phi_i\right) = \sum_{j \in \mathbf{I}} U_j a(\phi_j, \phi_i) = (f, \phi_i), \quad \text{for all } i \in \mathbf{I}.$$

Hence, the discrete problem B.20 reduces to the following problem:

Problem B.21 Given $f \in L^2(\Omega)$, find U_j, for $j \in \mathbf{I}$, such that

$$\sum_{j \in \mathbf{I}} U_j a(\phi_j, \phi_i) = (f, \phi_i), \quad \text{for all } i \in \mathbf{I}. \tag{B.43}$$

Thus, in order to find u_h, we have to solve the linear system (B.43). Let us now define the vectors:

$$\mathbf{U} = [U_1, \ldots, U_N]^T,$$
$$\hat{\mathbf{U}} = \mathbf{U}(\mathbf{I}) = [U_j]^T, \; j \in \mathbf{I},$$
$$\mathbf{F} = [F_1, \ldots, F_N]^T,$$
$$\hat{\mathbf{F}} = \mathbf{F}(\mathbf{I}) = [F_j]^T, \; j \in \mathbf{I},$$

with $F_i = (f, \phi_i)$, for $i = 1, \ldots, N$. Then define the matrix:

$$\hat{\mathrm{K}} = \mathrm{K}(\mathbf{I}, \mathbf{I}),$$

being K the $N \times N$ matrix such that

$$\mathrm{K} = \{k_{ij}\}_{i,j=1}^{N}, \quad k_{ij} = a(\phi_j, \phi_i), \quad \text{for all } i, j = 1, \ldots, N. \tag{B.44}$$

The matrix K is called *stiffness matrix*. Note that, with the notation

$$\mathrm{K}(\mathbf{I}, \mathbf{I}),$$

we indicate the submatrix of K constituted by the rows $i \in \mathbf{I}$ and the columns $j \in \mathbf{I}$.

With this notation, the linear system (B.43) in compact form becomes:

$$\hat{\mathrm{K}}\hat{\mathbf{U}} = \hat{\mathbf{F}}.$$

The crucial points of the implementation are therefore the assembling of the stiffness matrix K and of the right-hand side vector $\mathbf{F}$.

Assembling the Stiffness Matrix Observe that the generic element of matrix K is

$$k_{ij} = a(\phi_j, \phi_i) = \int_{\Omega} \nabla\phi_j \cdot \nabla\phi_i \, d\mathbf{x} = \sum_{l=1}^{N^{el}} \int_{E_l} \nabla\phi_j \cdot \nabla\phi_i \, d\mathbf{x}.$$

As a consequence, to assemble K, we can proceed with the following double loop algorithm:

$$\begin{aligned}
&\text{for } i = 1, \ldots, N \\
&\quad \text{for } j = 1, \ldots, N \\
&\quad\quad \text{for } l = 1, \ldots, N^{el} \\
&\quad\quad\quad k_{ij} = k_{ij} + \int_{E_l} \nabla\phi_j \cdot \nabla\phi_i \, d\mathbf{x} \\
&\quad\quad \text{end for} \\
&\quad \text{end for} \\
&\text{end for}
\end{aligned} \tag{B.45}$$

By exchanging the order of the for loops, we can re-write algorithm (B.45) as follows:

$$\begin{aligned}
&\text{for } l = 1, \ldots, N^{el}\\
&\quad \text{for } i = 1, \ldots, N\\
&\quad\quad \text{for } j = 1, \ldots, N\\
&\quad\quad\quad k_{ij} = k_{ij} + \int_{E_l} \nabla\phi_j \cdot \nabla\phi_i \, d\mathbf{x}\\
&\quad\quad \text{end for}\\
&\quad \text{end for}\\
&\text{end for}
\end{aligned} \tag{B.46}$$

Thanks to the tight support of the *hat basis functions* $\{\phi_i\}_{i=1}^N$, on the generic mesh element E_l only four basis functions do not vanish. Let $elem$ be the $N^{el} \times 4$ matrix, whose generic l-th row contains the indices of the four basis functions non-vanishing of element E_l. Thus algorithm (B.46) becomes:

$$\begin{aligned}
&\text{for } l = 1, \ldots, N^{el}\\
&\quad \text{for } i \in elem(l, :)\\
&\quad\quad \text{for } j \in elem(l, :)\\
&\quad\quad\quad k_{ij} = k_{ij} + \int_{E_l} \nabla\phi_j \cdot \nabla\phi_i \, d\mathbf{x}\\
&\quad\quad \text{end for}\\
&\quad \text{end for}\\
&\text{end for}
\end{aligned}$$

It remains to compute the integrals

$$\int_{E_l} \nabla\phi_j \cdot \nabla\phi_i \, d\mathbf{x}.$$

To this end, we assume that, up to a shift, the generic subinterval E_l is $\hat{E} = [0, h] \times [0, h]$. The four *hat basis functions* living on $\hat{E}$ are

$$\begin{aligned}
\hat{\phi}_1(x) &= 1 - \frac{x}{h} - \frac{y}{h} + \frac{xy}{h^2},\\
\hat{\phi}_2(x) &= \frac{x}{h} - \frac{xy}{h^2},\\
\hat{\phi}_3(x) &= \frac{y}{h} - \frac{xy}{h^2},\\
\hat{\phi}_4(x) &= \frac{xy}{h^2}.
\end{aligned}$$

Define the 4×4 matrix $\mathrm{K}^{loc} = \{k_{ij}^{loc}\}_{i,j=1}^4$ by

$$k_{ij}^{loc} = \int_{\hat{E}} \nabla\hat{\phi}_j \cdot \nabla\hat{\phi}_i \, d\mathbf{x}, \quad i, j = 1, \ldots, 4.$$

After some basic computations, we have:

$$\mathrm{K}^{loc} = \begin{bmatrix} \frac{2}{3} & -\frac{1}{6} & -\frac{1}{6} & -\frac{1}{3} \\ -\frac{1}{6} & \frac{2}{3} & -\frac{1}{3} & -\frac{1}{6} \\ -\frac{1}{6} & -\frac{1}{3} & \frac{2}{3} & -\frac{1}{6} \\ -\frac{1}{3} & -\frac{1}{6} & -\frac{1}{6} & \frac{2}{3} \end{bmatrix}.$$

Finally, the algorithm to assemble the *stiffness matrix* K can be written in compact form as follows:

for $l = 1, \ldots, N^{el}$
   $\mathrm{K}(elem(l, :), elem(l, :)) = \mathrm{K}(elem(l, :), elem(l, :)) + \mathrm{K}^{loc}$
end for

Assembling the Mass Matrix We call *mass matrix* M the $N \times N$ matrix such that

$$\mathrm{M} = \{m_{ij}\}_{i,j=1}^{N}, \quad m_{ij} = (\phi_j, \phi_i), \quad \text{for all } i, j = 1, \ldots, N. \tag{B.47}$$

This matrix is used to assemble the right-hand side vector **F** in this example and it will be used in the next examples to approximate reaction–diffusion and time-dependent problems.

Proceeding as in the assembling of the *stiffness matrix*, the algorithm to assemble the *mass matrix* reads:

for $l = 1, \ldots, N^{el}$
   for $i \in elem(l, :)$
      for $j \in elem(l, :)$
         $k_{ij} = k_{ij} + \int_{E_l} \phi_j \phi_i \, d\mathbf{x}$
      end for
   end for
end for

To compute the integrals

$$\int_{E_l} \phi_j \phi_i \, d\mathbf{x},$$

we focus again on the reference element $\hat{E} = [0, h] \times [0, h]$. Define the 4×4 matrix $\mathrm{M}^{loc} = \{m_{ij}^{loc}\}_{i,j=1}^{2}$ by

$$m_{ij}^{loc} = \int_{\hat{E}} \hat{\phi}_j \hat{\phi}_i \, dx, \quad i, j = 1, \ldots, 4.$$

If we apply again the *mass lumping* technique, the integrals above are approximated using the two-dimensional trapezoidal rule, thus the matrix M^{loc} becomes

$$\mathrm{M}^{loc} = \frac{h^2}{4} \begin{bmatrix} 1 & 0 & 0 & 0 \\ 0 & 1 & 0 & 0 \\ 0 & 0 & 1 & 0 \\ 0 & 0 & 0 & 1 \end{bmatrix}.$$

The algorithm to assemble M is then:

for $l = 1, \ldots, N^{el}$
  $\mathrm{M}(elem(l, :), elem(l, :)) = \mathrm{M}(elem(l, :), elem(l, :)) + \mathrm{M}^{loc}$
end for

Assembling the Right-Hand Side Vector The generic element F_i, for $i = 1, \ldots, N$, of the right-hand side vector $\mathbf{F}$ can be computed as follows:

$$\begin{aligned} F_i &= (f, \phi_i) \\ &\approx (\textstyle\sum_{j=1}^{N} f(x_j)\, \phi_j, \phi_i) \\ &= \textstyle\sum_{j=1}^{N} f(x_j)(\phi_j, \phi_i) \\ &= \textstyle\sum_{j=1}^{N} f(x_j)\, m_{ij}. \end{aligned}$$

Handling the Case of General Dirichlet Boundary Conditions In case of general Dirichlet boundary conditions, Problem B.18 becomes:

Problem B.22 Given $f : \Omega \to \mathbb{R}$ and $g : \partial\Omega \to \mathbb{R}$ find $u : \overline{\Omega} \to \mathbb{R}$ such that

$$\begin{cases} -\Delta u(\mathbf{x}) = f(\mathbf{x}), & \mathbf{x} \in \Omega, \\ u(\mathbf{x}) = g(\mathbf{x}), & \mathbf{x} \in \partial\Omega. \end{cases}$$

Thus, proceeding analogously as in the case of homogeneous Dirichlet boundary conditions, the discrete problem consists of finding $u_h \in V^h$ such that

$$a(u_h, \phi_i) = (f, \phi_i), \quad \text{for all } i \in \mathbf{I}, \tag{B.48}$$

with $u_h(\mathbf{x}_j) = g(\mathbf{x}_j)$, for all $j \in \mathbf{B}$. Expanding the unknown u_h with respect to the basis $\{\phi_i\}_{i=1}^N$, we have

$$u_h(\mathbf{x}) = \sum_{j\in\mathbf{I}} U_j\phi_j(\mathbf{x}) + \sum_{j\in\mathbf{B}} g(\mathbf{x}_j)\phi_j(\mathbf{x}), \tag{B.49}$$

where the coefficients U_j, $j \in \mathbf{I}$ have to be determined. Setting $U_j = g(\mathbf{x}_j)$, for all $j \in \mathbf{B}$, and substituting the expression (B.49) into equation (B.48), we obtain

$$a(u_h, \phi_i) = \sum_{j\in\mathbf{I}} U_j a(\phi_j, \phi_i) + \sum_{j\in\mathbf{B}} U_j a(\phi_j, \phi_i) = (f, \phi_i), \qquad \text{for all } i \in \mathbf{I}.$$

Hence, the discrete problem B.20, in case of general Dirichlet boundary conditions, reduces to the following linear system:

Problem B.23 Given $f \in L^2(\Omega)$, find U_j, for $j \in \mathbf{I}$, such that

$$\sum_{j\in\mathbf{I}} U_j a(\phi_j, \phi_i) = (f, \phi_i) - \sum_{j\in\mathbf{B}} U_j a(\phi_j, \phi_i), \qquad \text{for all } i \in \mathbf{I}. \tag{B.50}$$

Let us now define the vector $\mathbf{U}_b \in \mathbb{R}^N$ such that $\mathbf{U}_{b,j} = U_j$, for all $j \in \mathbf{B}$, and $\mathbf{U}_{b,j} = 0$, $textnormal forall$ $j \in \mathbf{I}$. Then, recalling the previous definitions, the linear system (B.50) in compact form becomes:

$$\hat{\mathrm{K}}\hat{\mathbf{U}} = \hat{\mathbf{F}} - \mathrm{K}(\mathbf{I}, 1:N)\mathbf{U}_b.$$

B.2.1.4 Numerical Tests

We report in this section the results obtained with the MATLAB code

```
run_stat_diff_2d.m,
```

available on the Birkhäuser website provided in the Preface and also at the link

```
http://www.mat.unimi.it/users/scacchi/...
ACS_codes/run_stat_diff_2d.m
```

As a numerical example, we consider the solution of equation (B.39), imposing as exact solution

$$u(\mathbf{x}) = u(x, y) = \sin(8\pi x)\, y^2 - e^{x-4y}.$$

We perform a convergence test by reducing the mesh size h, thus increasing the number of nodes N, as detailed in Table B.5. The behavior of the L^2 relative error,

Table B.5 Example 5, two-dimensional stationary diffusion equation. L^2 relative error $err = \frac{||u_h - u||_0}{||u||_0}$ as a function of the mesh size h, thus of the number of nodes N

$1/h$	N	err
16	289	7.83e-2
32	1089	1.86e-2
64	4225	4.69e-3
128	16641	1.19e-3
256	66049	2.99e-4
512	263169	7.50e-5

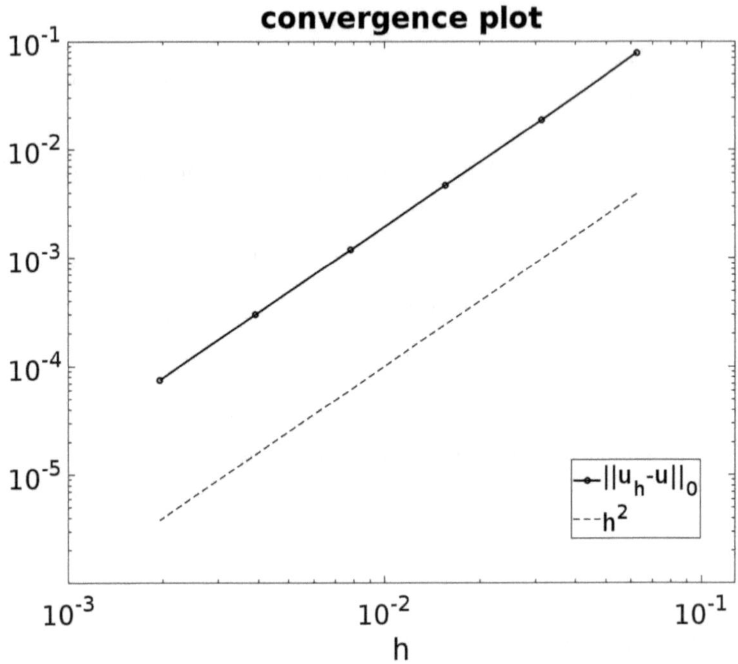

Fig. B.8 Example 5, two-dimensional stationary diffusion equation. Convergence plot

$$err := \frac{||u_h - u||_0}{||u||_0},$$

reported in Table B.5 and Figure B.8 shows that the method converges with order 2, since halving the mesh size h yields a reduction of the error of a factor 4. We also plot in Figure B.9 the approximate solution with $h = 1/128$ (left) and the exact solution (right).

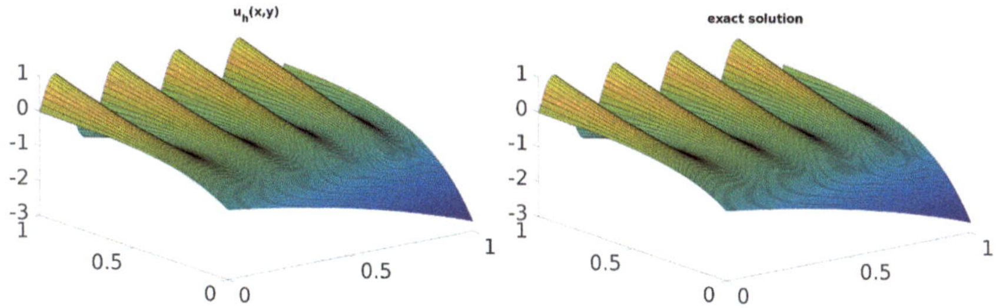

Fig. B.9 Example 5, two-dimensional stationary diffusion equation. Plot of approximate solution with $h = 1/128$ (left) and exact solution (right)

B.2.2 Example 6: The Two-Dimensional FitzHugh–Nagumo System

The last example proposed is the solution of the two-dimensional FitzHugh–Nagumo model [110], a parabolic nonlinear reaction–diffusion equation, coupled with an ODE, on $\Omega = (0, 1)^2$, during a time interval $(0, T)$, being $T > 0$, with homogeneous Neumann boundary conditions:

Problem B.24 Given $u_0,\ w_0 : \Omega \to \mathbb{R}$ and $I_{app} : \Omega \times (0, T) \to \mathbb{R}$, find $u,\ w : \overline{\Omega} \times [0, T] \to \mathbb{R}$ such that

$$\begin{cases} \dfrac{\partial u}{\partial t}(x,t) - \sigma \Delta u(x,t) + f(u(x,t), w(x,t)) = I_{app}\,(x,t), & \\ & (x,t) \in \Omega \times (0,T), \\ \dfrac{\partial w}{\partial t}(x,t) + g(u(x,t), w(x,t)) = 0, & (x,t) \in \Omega \times (0,T), \\ \dfrac{\partial u}{\partial \mathbf{n}}(x,t) = 0, & (x,t) \in \partial\Omega \times (0,T), \\ u(x,0) = u_0(x), & x \in \Omega, \\ w(x,0) = w_0(x), & x \in \Omega, \end{cases} \tag{B.51}$$

where $\mathbf{n}$ denotes the outward normal versor with respect to $\partial\Omega$ and

$$\begin{cases} f(u,w) = -bu(u-\beta)(\delta - u) + cw, \\ g(u,w) = -e(u - \gamma w), \end{cases}$$

being $\sigma,\ b,\ \beta,\ \delta,\ c,\ e,\ \gamma$ positive parameters.

The FitzHugh–Nagumo model describes the spread of the electrical impulse in a portion of excitable tissue, such as the cardiac tissue [91]. We introduce now a *weak formulation* of problem (B.51).

B.2.2.1 Weak Formulation

Proceeding as in the previous examples, we first multiply equations (B.51), for each $t > 0$, by a test function $v = v(\mathbf{x})$ and then we integrate on Ω. Setting $V = H^1(\Omega)$, denoting by $(\cdot, \cdot)$ the $L^2(\Omega)$ inner product and by $a(\cdot, \cdot)$ the bilinear form

$$a(u, v) := \int_\Omega \sigma \nabla u(\mathbf{x}) \cdot \nabla v(\mathbf{x})\, d\mathbf{x}, \quad \text{for all } u, v \in V,$$

the *weak formulation* of Problem B.24 reads:

Problem B.25 Given $u_0,\ w_0 \in L^2(\Omega)$, find $u,\ w : (0, T] \to V$ such that

$$\begin{cases} \left(\dfrac{\partial u}{\partial t}(t), v\right) + a(u(t), v) + (f(u(t), w(t)), v) = (I_{app}(t), v), & \text{for all } v \in V, \\ \left(\dfrac{\partial w}{\partial t}(t), v\right) + (g(u(t), w(t)), v) = 0, & \text{for all } v \in V, \end{cases}$$

with $u(0) = u_0,\ w(0) = w_0$.

B.2.2.2 Semi-discretization in Space with $\mathbf{Q}^1$ Finite Elements

Let $\mathcal{T}^h$ be a Cartesian triangulation of the domain $\Omega = (0, 1)^2$ into N^{el} square elements of equal edge h; let V^h be the associated $\mathbf{Q}^1$ finite element space defined in Definition B.3 and recall the notations of Example 5. The semi-discretization in space of Problem B.25 reads:

Problem B.26 Given $u_0,\ w_0 \in L^2(\Omega)$, find $u_h,\ w_h : (0, T] \to V^h$ such that

$$\begin{cases} \left(\dfrac{\partial u_h}{\partial t}(t), v_h\right) + a(u_h(t), v_h) + (f(u_h(t), w_h(t)), v_h) = (I_{app}(t), v_h), \\ \left(\dfrac{\partial w_h}{\partial t}(t), v_h\right) + (g(u_h(t), w_h(t)), v_h) = 0, \end{cases}$$

for all $v_h \in V^h$, with $u_h(0) = u_0^h$ and $w_h(0) = w_0^h$, being $u_0^h,\ w_0^h$ suitable approximations of $u_0,\ w_0$ in V^h.

B.2.2.3 Implementation of the Semi-discretization in Space

Denoting by $\{\phi_i\}_{i=1}^N$ the *hat functions* basis of V^h, Problem B.26 is equivalent to finding $u_h,\ w_h : (0, T] \to V^h$ such that

$$\begin{cases} \left(\dfrac{\partial u_h}{\partial t}(t), \phi_i\right) + a(u_h(t), \phi_i) + (f(u_h(t), w_h(t)), \phi_i) = (I_{app}(t), \phi_i), \\ \left(\dfrac{\partial w_h}{\partial t}(t), \phi_i\right) + (g(u_h(t), w_h(t)), \phi_i) = 0, \end{cases} \tag{B.52}$$

for all $i = 1, \ldots, N$, with $u_h(0) = u_0^h$, $w_h(0) = w_0^h$. Expanding the unknowns $u_h(t)$, $w_h(t)$ with respect to the basis $\{\phi_i\}_{i=1}^N$, we get

$$u_h(x,t) = \sum_{j=1}^N U_j(t)\phi_j(x), \quad w_h(x,t) = \sum_{j=1}^N W_j(t)\phi_j(x), \tag{B.53}$$

where $U_j(t)$, $W_j(t)$, for $j = 1, \ldots, N$, are $2N$ unknown scalar functions to be determined. Substituting now the expression (B.53) into the first equation in (B.52), we obtain

$$\begin{aligned} &\sum_{j=1}^N \frac{dU_j}{dt}(t)(\phi_j, \phi_i) + \sum_{j=1}^N U_j(t) a(\phi_j, \phi_i) \\ &+ \left(f\left(\sum_{j=1}^N U_j(t)\phi_j, \sum_{j=1}^N W_j(t)\phi_j\right), \phi_i \right) = (I_{app}(t), \phi_i), \end{aligned}$$

for all $i = 1, \ldots, N$.

Then, by approximating the nonlinear term as follows

$$\left(f\left(\sum_{j=1}^N U_j(t)\phi_j, \sum_{j=1}^N W_j(t)\phi_j\right), \phi_i \right) \approx \sum_{j=1}^N f(U_j(t), W_j(t))(\phi_j, \phi_i),$$

and proceeding analogously for the second equation of (B.52), the semi-discrete problem B.26 reduces to the following system of nonlinear ODEs:

Problem B.27 Given $u_0, w_0 \in L^2(\Omega)$, find $U_j, W_j : (0, T] \to \mathbb{R}$, for $j = 1, \ldots, N$, such that

$$\begin{cases} \displaystyle\sum_{j=1}^N \frac{dU_j}{dt}(t)(\phi_j, \phi_i) + \sum_{j=1}^N U_j(t) a(\phi_j, \phi_i) \\ \qquad \displaystyle + \sum_{j=1}^N f(U_j(t), W_j(t))(\phi_j, \phi_i) = (I_{app}(t), \phi_i), \\ \displaystyle\sum_{j=1}^N \frac{dW_j}{dt}(t)(\phi_j, \phi_i) + \sum_{j=1}^N g(U_j(t), W_j(t))(\phi_j, \phi_i) = 0, \end{cases} \tag{B.54}$$

for all $i = 1, \dots, N$, with $U_j(0) = u_0(x_j)$ and $W_j(0) = W_0(x_j)$, for all $j = 1, \dots, N$.

Let us now denote by $\mathbf{U}(t), \mathbf{W}(t), \mathbf{F}(\mathbf{U}(t), \mathbf{W}(t)), \mathbf{G}(\mathbf{U}(t), \mathbf{W}(t)), \mathbf{I}_{app}(t) \in \mathbb{R}^N$ the vectors

$$
\begin{aligned}
\mathbf{U}(t) &= [U_1(t), \dots, U_N(t)]^T, \\
\mathbf{W}(t) &= [W_1(t), \dots, W_N(t)]^T, \\
\mathbf{F}(\mathbf{U}(t), \mathbf{W}(t)) &= [f(U_1(t), W_1(t)), \dots, f(U_N(t), W_N(t))]^T, \\
\mathbf{G}(\mathbf{U}(t), \mathbf{W}(t)) &= [g(U_1(t), W_1(t)), \dots, g(U_N(t), W_N(t))]^T, \\
\mathbf{I}_{app}(t) &= [I_{app}(\mathbf{x}_1, t), \dots, I_{app}(\mathbf{x}_N, t)]^T,
\end{aligned}
$$

and by $\mathrm{K} = \{k_{ij}\}$ and $\mathrm{M} = \{m_{ij}\}$ the *stiffness* and *mass* matrices defined in (B.44) and (B.47), respectively. Then, the ODEs system (B.54) in compact form becomes:

$$
\begin{cases}
\mathrm{M}\dfrac{d\mathbf{U}}{dt}(t) + \mathrm{A}\mathbf{U}(t) + \mathrm{M}\mathbf{F}(\mathbf{U}(t), \mathbf{W}(t)) = M\mathbf{I}_{app}(t), & t \in (0, T), \\
\mathrm{M}\dfrac{d\mathbf{W}}{dt}(t) + \mathrm{M}\mathbf{G}(\mathbf{U}(t), \mathbf{W}(t)) = 0, & t \in (0, T), \\
\mathbf{U}(0) = \mathbf{U}^0, \\
\mathbf{W}(0) = \mathbf{W}^0,
\end{cases}
\tag{B.55}
$$

with

$$
\begin{cases}
\mathbf{U}^0 = [u_0(\mathbf{x}_1), \dots, u_0(\mathbf{x}_N)]^T, \\
\mathbf{W}^0 = [w_0(\mathbf{x}_1), \dots, w_0(\mathbf{x}_N)]^T.
\end{cases}
\tag{B.56}
$$

B.2.2.4 Time Discretization

System (B.55) can be approximated by any numerical method for ODEs [64]. Here we adopt a semi-implicit method, where the linear diffusion term is treated implicitly, whereas the nonlinear reaction term is treated explicitly. Let us introduce a partition of the time interval $(0, T)$ into M subintervals $[t_n, t_{n+1}]$, for $n = 0, \dots, M-1$, of equal size $\tau = \frac{T}{M}$. Thus $t_0 = 0$ and $t_M = T$. A variable time step τ could be used as well. Set now $\mathbf{U}^n = \mathbf{U}(t_n)$, $\mathbf{W}^n = \mathbf{W}(t_n)$ and $\mathbf{I}^n_{app} = \mathbf{I}_{app}(t_n)$.

Then, starting from $\mathbf{U}^0$, $\mathbf{W}^0$ given in (B.56), our semi-implicit method consists of computing the solution $\mathbf{U}^{n+1}$, $\mathbf{W}^{n+1}$ at the generic time step t_{n+1}, for $n = 0, \dots, M-1$, as follows:

$$\begin{cases} \mathrm{M}\dfrac{\mathbf{U}^{n+1}-\mathbf{U}^{n}}{\tau}+\mathrm{A}\mathbf{U}^{n+1}+\mathrm{M}\mathbf{F}(\mathbf{U}^{n},\mathbf{W}^{n})=\mathrm{M}\mathbf{I}_{app}^{n+1}, \\ \mathrm{M}\dfrac{\mathbf{W}^{n+1}-\mathbf{W}^{n}}{\tau}+\mathrm{M}\mathbf{G}(\mathbf{U}^{n},\mathbf{W}^{n})=0. \end{cases}$$

Thus, at the generic time step t_{n+1}, given $\mathbf{U}^n$, $\mathbf{W}^n$ computed at the previous time step, we have to find $\mathbf{U}^{n+1}$, $\mathbf{W}^{n+1}$ by solving the linear system

$$\begin{bmatrix} \dfrac{1}{\tau}\mathrm{M}+\mathrm{A} & 0 \\ 0 & \dfrac{1}{\tau}\mathrm{M} \end{bmatrix} \begin{bmatrix} \mathbf{U}^{n+1} \\ \mathbf{W}^{n+1} \end{bmatrix} = \begin{bmatrix} \dfrac{1}{\tau}\mathrm{M}\mathbf{U}^{n}-\mathrm{M}\mathbf{F}(\mathbf{U}^{n},\mathbf{W}^{n})+\mathrm{M}\mathbf{I}_{app}^{n+1} \\ \dfrac{1}{\tau}\mathrm{M}\mathbf{W}^{n}-\mathrm{M}\mathbf{G}(\mathbf{U}^{n},\mathbf{W}^{n}) \end{bmatrix}.$$

B.2.2.5 Numerical Tests

We report in this section the results obtained with the MATLAB code

```
run_fhn_2d.m,
```

available on the Birkhäuser website provided in the Preface and also at the link

```
http://www.mat.unimi.it/users/scacchi/ACS_codes/run_
fhn_2d.m
```

As a numerical example, we consider the solution of the two-dimensional FitzHugh–Nagumo system, with $\Omega = (0,1)^2$, $T = 30$, initial data $u_0(x) = 0$, $w_0(x) = 0$, *applied current*

$$I_{app}(\mathbf{x},t) = \begin{cases} 5, & \text{for } |\mathbf{x}| \le 0.04,\ 0 \le t \le 1, \\ 0, & \text{otherwise,} \end{cases}$$

$\sigma = 2e-3$, $b = 5$, $\beta = 0.1$, $\delta = 1$, $c = 1$, $\gamma = 0.25$, and $e = 0.1$. We discretize the unit square domain with a grid of 100×100 Q1 finite elements, yielding a total amount of 10201 discretization nodes, and we set the time step size $\tau = 0.01$. Figure B.10 reports the spatial distribution of the potential $u(\mathbf{x},t)$ at 12 time frames, where we observe the propagation of the traveling wave from the center of the domain, where the current is applied at the beginning of the simulation, and the subsequent recovery phase, where the potential returns to the resting state. In Figure B.11, we show instead the time evolution of the potential $u(\mathbf{x},t)$ and of the gating variable $w(\mathbf{x},t)$ at the point $\mathbf{x} = (x,y) = (0.25, 0.25)$.

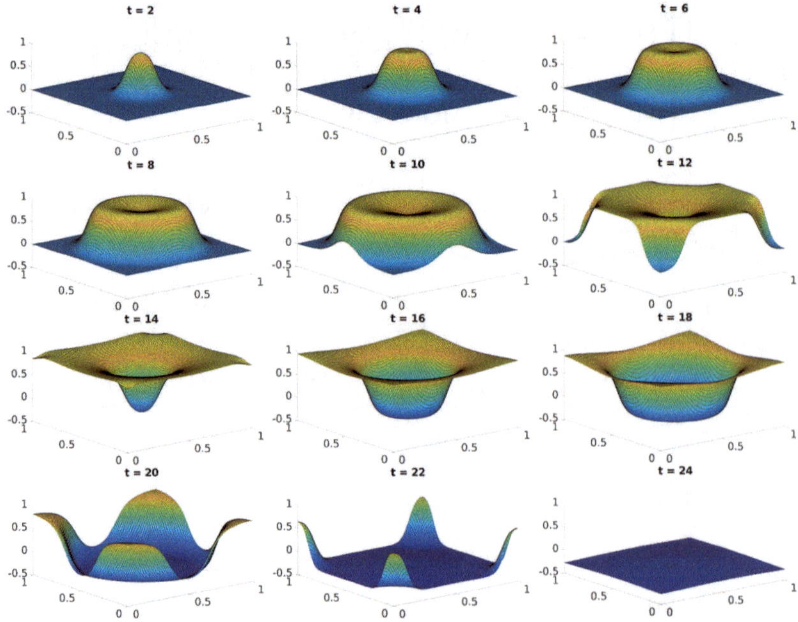

Fig. B.10 Example 6, two-dimensional FitzHugh–Nagumo system on $\Omega = (0, 1)^2$, $T = 30$. Spatial distribution of the potential $u(\mathbf{x}, t)$ at 12 time frames

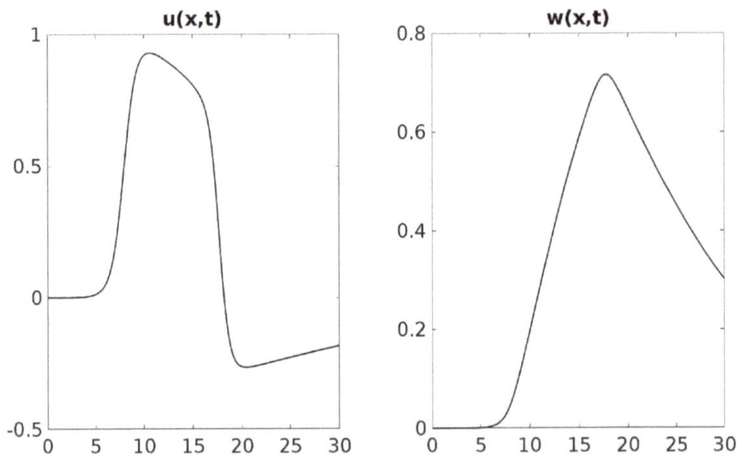

Fig. B.11 Example 6, two-dimensional FitzHugh–Nagumo system on $\Omega = (0, 1)^2$, $T = 30$. Time evolution of the potential $u(\mathbf{x}, t)$ and of the gating variable $w(\mathbf{x}, t)$ at the point $\mathbf{x} = (x, y) = (0.25, 0.25)$

References

1. Ainseba, B., & Aniţa, S. (2005). Internal stabilizability for a reaction-diffusion problem modeling a predator-prey system. *Nonlinear Analysis, Theory, Methods and Applications, 61*, 491–501.
2. Almeida, R. P. P., Blua, M. J., Lopes, J. R. S., & Purcell, A. H. (2005). Vector transmission of *Xylella fastidiosa*: Applying fundamental knowledge to generate disease management strategies. *Annals of the Entomological Society of America, 98*, 775–786.
3. Amann, H. (1976). Fixed point equations and nonlinear eigenvalue problems in ordered Banach spaces. *SIAM Review, 18*, 620–709.
4. Aniţa, L.-I., & Aniţa, S. (2005). Note on the stabilization of a reaction-diffusion model in epidemiology. *Nonlinear Analysis: Real World Applications, 6*, 537–544.
5. Aniţa, L.-I., Aniţa, S., & Arnăutu, V. (2008). Global behavior for an age-dependent model with logistic term and periodic vital rates. *Applied Mathematics and Computation, 206*, 368–379.
6. Aniţa, L.-I., Aniţa, S., & Arnăutu, V. (2009). Optimal harvesting for periodic age-dependent population dynamics with logistic term. *Applied Mathematics and Computation, 215*, 2701–2715.
7. Aniţa, L.-I., Aniţa, S., & Moroşanu, C. (2015). Optimal strategies to diminish a pest population via bilinear controls. *Applied Mathematics Letters, 40*, 7–12.
8. Aniţa, S. (2000). *Analysis and control of age-dependent population dynamics.* Kluwer Acad. Publ.
9. Aniţa, S., Arnăutu, V., & Capasso, V. (2011). *An introduction to optimal control problems in life sciences and economics.* Birkhäuser.
10. Aniţa, S., & Capasso, V. (2002). A stabilizability problem for a reaction-diffusion system modelling a class of spatially structured epidemic systems. *Nonlinear Analysis: Real World Applications, 3*, 453–464.
11. Aniţa, S., & Capasso, V. (2009). A stabilization strategy for a reaction-diffusion system modelling a class of spatially structured epidemic systems (*think globally, act locally*). *Nonlinear Analysis: Real World Applications, 10*, 2026–2035.
12. Aniţa, S., & Capasso, V. (2010). On the stabilization of reaction-diffusion systems modeling a class of man-environment epidemics: A review. *Mathematical Methods in the Applied Sciences, 33*, 1235–1244.
13. Aniţa, S., & Capasso, V. (2011). Stabilization for a reaction-diffusion system modelling a class of spatially structured epidemic systems. The periodic case. In S. Sivasundaram & A. V. Balakrishnan (Eds.), *Advances in dynamics and control: Theory, methods, and applications* (pp. 181–193). Cambridge Sci. Publ. Ltd.

S. Aniţa et al., *Mathematical Modeling and Control in Life and Environmental Sciences*, Modeling and Simulation in Science, Engineering and Technology,
https://doi.org/10.1007/978-3-031-49971-5

14. Aniţa, S., & Capasso, V. (2012). Stabilization of a reaction-diffusion system modelling a class of spatially structured epidemic systems via feedback control. *Nonlinear Analysis: Real World Applications, 13*, 725–735.
15. Aniţa, S., & Capasso, V. (2012). Stabilization of a reaction-diffusion system modelling malaria transmission. *Discrete and Continuous Dynamical Systems - B, 17*, 1673–1684.
16. Aniţa, S., & Capasso, V. (2021). Regional control for spatially structured mosquito borne epidemics. Malaria as a working example. *Vietnam Journal of Mathematics, 49*, 21–35.
17. Aniţa, S., Capasso, V., & Dimitriu, G. (2019). Controlling an alien predator population by regional controls. *Nonlinear Analysis: Real World Applications, 46*, 82–97.
18. Aniţa, S., Capasso, V., Kunze, H., & La Torre, D. (2013). Optimal control and long-run dynamics for a spatial economic growth model with physical capital accumulation and pollution diffusion. *Applied Mathematics Letters, 26*, 908–912.
19. Aniţa, S., Capasso, V., Kunze, H., & La Torre, D. (2015). Dynamics and optimal control in a spatially structured economic growth model with pollution diffusion and environmental taxation. *Applied Mathematics Letters, 42*, 36–40.
20. Aniţa, S., Capasso, V., Kunze, H., & La Torre, D. (2015). Dynamics and control of an integro-differential system of geographical economics. *Annals of the Academy of Romanian Scientists: Series on Mathematics and its Applications, 7*, 8–26.
21. Aniţa, S., Capasso, V., Kunze, H., & La Torre, D. (2017). Optimizing environmental taxation on physical capital for a spatially structured economic growth model including pollution diffusion. *Vietnam Journal of Mathematics, 45*, 199–206.
22. Aniţa, S., Capasso, V., Montagna, M., & Scacchi, S. (2022). Predators as a possible strategy for controlling a *Xylella* epidemic? *Mathematical Modelling of Natural Phenomena, 7*, 279–305.
23. Aniţa, S., Capasso, V., & Moşneagu, A.M. (2016). Regional control in optimal harvesting of population dynamics. *Nonlinear Analysis: Theory, Methods & Applications, 147*, 191–212.
24. Aniţa, S., Capasso, V., & Scacchi, S. (2021). Regional control for spatially structured mosquito borne epidemics. Part II: Computational issues. *Vietnam Journal of Mathematics, 49*, 189–206.
25. Aniţa, S., Capasso, V., & Scacchi, S. (2021). Controlling the spatial spread of a Xylella epidemic. *Bulletin of Mathematical Biology, 83*, 32.
26. Aniţa, S., Casas, J., & Suppo, C. (2014). Impulsive spatial control of invading pests by generalist predators. *Mathematical Medicine and Biology, 31*, 284–301.
27. Aniţa, S., Fitzgibbon, W. E., & Langlais, M. (2009). Global existence and internal stabilization for a class of predator-prey systems posed on non coincident spatial domains. *Discrete and Continuous Dynamical Systems - B, 11*, 805–822.
28. Arendt, W., ter Elst, A.F.M. & Glück, J. (2020). Strict positivity for the principal eigenfunction of elliptic operators with various boundary conditions. *Advanced Nonlinear Studies, 20*, 633–650.
29. Arnăutu, V., Barbu, V. & Capasso, V. (1989). Controlling the spread of a class of epidemics. *Applied Mathematics & Optimization, 20*, 297–317.
30. Arnăutu, V., & Neittaanmäki, P. (2003). *Optimal control from theory to computer programs*. Kluwer Acad. Publ.
31. Arns, C., Knackstedt, M., & Mecke, K. (2002). Characterising the morphology of disordered materials. In L. Mecke & D. Stoyan (Eds.), *Morphology of condensed matter* (pp. 37–74). Springer.
32. Aron, J. L., & May, R. M. (1982). The population dynamics of malaria. In R. M. Anderson (Ed.), *Population dynamics of infectious diseases. Theory and applications* (pp. 139–179). Chapman & Hall.
33. Aronson, D. G. (1977). The asymptotic speed of propagation of a simple epidemic. In W.R. Fitzgibbon & A. F. Walker (Eds), *Nonlinear diffusion* (pp. 1–23). Pitman.

34. Athanassoglou, S., & Xepapadeas, A. (2012). Pollution control with uncertain stock dynamics: When, and how, to be precautious. *Journal of Environmental Economics and Management, 63*, 304–320.
35. Bacaer, N., & Sokhna, C. (2005). A reaction-diffusion system modeling the spread of resistance to an antimalarial drug. *Mathematical Biosciences and Engineering, 2*, 227–238.
36. Bai, Z. (2017). A periodic reaction-diffusion system modelling man-environment-man epidemics. *International Journal of Biomathematics, 10*, 1750074.
37. Baras, P. (1978). Compacité de l'opérateur $f \mapsto u$ solution d'une équation non linéaire $(du/dt) + Au \ni f$. *Comptes rendus de l'Académie des Sciences Paris, Série A, 286*, 1113–1116.
38. Barbu, V. (1994). *Mathematical methods in optimization of differential systems*. Kluwer Acad. Publ.
39. Barbu, V. (1998). *Partial differential equations and boundary value problems*. Kluwer Acad. Publ.
40. Barbu, V. (2018). *Controllability and stabilization of parabolic equations*. Birkhäuser.
41. Barbu, V., & Iannelli, M. (1999). Optimal control of population dynamics. *Journal of Optimization Theory and Applications, 102*, 1–14.
42. Barbu, V., & Precupanu, T. (1986). *Convexity and optimization in Banach spaces*. D. Reidel Publ. Comp.
43. Basile, N., & Mininni, M. (1990). An extension of the maximum principle for a class of optimal control problems in infinite dimensional spaces. *SIAM Journal on Control and Optimization, 28*, 1113–1135.
44. Basile, N., & Mininni, M. (1991). A vector valued optimization approach to the study of a class of epidemics. *Journal of Mathematical Analysis and Applications, 155*, 485–498.
45. Beckerman, W. (1992). Economic growth and the environment: Whose growth? Whose environment? *World Development, 20*, 481–496.
46. Belyakov, A. O., & Veliov, V. (2014). Constant versus periodic fishing: Age structured optimal control approach. *Mathematical Modelling of Natural Phenomena, 9*, 20–37.
47. Belyakov, A. O., & Veliov, V. M. (2015). On optimal harvesting in age-structured populations. *Research Report 2015-08, ORCOS, TU Wien.*
48. Beretta, E., Capasso, V., Scacchi, S., Brunetti, M., & Montagna, M. (2022). Prevention and control of OQDS (olive quick decline syndrome) outbreaks caused by *Xylella fastidiosa*. *Journal of Theoretical Biology, 542*, 111118.
49. Bertinelli, L., Strobl, E., & Zou, B. (2012). Sustainable economic development and the environment: Theory and evidence. *Energy Economics, 34*, 1105–1114.
50. Boscia, D., Altamura, G., Saponari, M., Tavano, D., Zicca, S., Pollastro, P., Silletti, M. R., Savino, V. N., Martelli, G. P., Delle Donne, A., Mazzotta, S., Signore, P. P., Troisi, M., Drazza, P., Conte, P., D'Ostuni, V., Merico, S., Perrone, G., Specchia, F., Stanca, A., & Tanieli, M. (2017). Incidenza di Xylella in oliveti con disseccamento rapido. *Informatore Agrario, 27*, 47–50.
51. Boucekkine, R., Camacho, C., & Zou, B. (2009). Bridging the gap between growth theory and the new economic geography: The spatial Ramsey model. *Macroeconomic Dynamics, 13*, 20–45.
52. Braess, D. (2007). *Finite elements*. Cambridge Univ. Press.
53. Bressan, A., Coclite, G. M., & Shen, W. (2013). A Multidimensional optimal-harvesting problem with measure-valued solutions. *SIAM Journal on Control and Optimization, 51*, 1186–1202.
54. Bressan, A., & Zhang, M. (2023). Controlled traveling profiles for models of invasive biological species. arXiv:2302.09728v1 [math.OC] 20 Feb 2023.
55. Brezis, H. (2011). *Sobolev spaces and partial differential equations*. Springer.
56. Brito, P. (2004). The dynamics of growth and distribution in a spatially heterogeneous world. UECE-ISEG, Technical University of Lisbon.
57. Brock, W. A., & Taylor, M. S. (2010). The green Solow model. *Journal of Economic Growth, 15*, 127–153.

58. Brock, W., & Xepapadeas, A. (2008). Diffusion-induced instability and pattern formation in infinite horizon recursive optimal control. *Journal of Economic Dynamics and Control, 32*, 2745–2787.
59. Brock, W., & Xepapadeas, A. (2010). Pattern formation, spatial externalities and regulation in coupled economic-ecological systems. *Journal of Environmental Economics and Management, 59*, 149–164.
60. Brokate, M. (1985). Pontryagin's principle for control problems in age-dependent population dynamics. *Journal of Mathematical Biology, 23*, 75–101.
61. Brokate, M. (1987). On a certain optimal harvesting problem with continuous age structure. In K.-H. Hoffmann & W. Krabs (Eds.), *Optimal control of partial differential equations II* (pp. 29–42). Birkhäuser.
62. Brunetti, M., Capasso, V., Montagna, M., & Venturino, E. (2020). A mathematical model for *Xylella fastidiosa* epidemics in the Mediterranean regions. Promoting good agronomic practices for their effective control. *Ecological Modelling, 432*, 109–204.
63. Bucur, D., & Buttazzo, G. (2002). *Variational methods in some shape optimization problems*. Notes of Courses Given by the Teachers at the School, Scuola Normale Superiore. Pisa.
64. Butcher, J. C. (2016). *Numerical methods for ordinary differential equations*. Wiley.
65. Camacho, C., & Zou, B. (2004). The spatial Solow model. *Economics Bulletin, 18*, 1–11.
66. Capasso, V. (1978). Global solution for a diffusive nonlinear deterministic epidemic model. *SIAM Journal on Applied Mathematics, 35*, 274–284.
67. Capasso, V. (1984). Asymptotic stability for an integro-differential reaction-diffusion system. *Journal of Mathematical Analysis and Applications, 103*, 575–588.
68. Capasso, V. (1993). *Mathematical structures of epidemic systems*. Springer.
69. Capasso, V. (2008). *Mathematical structures of epidemic systems. 2nd corrected printing*. Lecture Notes in Biomathematics (Vol. 97.) Springer.
70. Capasso, V., & Aniţa, S. (2018). The interplay between models and public health policies: Regional control for a class of spatially structured epidemics (Think globally, act locally). *Mathematical Biosciences and Engineering, 15*, 1–20.
71. Capasso, V., Engbers, R., & La Torre, D. (2010). On the spatial Solow model with technological diffusion and nonconcave production function. *Nonlinear Analysis: Real World Applications, 11*, 3858–3876.
72. Capasso, V., & Kunisch, K. (1988). A reaction-diffusion system arising in modelling man-environment diseases. *Quarterly of Applied Mathematics, 46*, 431–450.
73. Capasso, V., & Maddalena, L. (1981). Convergence to equilibrium states for a reaction-diffusion system modelling the spatial spread of a class of bacterial and viral diseases. *Journal of Mathematical Biology, 13*, 173–184.
74. Capasso, V., & Maddalena, L. (1983). Periodic solutions for a reaction-diffusion system modelling the spread of a class of epidemics. *SIAM Journal on Applied Mathematics, 43*, 417–427.
75. Capasso, V., & Maddalena, L. (1987). On a degenerate nonlinear diffusion problem with boundary feedback. *Application Analysis, 24*, 256–298.
76. Capasso, V., & Paveri-Fontana, S.L. (1979). A mathematical model for the 1973 cholera epidemic in the European Mediterranean region. *Revue d'Epidemiologie et de la Santé Publique, 27*, 121–132. Errata corrige (1980) *Ibidem, 28*, 390.
77. Capasso, V., & Serio, G. (1978). A generalization of the Kermack-McKendrick deterministic epidemic model. *Mathematical Biosciences, 42*, 43–61.
78. Carlucci, A., Lops, F., Marchi, G., Mugna, L., & Surico, G. (2013). Has *Xylella fastidiosa* "chosen" olive trees to establish in the Mediterranean basin? *Phytopathologia Mediterranea, 52*, 541–544.
79. Chamchod, F., & Britton, N.F. (2011). Analysis of a vector-bias model on malaria transmission. *Bulletin of Mathematical Biology, 73*, 639–657.
80. Chan, T. F., & Vese, L. A. (2001). Active contours without edges. *IEEE Transactions on Image Processing, 10*, 266–277.

81. Chang, K.-C., Wang, X., & Wu, X. (2020). On the spectral theory of positive operators and PDE applications. *Discrete and Continuous Dynamical Systems, 40*, 3171–3200.
82. Charlesworth, B. (1980). *Evolution in age-structured populations*. Cambridge Univ. Press.
83. Chasseigne, E., Chavez, M., & Rossi, J.D. (2006). Asymptotic behaviour for nonlocal diffusion equations. *Journal de Mathématiques Pures et Appliquées, 86*, 271–291.
84. Chen, F. (2002). Almost periodic traveling waves of nonlocal evolution equations. *Nonlinear Analysis, 50*, 807–838.
85. Chen, X. (1997). Existence, uniqueness, and asymptotic stability of traveling waves in nonlocal evolution equations. *Advances in Difference Equations, 2*, 125–160.
86. Chitnis, N., Cushing, J. M., & Hyman, J. M. (2006). Bifurcation analysis of a mathematical model for malaria transmission. *SIAM Journal on Applied Mathematics, 67*, 24–45.
87. Chitnis, N., Smith, T. A., & Steketee, R. W. (2008). A mathematical model for the dynamics of malaria in mosquitoes feeding on a heterogeneous host population. *Journal of Biological Dynamics, 2*, 259–285.
88. Coclite, G. M., & Garavello, M. (2017). A time-dependent optimal harvesting problem with measure-valued solutions. *SIAM Journal on Control and Optimization, 55*, 913–935.
89. Coclite, G. M., Garavello, M., & Spinolo, L. V. (2018). Optimal strategies for a time-dependent harvesting problem. *Discrete and Continuous Dynamical Systems - S, 11*, 865–900.
90. Codeço, C. T. (2001). Endemic and epidemic dynamics of cholera: The role of the aquatic reservoir. *BMC Infectious Diseases, 1*, 1–14.
91. Colli Franzone, P., Pavarino, L. F., & Scacchi, S. (2014). *Mathematical cardiac electrophysiology*. Springer.
92. Cornara, D., Cavalieri, V., Dongiovanni, C., Altamura, G., Palmisano, F., Bosco, D., Porcelli, F., Almeida, R. P. P., & Saponari, M. (2017). Transmission of *Xylella fastidiosa* by naturally infected *Philaenus spumarius (Hemiptera, Aphrophoridae)* to different host plants. *Journal of Applied Entomology, 141*, 80–87.
93. Cushing, J. (1998). *An introduction to structured population dynamics*. SIAM.
94. Davranoglou, L. R. (2011). *Zelus renardii* (Kolenati, 1856), a New World reduviid discovered in Europe (Hemiptera: Reduviidae: Harpactorinae). *Entomologist's Monthly Magazine, 147*, 157–162.
95. Deimling, K. (1985). *Nonlinear functional analysis*. Springer.
96. Delfour, M. C., & Zolesio, J.-P. (2011). *Shapes and geometries. Metrics, analysis, differential calculus and optimization* (2nd ed). SIAM.
97. Dietz, K. (1988). Mathematical models for transmission and control of malaria. In W. Wernsdorfer & Y. McGregor (Eds.), *Principles and practice of malariology* (pp. 1091–1133). Churchill Livingstone.
98. Dietz, K., Molineaux, T., & Thomas, A. (1974). A malaria model tested in the African savannah. *Bulletin of the World Health Organization, 50*, 347–357.
99. d'Onofrio, A., & Manfredi, P. (2009). Information-related changes in contact patterns may trigger oscillations in the endemic prevalence of infectious diseases. *Journal of Theoretical Biology, 256*, 473–478.
100. d'Onofrio, A., Manfredi, P., & Salinelli, E. (2007). Vaccination behaviour, information, and the dynamics of SIR vaccine preventable diseases. *Theoretical Population Biology, 71*, 301–317.
101. Dryden I. L., & Mardia, K. V. (2016). *Statistical shape analysis: With applications in R* (2nd Ed). Wiley.
102. Du, Y. (2006). *Order structure and topological methods in nonlinear partial differential equations. Vol. 1: Maximum principles and applications*. World Scientific.
103. Evans, L.C. (1998). *Partial differential equations*. AMS.
104. Feichtinger, G., Tragler, G., & Veliov, V. M. (2003). Optimality conditions for age-structured control systems. *Journal of Mathematical Analysis and Applications, 288*, 47–68.
105. Feichtinger, G., Prskawetz, A., & Veliov, V. M. (2004). Age-structured optimal control in population economics. *Theoretical Population Biology, 65*, 373–387.

106. Fierro, A., Liccardo, A., & Porcelli, F. (2019). A lattice model to manage the vector and the infection of the *Xylella fastidiosa* on olive trees. *Scientific Reports, 9*, 8723.
107. Fife, P. (1996). An integrodifferential analogue of semilinear parabolic PDEs. In *Partial differential equations and applications*. Lecture Notes in Pure and Applied Mathematics 177 (pp. 137–145). Dekker.
108. Fife, P., (2003). Some nonclassical trends in parabolic and parabolic-like evolutions. In *Trends in nonlinear analysis* (pp. 153–191). Springer.
109. Fister, K. R., & Lenhart, S. (2006). Optimal harvesting in an age-structured predator-prey model. *Applied Mathematics & Optimization, 54*, 1–15.
110. FitzHugh, R. (1969). Mathematical models of excitation and propagation in nerve. In H. P. Schwan (Ed.), *Biological engineering* (pp. 1–85). MacGraw-Hill.
111. Friedman, A. (1964). *Partial differential equations of parabolic type*. Prentice-Hall.
112. Friedman, A. (1969). *Partial differential equations*. Holt, Rinehart, & Winston.
113. Friedman, A. (2008). *Partial differential equations of parabolic type*. Dover Publ.
114. Fujita, M., Krugman, P., & Venables, A. (1999). *The spatial economy. Cities, regions and international trade*. MIT Press.
115. Genieys, S., Volpert, V., & Auger, P. (2006). Pattern and waves for a model in population dynamics with nonlocal consumption of resources. *Mathematical Modelling of Natural Phenomena, 1*, 65–82.
116. Getreuer, P. (2012). Chan-Vese segmentation. *Image Processing On Line, 2*, 214–224.
117. Goldstein Ruiz, G. (2006). Derivation and physical interpretation of general boundary conditions. *Advances in Difference Equations, 11*, 457–480.
118. Greenman, J. V., & Norman, R. A. (2007). Environmental forcing, invasion and control of ecological and epidemiological systems. *Journal of Theoretical Biology, 247*, 492–506.
119. Grisvard, P. (1985). *Elliptic problems in nonsmooth domains*. Pitman Publ.
120. Grossman, G. M., & Krueger, A. B. (1995). Economic growth and the environment. *The Quarterly Journal of Economics, 110*, 353–377.
121. Gurtin, M. E., & Murphy, L. F. (1981). On the optimal harvesting of age-structured populations: Some simple models. *Mathematical Biosciences, 55*, 115–136.
122. Gurtin, M. E., & Murphy, L. F. (1981). On the optimal harvesting of persistent age-structured populations. *Journal of Mathematical Biology, 13*, 131–148.
123. Hadwiger, H. (1957). *Vorlesungen über Inhalt, Oberfläche, und Isoperimetrie*. Springer-Verlag.
124. He, Z.-R. (2006). Optimal harvesting of two competing species with age dependence. *Nonlinear Analysis: Real World Applications, 7*, 769–788.
125. He, Z.-R., & Liu, R. (2014). Theory of optimal harvesting for a nonlinear size-structured population in periodic environments. *International Journal of Biomathematics, 7*, 18.
126. Henrot, A., & Pierre, M. (2005). *Variation et Optimisation de Formes, Mathématiques et Applications*. Springer-Verlag.
127. Hethcote, H. W., & van den Driessche, P. (1991). Some epidemiological models with nonlinear incidence. *Journal of Mathematical Biology, 29*, 271–287.
128. Hoppensteadt, F. (1975). *Mathematical theories of populations: Demographics, genetics and epidemics*. SIAM Reg. Conf. Appl. Math., 20. SIAM.
129. Hritonenko, N., & Yatsenko, Y. (2005). Optimization of harvesting age in integral age-dependent model of population dynamics. *Mathematical Biosciences, 195*, 154–167.
130. Hritonenko, N., & Yatsenko, Y. (2012). Bang-bang, impulse, and sustainable harvesting in age-structured populations. *Journal of Biological Systems, 20*, 133–154.
131. Hritonenko, N., & Yatsenko, Y. (2013). *Mathematical modeling in economics, ecology and the environment. 2nd updated ed.* Springer.
132. Iannelli, M. (1995). *Mathematical theory of age-structured population dynamics*. Giardini Editori e Stampatori.
133. Iannelli, M., & Milner, F. (2017). *The basic approach to age-structured population dynamics. Models, methods and numerics*. Springer.

134. Inaba, H. (2017). *Age-structured population dynamics in demography and epidemiology*. Springer.
135. Janse, J. D., & Obradovic, A. (2010). *Xylella fastidiosa*: Its biology, diagnosis, control and risks (Minireview). *Journal of Plant Pathology, 92*, S1.35–S1.48.
136. Johnson, C. (1988). *Numerical solutions of partial differential equations by the finite element method.* Cambridge Univ. Press.
137. Kendall, D. G. (1965). Mathematical models for the spread of infections. In *Mathematics and computer science in biology and medicine* (pp. 213–225). H.M.S.O.
138. Killeen, G. F., McKenzie, F. F., Foy, B. D., Schieffelin, C., Billingsley, P. F., & Beier, J. C. (2000). A simplified model for predicting malaria entomological inoculation rates based on entomologic and parasitologic parameters relevant to control. *American Journal of Tropical Medicine and Hygiene, 62*, 535–544.
139. Klain, D. A., & Rota, G. C. (1997). *Introduction to geometric probability*. Cambridge Univ. Press.
140. Klamka, J., Maurer, H., & Swierniak, A. (2017). Local control lability and optimal control for a model of combined anticancer therapy with control delays. *Mathematical Biosciences and Engineering, 14*, 195–216.
141. Krugman, P. (1991). Increasing returns and economic geography. *Journal of Political Economy, 99*, 483–499.
142. Lago, C., Garzo, E., Moreno, A., et al. (2021). Flight performance and the factors affecting the flight behaviour of *Philaenus spumarius* the main vector of *Xylella fastidiosa* in Europe. *Scientific Reports, 11*, 17608.
143. Lahbib, N., et al. (2022). *Zelus renardii* roaming in Southern Italy. *Insects, 13*, 158.
144. Leitmann, G. (1981). *The calculus of variations and optimal control*. Plenum.
145. Lenhart, S., & Workman, J. T. (2007). *Optimal control applied to biological models*. Chapman, & Hall/CRC.
146. Lewis, M. A., Petrovskii, S. V., & Potts, J. R. (2016). *The mathematics behind biological invasions*. Springer.
147. Li, X.-Z., Yang, J., & Martcheva, M. (2020) *Age structured epidemic modeling*. Springer.
148. Liccardi, A., Fierro, A., Garganese, F., Picciotti, U., & Porcelli, F. (2020). A biological control model to manage the vector and the infection of *Xylella fastidiosa* on olive trees. *PLoS ONE, 15*, e0232363.
149. Lions, J. L. (1988). *Controlabilité exacte, stabilisation et perturbation de systèmes distribués*, RMA 8. Masson.
150. Luo, Z. (2007). Optimal harvesting problem for an age-dependent n-dimensional food chain diffusion model. *Applied Mathematics and Computation, 186*, 1742–1752.
151. Macdonald, G. (1952). The analysis of equilibria in malaria. *Tropical Diseases Bulletin, 49*, 813–829.
152. Macdonald, G. (1957). *The epidemiology and control of malaria*. Oxford Univ. Press.
153. Magal, P., & Ruan, S. (2018). *Theory and applications of abstract semilinear Cauchy problems*. Springer.
154. Martin, R. H. Jr. (1978). Asymptotic stability and critical points for nonlinear quasimonotone parabolic systems. *Journal of Differential Equations, 30*, 391–423.
155. Martin, R. H. Jr, & Smith, H. L. (1990). Abstract functional differential equations and reaction-diffusion systems. *Transactions of the American Mathematical Society, 321*, 1–44.
156. Matricardi, P. M., Dal Negro, R. W., & Nisini, R. (2020). The first holistic immunological model of COVID-19: Implications for prevention, diagnosis, control and public health measures. *Pediatric Allergy and Immunology, 31*, 454–470.
157. Modica, L., & Mortola, S. (1977). Un esempio di Γ-cosnvergenza (in Italian). *Bollettino dell'Unione Matematica Italiana, 14-B*, 285–299.
158. Molineaux, L., & Gramiccia, G. (1980). *The Garki project. Research on the epidemiology and control of malaria in the Sudan Savanna of West Africa*. World Health Org.
159. Murphy, L. F., & Smith, S. J. (1990). Optimal harvesting of an age-structured population. *Journal of Mathematical Biology, 29*, 77–90.

160. Näsell, I., & Hirsch, W. M. (1973). The transmission dynamics of schistosomiasis. *Communications on Pure and Applied Mathematics, 26*, 395–453.
161. Norregard, J., & Reppelin-Hill, V. (2000). Taxes and tradable permits as instruments for controlling pollution: Theory and practice. *IMF Working paper WP/00/13*. http://www.imf.org/external/pubs/ft/wp/2000/wp0013.pdf
162. Ohser, J., & Mücklich, F. (2000). *Statistical analysis of microstructures in materials science. Statistics in practice*. John Wiley, & Sons.
163. Osher, S., & Fedkiw, R. (2003). *Level set methods and dynamic implicit surfaces*. Springer.
164. Pazy, A. (1983). *Semigroups of linear operators and applications to partial differential equations*. Springer.
165. Protter, M. H., & Weinberger, H. F. (1984). *Maximum principles in differential equations*. Springer.
166. Puu, T. (1982). Outline of a trade cycle model in continuous space and time. *Geographical analysis, 14*, 1–9.
167. Quarteroni, A. (2009). *Numerical models for differential problems*. Springer-Verlag Italia.
168. Quarteroni, A., & Valli, A. (1994). *Numerical approximation of partial differential equations*. Springer.
169. Ramsey, F. P. (1928). A mathematical theory of saving. *The Economic Journal, 38*, 543–559.
170. Redak, R. A., Purcell, A. H., Lopes, J. R. S., Blua, M. J., Mizell, R. F. III, & Andersen, P. C. (2004). The biology of xylem fluid-feeding insect vectors of *Xylella fastidiosa* and their relation to disease epidemiology, applying fundamental knowledge to generate disease management. *Annual Review of Entomology, 49*, 243–270.
171. Ross, R. (1911). *The prevention of malaria* (2nd ed.). Murray.
172. Rossi, J. D. (2009). *Asymptotic behaviour of solutions to evolution problems with nonlocal diffusion*. Lecture notes SIMUMAT Summer School 2009, CIEM, Cantabria, Spain.
173. Ruan, S., Xiao, D., & Beier, J. C. (2008). On the delayed Ross-Macdonald model for malaria transmission. *Bulletin of Mathematical Biology, 70*, 1098–1114.
174. Saponari, M., Boscia, D., Altamura, G., Loconsole, G., Zicca, S., D'Attoma, G., Morelli, M., Palmisano, F., Saponari, A., Tavano, D., Savino, V. N., Dongiovanni, C., & Martelli, G. P. (2017). Isolation and pathogenicity of *Xylella fastidiosa* associated to the olive quick decline syndrome in southern Italy. *Scientific Reports, 7*, 17723.
175. Saponari, M., Giampetruzzi, A., Loconsole, G., Boscia, D., & Saldarelli, P. (2018). *Xylella fastidiosa* in olive in Apulia: Where we stand. *Phytopathology, 109*, 175–186.
176. Schättler, H., & Ledzewicz, U. (2015). *Optimal control for mathematical models of cancer therapies. An application of geometric methods*. Springer.
177. Schneider, K., van der Werf, W., Cendoya, M., Maurits, M., & Navas-Cortes, J. A. (2020). Impact of *Xylella fastidiosa* subspecies *pauca* in European olives. *PNAS, 117*, 9250–9259.
178. Sethian, J. A. (1999). *Level set methods and fast marching methods. Evolving interfaces in computational geometry, fluid mechanics, computer vision, and materials science*. Cambridge Univ. Press.
179. Shcherbacheva, A., Haario, H., & Killeen, G. F. (2018). Modeling host-seeking behavior of African malaria vector mosquitoes in the presence of long-lasting insecticidal nets. *Mathematical Biosciences, 295*, 36–47.
180. Shi, J. P., & Wang, X. F. (2009). On global bifurcation for quasilinear elliptic systems on bounded domains. *Journal of Differential Equations, 246*, 2788–2812.
181. Silva, S. E., Rodrigues, A. S. B., Marabuto, E., Yurtsever, S., Borges, P. A. V., Quartau, J. A., Paulo, O. S., & Seabra, S. G. (2015). Differential survival and reproduction in colour forms of *Philaenus spumarius* give new insights to the study of its balanced polymorphism. *Ecological Entomology, 40*, 759–766.
182. Smith, T., Killeen, G., Maire, N., Ross, A., Molineaux, L., Tediosi, F., Hutton, G., Utzinger, J., Dietz, K., & Tanner, M. (2006). Mathematical modeling of the impact of malaria vaccines on the clinical epidemiology and natural history of *Plasmodium Falciparum* malaria: Overview. *American Journal of Tropical Medicine and Hygiene, 75*, 1–10.

183. Sochantha, T., Hewitt, S., Nguon, C., Okell, L., Alexander, N., Yeung, S., Vannara, H., Rowland, M., & Socheat, D. (2006). Insecticide-treated bednets for the prevention of Plasmodium falciparum malaria in Cambodia: A cluster-randomized trial. *Tropical Medicine & International Health, 11*, 1166–1177.
184. Sokolowski, J., & Zolesio, J.-P. (1992). *Introduction to shape optimization*. Springer.
185. Solow, R. M. (1956). A contribution to the theory of economic growth. *The Quarterly Journal of Economics, 70*, 65–94.
186. Swierniak, A., Kimmel, M., Smieja, J., Puszynski, K., & Psiuk-Maksymowicz, K. (2016). *System engineering approach to planning anticancer therapies*. Springer Internat. Publ.
187. Thieme, H. R. (1984). Renewal theorems for linear periodic Volterra integral equations. *The Journal of Integral Equations, 7*, 253–274.
188. Thieme, H. R., & Zhao X-Q. (2003). Asymptotic speeds of spread and traveling waves for integral equations and delayed reaction-diffusion models. *Journal of Differential Equations, 195*, 430–470.
189. Thomee, V. (1984). *Galerkin finite element methods for parabolic problems*. Springer.
190. Turchin, P. (2003). *Complex population dynamics: A theoretical/empirical synthesis*. Monographs in Population Biology. Princeton Univ. Press.
191. Villalobos, F. J., Testi, L., Hidalgo, J., Pastor, M., & Orgaz, F. (2006). Modelling potential growth and yield of olive (*Olea europaea L.*) canopies. *European Journal of Agronomy, 24*, 296–303.
192. Volpert, V., & Petrovskii, S. (2009). Reaction-diffusion waves in biology. *Physics of Life Reviews, 6*, 267–310.
193. Webb, G. (1985). *Theory of nonlinear age-dependent population dynamics*. Marcel Dekker.
194. WHO-UNICEF (2003). The Africa Malaria Report.
195. Xepapadeas, A. (2010). The spatial dimension in environmental and resource economics. *Environment and Development Economics, 15*, 747–758.
196. Xu, D., & Zhao, X-Q. (2005). Asymptotic speed of spread and traveling waves for a nonlocal epidemic model. *Discrete and Continuous Dynamical Systems - B, 5*, 1043–1056.
197. Yurtsever, S. (2000). On the polymorphic meadow spittlebug, *Philaenus spumarius (L.) (Homoptera: Cercopidae)*. *Turkish Journal of Zoology, 24*, 447–460.
198. Zhao, C., Wang, M., & Zhao, P. (2005). Optimal harvesting problems for age-dependent interacting species with diffusion. *Applied Mathematics and Computation, 163*, 117–129.
199. Zhao, C., Zhao, P., & Wang, M. (2006). Optimal harvesting for nonlinear age-dependent population dynamics. *Mathematical and Computer Modelling, 43*, 310–319.
200. Zorich, V. A. (2015). *Mathematical analysis I* (2nd ed.). Springer.

Index

S. Aniţa et al., *Mathematical Modeling and Control in Life and Environmental Sciences*, Modeling and Simulation in Science, Engineering and Technology,
https://doi.org/10.1007/978-3-031-49971-5

MIX
Papier aus verantwortungsvollen Quellen
Paper from responsible sources
FSC® C105338

If you have any concerns about our products,
you can contact us on
ProductSafety@springernature.com

In case Publisher is established outside the EU,
the EU authorized representative is:
Springer Nature Customer Service Center GmbH
Europaplatz 3, 69115 Heidelberg, Germany

Printed by Libri Plureos GmbH
in Hamburg, Germany